MAGNETIC PROPERTIES
OF AMORPHOUS METALS

Proceedings of the Symposium on
MAGNETIC PROPERTIES OF AMORPHOUS METALS

held at Benalmádena, Spain
May 25–29, 1987

Edited by

Antonio HERNANDO, Vicente MADURGA,
Mari C. SÁNCHEZ-TRUJILLO and Manuel VÁZQUEZ

Departamento Física de Materiales,
Facultad de Ciencias Físcas,
Universidad Complutense,
Madrid,
Spain

1987

NORTH-HOLLAND
AMSTERDAM · OXFORD · NEW YORK · TOKYO

© Elsevier Science Publishers B.V., 1987

All rights reserved. No part of this publication may be reproduced, stored in a retrieval system, or transmitted, in any form or by any means, electronic, mechanical, photocopying, recording or otherwise, without the prior permission of the publisher, Elsevier Science Publishers B.V. (North-Holland Physics Publishing Division), P.O. Box 103, 1000 AC Amsterdam, The Netherlands.
Special regulations for readers in the USA: This publication has been registered with the Copyright Clearance Center Inc. (CCC), Salem, Massachusetts. Information can be obtained from the CCC about conditions under which photocopies of parts of this publication may be made in the USA.
All other copyright questions, including photocopying outside of the USA, should be referred to the publisher.

ISBN: 0 444 87086 5

Published by:

North-Holland Physics Publishing
a division of
Elsevier Science Publishers B.V.
P.O. Box 103
1000 AC Amsterdam
The Netherlands

Sole distributors for the U.S.A. and Canada:

Elsevier Science Publishing Company, Inc.
52 Vanderbilt Avenue
New York, N.Y. 10017
U.S.A.

Library of Congress Cataloging-in-Publication Data

Symposium on Magnetic Properties of Amorphous Metals
 (1987 : Benalmádena, Spain)
 Symposium on Magnetic Properties of Amorphous Metals.

 Includes index.
 1. Amorphous sbustances—-Magnetic properties--
Congresses. I. Hernando, Antonio, 1947-   .
II. Title.
QC 176.8.A44S96   1987   530.4'1   87-28209
 ISBN 0-444-87086-5 (U.S.)

PRINTED IN THE NETHERLANDS

ADVISORY COMMITTEE

P. ALLIA	– Italy	H.K. LACHOWICZ	– Poland
J. DURAND	– France	T. MASUMOTO	– Japan
A.L. GREER	– U.K.	O.V. NIELSEN	– Denmark
A. HERNANDO	– Spain	K.V. RAO	– Sweden
L.T. KABAKOFF	– U.S.A.	H.T. SAVAGE	– U.S.A.
H. KRONMÜLLER	– F.R.G.	A. SEREBRIAKOV	– U.S.S.R.

ORGANIZING COMMITTEE

A. HERNANDO   V. MADURGA   M.C. SÁNCHEZ-TRUJILLO   M. VÁZQUEZ

NATIONAL SCIENTIFIC COMMITTEE SPAIN

J.M. ALAMEDA — (Universidad de Oviedo)
C. AROCA — (Universidad de Madrid)
J.M. BARANDIARÁN — (Universidad de Bilbao)
F. BRIONES — (C.S.I.C. Madrid)
A. CONDE — (Universidad de Sevilla)
J.C. GÓMEZ — (Universidad de Santander)
D. GONZÁLEZ — (Universidad de Zaragoza)
E. LÓPEZ — (Universidad de Madrid)
V. MADURGA — (Universidad de Madrid)
M.T. MORA — (Universidad Autónoma de Barcelona)
A. DEL MORAL — (Universidad de Zaragoza)
J. MUÑOZ — (Universidad Autónoma de Barcelona)
R. RIVAS — (Universidad de Santiago de Compostela)
J.M. RIVEIRO — (Universidad de Madrid)
M.C. SÁNCHEZ — (Universidad de Madrid)
J.M. SERRATOSA — (C.S.I.C. Madrid)
J. TEJADA — (Universidad Central de Barcelona)
M. TEJEDOR — (Universidad de Oviedo)
M. VÁZQUEZ — (Universidad de Madrid)
J.L. VICENT — (Universidad de Madrid)

SPONSORS

Comisión Asesora de Investigación Científica y Técnica
Consejería de Educacion y Ciencia de la Junta de Andalucía
Fundación Banco Exterior de España
Universidad Complutense de Madrid
Swedish Companies:— Alfa Laval, Atlas Copco, Bohin Reologi, Ecupan, Enpece AB, Sandvik Electronics, Sandvik Coromant/Proengco, SKF, Volvo, Volvo Flygmotor
European Office of the Navy (U.S.A.)

Dedicated to the memory of Professor Jacques Durand (1938–1987)

OBITUARY

Jacques Durand, 1938 – 1987

With a deep sorrow, the scientific world has learned about the death of *Professor Jacques Durand* who died suddenly on April 20th, 1987 in Nancy. He was 49 years old.

Jacques was born in 1938 in Vendée and started his scientific career studying philosophy at the University of Poitiers where he received his M.Sc. in 1964.

Afterwards, he changed his interests to physics and he obtained in 1969 his M.Sc. in physics from the University of Strasbourg where he completed his Ph.D. in 1973 on the properties of transition metal based alloys studied by means of NMR. In 1974, he joined the group of Professor Paul Duwez at Caltech. in Pasadena (U.S.A.) where he stayed for over three years and performed studies of magnetic properties of amorphous alloys. His outstanding contributions quickly gained world wide scientific attention. He returned to France in 1977 and created a very active group studying the rare-earth amorphous systems at the University of Strasbourg.

In 1981, he was appointed as a Professor at the University of Nancy, joining the Solid state Physics Laboratory, there.

In 1986, he created a new joint research laboratory of C.N.R.S. (Centre National de la Recherche Scientifique) and Saint-Gobain Co. in Pont-à-Mousson near Nancy. In the same year, he became a director of this laboratory.

Durand's contributions to science were widely recognized. He was a member of the French Physical Society and of numerous scientific advisory committees of the most important international magnetic conferences. He authored a large number of original research papers and several important review articles on magnetism in amorphous metallic alloys.

His colleagues and friends from the Solid State Physics Laboratory,
University of Nancy

Participants at Benalmádena, 29 May 1987

PREFACE

" The Symposium on Magnetic Properties of Amorphous Metals ", was held in Benalmádena (Malaga), Spain from 25 to 29 May, 1987. More than 150 people arriving from 20 different countries all over the World joined this meeting.

The book contains 18 invited papers and 80 contributed papers covering some of the most interesting and very novel aspects concerning metallic glasses. These include in particular: different ways of production, basic magnetic properties, magnetoelastic phenomena, structural relaxation, crystallization, technological applications.

All contributions were reviewed by specialists on each subject and revisions were made when necessary.

We wish to thank the publishers for their advice and help in planning the book.

The Editors
A. HERNANDO
V. MADURGA
M.C. SÁNCHEZ-TRUJILLO
M. VÁZQUEZ

CONTENTS

Committees v
Dedication to Professor Jacques Durand vi
Preface ix

Ultrafine amorphous alloy particles prepared by chemical methods
S. MØRUP and J. VAN WONTERGHEM 1
Laboratory scale production of soft magnetic wide amorphous tapes: Optimum production conditions for improved properties
A.R. YAVARI and P. DESRE 6
Synthesis of amorphous materials by mechanical alloying
W. KRAUSS, C. POLITIS and J.R. THOMPSON 15
What is wrong with magnetic domains in amorphous thin films?
I.B. PUCHALSKA 21
Effect of both the conventional heat treatment and the magnetic annealing on magnetoelastic coupling coefficient of amorphous metal ribbons
L. LANOTTE and F. PORRECA 28
Study of the preparation dependence of the quality of different soft magnetic amorphous ribbons
H. SASSIK and R. WEZULEK 32
Magnetic properties of amorphous $Fe_{80}Sb_{20}$ thin films
I. GOSCIANSKA, H. RATAJCZAK, M. KONC and P. SPISAK 35
Magnetic properties and induced anisotropy in $(Fe_{0.79}Co_{0.21})_{75+x}Si_{15-1.4x}B_{10+0.4x}$ metallic glass ribbons
M.L. FDEZ-GUBIEDA, O.V. NIELSEN and P. VOITANIK 38
Magnetic anisotropy in modulated Nd-Fe multilayers
L.T. BACZEWSKI, M. PIECUCH, J. DURAND, P. DELCROIX and G. MARCHAL 41
The Mössbauer study on magnetic anisotropy of the amorphous alloy $Fe_{71}Ni_{10}B_{13}Si_4C_2$
XU ZUXIONG, MA RUZHANG and HAN YONGZHU 45
Induced anisotropy and power losses of cobalt-based amorphous alloys
A.A. GLAZER, A.P. POTAPOV, I.E. STARTSEWA and V.V. SHULIKA 48
Near surface magnetization process of metglas 2605 Sc amorphous ribbons
M. TEJEDOR, A. FERNANDEZ and H. RUBIO 51
Measurement of saturation magnetization and anisotropies in amorphous ribbons using torque magnetometers
M. TEJEDOR, B. HERNANDO, J.A. GARCÍA and J. CARRIZO 54
Magnetic properties of amorphous Fe-B-Sc alloys
M. GHAFARI, R.K. DAY and J.B. DUNLOP 58
Static critical properties of amorphous ferromagnets from Mössbauer measurements
K. BRZÓZKA, K. JEZUITA and J. SZLANTA 61
Amorphization by solid state reaction of Co-Sn multilayers. Mössbauer spectroscopy and magnetic measurements
P. GUYOT, P. GUILMIN and P. DELCROIX 64
FMR study of amorphous $(Sm_xFe_{1-x})_{80}B_{20}$ thin films
W.A. KACZMAREK, J. PIETRZAK and H. RATAJCZAK 67
Surface magnetic properties of amorphous terbium-iron thin films
M. AESCHLIMANN, G.L. BONA, F. MEIER, M. STAMPANONI, A. VATERLAUS and H.C. SIEGMANN 70

Mechanism of chemical short range ordering during structural relaxation in Co–Fe based amorphous alloys
T. KOMATSU and K. MATUSITA 74
Medium range order and magnetic properties of amorphous $Tb_{0.50}Cu_{0.50}$
D. GUIMARAES, M. SANQUER, R. TOURBOT, M.C. BELLISSENT-FUNEL and B. BOUCHER 77
Influence of quench rate on the structural relaxation in Fe–B glasses
Z. ALTOUNIAN 80
Magnetic domain structure by measuring magnetic forces
J.J. SÁENZ, N. GARCÍA, P. GRUÜTTER, E. MEYER, H. HEINZELMANN, R. WIESENDANGER, L. ROSENTHALER, H.R. HIBDER and H.-J. GÜNTHERODT 83
Amorphous to crystalline transition in fine iron–cobalt alloy particles
S. WELLS and S.W. CHARLES 86
Flash annealing of cold-drawn amorphous magnetostrictive wires
R. MALMHÄLL, K. MOHRI, F.B. HUMPHREY, T. MANABE, H. KAWAMURA and J. YAMASAKI 89
Amorphous Tb(Fe, Co) thin films by facing targets sputtering
N. KITAMURA, T. HIRATA and M. NAOE 92
Temperature dependence of bistable magnetic properties in as-quenched and cold-drawn amorphous Fe–B–Si wires
R. MALMHÄLL, K. MOHRI, F.B. HUMPHREY, T. MANABE, H. KAWAMURA, J. YAMASAKI and D-X CHEN 95
Random magnetism and phase transitions in homogeneous and multilayered amorphous systems
D.J. SELLMYER 98
Magnetic properties of sputtered amorphous solids and modulated solids
C.L. CHIEN 104
Re-entrant magnetic flux reversal in amorphous wires
F.B. HUMPHREY, K. MOHRI, J. YAMASAKI, H. KAWAMURA, R. MALMHÄLL and I. OGASAWARA 110
High fatigue strength and unique magnetic properties of amorphous alloy wires
A. INOUE, T. MASUMOTO, I. OGASAWARA and M. HAGIWARA 117
Magnetic and crystallization studies in amorphous R–Fe and R–Co alloys
G. HADJIPANAYIS and A. NAZARETH 123
$Co_{1-x}P_x$ electrodeposited alloys: Structural and magnetic properties
L. LANOTTE and F. PORRECA 129
Low angle diffuse X-ray scattering in glassy $(Fe-Ni)_{1-c}B_c$ alloys before and after structural relaxation and embrittlement
D. BIJAOUI, A.R. YAVARI and F. LIVET 132
Elasticity and magnetomechanical coupling dependences of the $Fe_{79}Si_{12}B_9$ metallic glasses on the magnetic field
Z. KACZKOWSKI 136
Magnetic field dependence of the magnetomechanical coupling of the $Fe_{78}B_{12}Si_{10}$ metallic glasses annealed in perpendicular magnetic field
Z. KACZKOWSKI and HO SU NAM 139
Short range order and magnetoelastic properties of some Fe-rich metallic glasses
J.M. BARANDIARÁN, J. GUTIERREZ, F. PLAZAOLA and I. ZABALA 142
Enthalpy and Curie temperature relaxation in $Fe_{40}Ni_{38}Mo_4B_{18}$ glasses
J.M. BARANDIARÁN, A.L. GREER and I. TELLERIA 145
Magnetic stability of water quenched Co-based amorphous alloys
HAN ZHENGE, WANG XINLIN and KE CHENG 148
On diffusion in amorphous multilayer films $Fe_{1-x}B_x/Fe_{1-y}B_y$

F. STOBIECKI 151
    Coercive force and losses in FeBSi metallic glass ribbons
B. SZYMAŃSKI, M. SOIŃSKI and I. ŠKORVANEK 154
    Nonexponential decay and magnetic relaxation in metallic glasses
J. COLMENERO, I. TELLERIA and A. ALEGRIA 157
    Annealing effect of Pr–Co films deposited by ion beam sputtering
Y. HOSHI and M. NAOE 160
    Observation of crossover between reversible and irreversible magnetic after-effects in amorphous $Fe_{40}Ni_{40}P_{14}B_6$
J. RIVAS, F. WALZ and H. KRONMÜLLER 164
    Magnetostriction of Fe-based amorphous thin films
K. TWAROWSKI and J. WENDA 167
    Forced volume magnetostriction of Zr-containing Fe-rich metallic glasses
L.T. BACZEWSKI and A. WEGRZYN 170
    Stress induced anisotropy, magnetostriction and creep of current heated Co-based amorphous alloy
J. KRZYWINSKI, L. ZALUSKI and A. SIEMKO 173
    Structural relaxation and surface induced aging effects of amorphous GdTbFe films
S. KLAHN, M. HARTMANN, K. WITTER and H. HEITMANN 176
    Amorphous iron-base alloys. Electrochemical growth of oxide layers
T.F. OTERO and A.R. PIERNA 179
    Magnetoresistivity of random anisotropy a-$Dy_xGd_{1-x}Ni$ alloys
J.M. MOREIRA, V.S. AMARAL, M.M. AMADO, J.B. SOUSA, B. BARBARA and B. DIENY . 182
    Thermal dependence of electrical resistivity and crystallization of $(Co_{1-x}Ni_x)_{75}Si_{15}B_{10}$ amorphous ribbons
J.C. GOMEZ SAL, J. RODRIGUEZ FERNANDEZ, L. FERNANDEZ BARQUIN, J.M. BARANDIARÁN and F. PLAZAOLA 185
    Electrical and magnetic properties of amorphous FeZr and CoZr films
T. STOBIECKI, G. BAYREUTHER and H. HOFFMANN 188
    Anisotropic magnetoresistance within the NFE model
E. PILPCZUK and H. MATYJA 191
    Localization in amorphous Nb/Ni multilayers
M.T. PÉREZ FRIAS and J.L. VICENT 194
    Magnetoresistance in amorphous $Fe_{81.5}B_{14.5}Si_4$
J. FLORES and J.L. VICENT 197
    Amorphous melt spun Co-based alloys with high metalloid content: Thermal stability, magnetic and electrical properties studies
M. PONT, K.V. RAO and A. INOUE 200
    Study of the stress dependence of the hysteresis loop and magnetostriction of different 'zero-magnetostrictive' amorphous ribbons
R. GRÖSSINGER, A. PÖNNINGER and G. HERZER 203
    Creep-induced magnetic anisotropy of metallic glasses
L. KRAUS, N. ZÁRUBOVÁ, K. ZÁVĚTA and P. DUHAJ 206
    Magnetic after-effects in binary amorphous alloys
P. VOJTANÍK 209
    Elastic properties of amorphous ferromagnets
K. KAMIGAKI, S. ABE and H. FUJIMORI 212
    Electronic structure, bonding and magnetism in amorphous alloys
R.C. O'HANDLEY, A. COLLINS and M.E. McHENRY 215
    Investigation of domains walls in amorphous materials using scanning electron microscopy with spin polarization analysis
J. UNGURIS, G. HEMBREE, C. AROCA, R.J. CELOTTA and D.T. PIERCE 221
    Evaluation of perpendicular anisotropy of magnetic thin film using the spontaneous Hall effect

K. OKAMOTO 227
Stress dependence of saturation magnetostriction in metallic glasses
H.K. LACHOWICZ and H. SZYMCZAK 232
Magnetic and structural relaxation in amorphous metals
J.A. LEAKE 238
Magnetic properties of amorphous iron-rich binary alloys
D.H. RYAN 244
Correlation properties of inhomogeneities of amorphous magnets
V.A. IGNATCHENKO and R.S. ISKHAKOV 250
Thermomagnetic analysis of the crystallisation of $Fe_{77}B_{15}Sc_8$
C.P. FOLEY, R.K. DAY, J.B. DUNLOP and R.B. ROBERTS 256
Surface effects and magnetic properties of amorphous Co_xY_{1-x} thin films
J.M. ALAMEDA, M.C. CONTRERAS and A.R. LAGUNAS 259
Transverse susceptibility in amorphous NdFeB thin films
J.M. ALAMEDA, F. BRIONES, M.C. CONTRERAS, F. FUERTES, D. GIVORD and A. LIENARD 262
On the crystallisation of $Pd_{40}Ni_{40}P_{20}$
A. GARCÍA ESCORIAL, M.C. CRISTINA, P. ADEVA and A.L. GREER 265
Crystallization of thin amorphous $Fe_{1-x}Si_x$ films
T. LUCINSKI 268
Magnetic properties of ring-shaped electrodeposited Co–P amorphous alloys
G. RIVERO 271
Magnetization noise in amorphous ferromagnetic alloys
M. CELASCO, A. MASOERO and A. STEPANESCU 274
Application of amorphous alloys in ground fault interrupters
J. GUZ, G. MATRAS, A. NAFALSKI and A. WAC-WŁODARCZYK 278
Experimental study of the domain wall
E. LÓPEZ, L. DE PEDRO, P. SÁNCHEZ, C. AROCA, C. MUÑOZ and M.C. SÁNCHEZ 281
'Cluster-in-shell' modification of the micro-crystalline model of amorphous alloy structure
A.V. SEREBRYAKOV 284
Low and high field magnetic studies of heat treated amorphous $(Fe_{0.5}Ni_{0.5})_{83}P_{17}$
R. PUŹNIAK, J.S. MUÑOZ and K.V. RAO 287
Crystallization behaviour of the $Ni_{81}Cr_{15}B_4C$ (wt. %) alloy
A. CRIADO, M. MILLÁN, A. CONDE and R. MÁRQUEZ 291
Magnetoelastic properties of multilayered Co–P amorphous alloys
G. RIVERO, M. LINIERS, E. ASCASIBAR and J.M. GONZALEZ 294
Induced anisotropy by laser annealing
C. AROCA, M.C. SÁNCHEZ, M. GARCIA, E. LÓPEZ and P. SÁNCHEZ 297
The resonance magnetic field shift induced by dc current in amorphous alloys
A. WADAS 300
A semi-infinite transverse Ising model with surface amorphization: Phase diagrams
T. KANEYOSHI and E.F. SARMENTO 303
Distribution and evolution of temperature during flash-annealing of metallic glasses
J.M. BARANDIARÁN and N. ZABALA 306
Laser surface treatment of amorphous and crystalline $Fe_{40}Ni_{38}Mo_4B_{18}$
J.V. ARMSTRONG, J.M.D. COEY and J.G. LUNNEY 309
Magnetovolume effects in amorphous transition metal based alloys
J. SCHNEIDER, A. HANDSTEIN, J. ARNOLD and J. KAMARAD 312
Magnetic and structural characterization of $TbDyFe_2$ sputtered films
E.T.M. LACEY, A.D. BIRCHENOUGH, D.G. LORD and P.J. GRUNDY 315
Positron lifetime study and magnetic properties of $Fe_{81.5}B_{14.5}Si_4$ glass
S. LINDEROTH, C. HIDALGO, J.M. GONZALEZ, M. LINIERS and J.L. VICENT 318

Irreversible changes of resistivity induced by stress, applied at room temperature, in some as quenched amorphous ribbons
A. HERNANDO, V. MADURGA, M. VÁZQUEZ, E. ASCASIBAR and A. GARCÍA ESCORIAL ... 321
   Stress-field induced magnetic anisotropy in Co–Fe–Ni metallic glasses
M. VÁZQUEZ, J. GONZÁLEZ, V. MADURGA, J.M. BARANDIARÁN, A. HERNANDO and O.V. NIELSEN ... 324
   Temperature dependence and critical exponents of the magnetostriction of metallic glass ribbons
M. VÁZQUEZ, C. NUÑEZ DE VILLACICENCIO, V. MADURGA, J.M. BARANDIARÁN, A. HERNANDO and H. KRONMÜLLER ... 327
   Magnetostriction of a-rare earth random magnetic anisotropy spin glasses
A. DEL MORAL and J.I. ARNAUDAS ... 330
   Specific heat and electronic structure of ferromagnetic $Fe_{90-x}Co_xZr_{10}$ amorphous alloys
M. ROSENBERG and R. WERNHARDT ... 333
   Relaxation and embrittlement of Fe–Ni–P amorphous alloys studied by small-angle neutron scattering
A.R. YAVARI, D. BIJAOUI, M.C. BELLISSENT, P. CHIEUX and M. HARMELIN ... 336
   Galvano-magnetic properties of ion-beam mixed Fe–Zr
N. KARPE, K.V. RAO, B. TORP and J. BØTTIGER ... 340
   Investigation of magnetic properties in amorphous FeZr alloys
A. FORKL, R. REISSER and H. KRONMÜLLER ... 344
   Losses, aftereffect and disaccomodations in amorphous ferromagnetic alloys
P. ALLIA and F. VINAI ... 347
   Industrial applications of metallic glass ribbons
G. HERZER and H.R. HILZINGER ... 354
   Amorphous metal sensor
K. MOHRI ... 360
   SWR linewidth in thin films of Fe-B and long range fluctuation of exchange parameter
R.S. ISKHAKOV, L.J. MAKSYMOWICZ, D. TEMPLE and R. ŻUBEREK ... 367

Author Index ... 371
Subject Index ... 375
Materials Index ... 379

ULTRAFINE AMORPHOUS ALLOY PARTICLES PREPARED BY CHEMICAL METHODS

Steen Mørup and Jacques van Wonterghem

Laboratory of Applied Physics II, Technical University of Denmark, DK-2800 Lyngby, Denmark

The chemical preparation and the properties of ultrafine amorphous alloy particles is discussed. $Fe_{100-x}C_x$ particles can be prepared by thermal decomposition of $Fe(CO)_5$ in an organic liquid. In this case particles with diameters in the range 4-8 nm have been obtained. By reduction of metal ions in aqueous solution by KBH_4 we have prepared alloy particles containing Fe, Co, Ni, and B. In this case the particle size was in the range 10-100 nm.

1. INTRODUCTION

Amorphous alloys are normally prepared as thin ribbons or films by the liquid quench technique or by vapor deposition. Recently, we have shown that ultrafine particles of amorphous alloys can be prepared by chemical reactions at a temperature below the glass transition temperature T_g of the alloys[1,2].

The principles for formation of amorphous alloys by chemical methods and by the liquid quench technique are illustrated in Figure 1. When using the liquid quench technique, the material has to be cooled very rapidly ($\gtrsim 10^6$ Ks^{-1}) from the liquid state to a temperature below T_g. At a given cooling rate, formation of the amorphous phase is favored if the temperature interval ΔT between the liquidus curve T_L and T_g is small. This is probably one of the reasons that compositions near the eutectic are favorable for formation of an amorphous alloy, whereas compositions far from the eutectic may not lead to amorphous phases when using the presently available liquid quench technique. However, when the alloy is formed by a chemical method at a temperature T_k well below T_g, it is in principle possible to form alloys with compositions in the whole range indicated by the horizontal line at T_k in

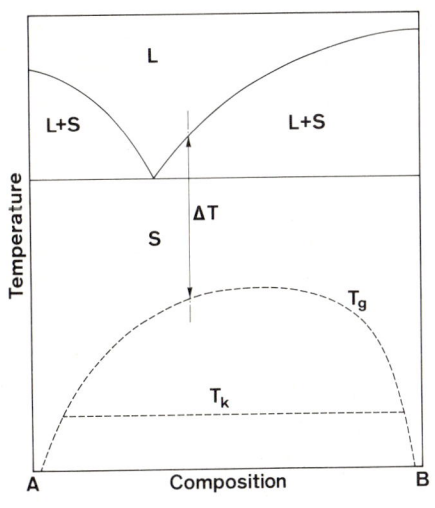

FIGURE 1
Phase diagram illustrating the principal features of formation of amorphous alloys by the liquid quench technique and the chemical methods.

Figure 1. Another advantage of the chemical methods is that they can easily be used for production of ultrafine amorphous alloy particles instead of ribbons or films.

2. THERMAL DECOMPOSITION OF Fe(CO)$_5$

Ultrafine particles of amorphous $Fe_{100-x}C_x$ can be prepared by thermal decomposition of Fe(CO)$_5$ in an organic liquid[1,3,4]. During the chemical reaction the temperature cannot exceed the boiling point of the liquid, and if this temperature is lower than the glass transition temperature, an amorphous phase may be formed. In order to produce ultrafine particles with dimensions below 10 nm, it is necessary to add a surfactant to the liquid before the reaction takes place. The preparation then leads to formation of a colloidal suspension of ultrafine ferromagnetic particles, i.e. a ferrofluid[5].

van Wonterghem et al.[1] have prepared such a ferrofluid by mixing a surfactant, Sarkosyl-O (n-oleyoyl sarcosine) and iron pentacarbonyl in Decalin (decahydronaphthalene). The mixture was refluxed and stirred for 5-6 h while the temperature was increased up to 460 K. The particles produced in this way were investigated by Mössbauer spectroscopy, X-ray diffraction, and electron microscopy. Figure 2 shows a Mössbauer spectrum of the sample, obtained at 80 K. The spectrum consists of a magnetically split component with very broad lines. The average hyperfine field is $B_{hf} = 26.0$ T, and the isomer shift is $\delta \approx 0.17$ mm s^{-1} relative to α-Fe at room temperature. It was shown that the line broadening is due to a broad distribution of hyperfine parameters, and this suggests that the particles are amorphous. This was confirmed by X-ray diffraction studies. Moreover, it was shown that the particles crystallize into a mixture of α-Fe and iron carbides during annealing at 523 K. The Mössbauer spectra are quite similar to those of amorphous iron-carbon alloys prepared by sputtering[6].

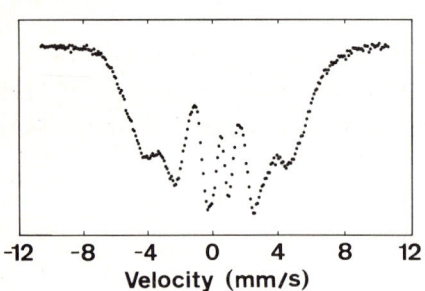

FIGURE 2
Mössbauer spectrum of amorphous particles of $Fe_{100-x}C_x$ in a ferrofluid at 80 K.

The stability of the amorphous phase is probably due to the presence of the carbon atoms. It is likely that the carbon atoms enter the particles during the preparation as a result of chemisorption and disintegration of carbon monoxide at the surface of the particles[1].

The chemical reactions leading to the formation of the particles have later been studied by use of Mössbauer spectroscopy[4]. It was found that after some time of reaction, an intermediate carbonyl complex was formed. This complex seems to be a precursor phase for the formation of the amorphous alloy particles. When the particles start to form, they have a diameter of about 5 nm. Later, the particle size slowly increases to about 6.8 nm.

In a more recent study, Mørup et al.[7] have prepared similar particles with oleic acid as a surfactant. In this case Mössbauer studies showed that the particles were superparamagnetic above 80 K. Mössbauer spectra obtained at 161 K in various applied magnetic fields, B, are shown in Figure 3. At B = 0 the alloy particles yield a paramagnetic spectrum. The shoulder at about 2 mm s^{-1} is due to an Fe^{2+} impurity. When the magnetic field is applied, a substantial magnetic splitting of the alloy component appears. This result shows that the particles are superparamagnetic. From the field dependence of the magnetic hyperfine splitting a particle size of about 4.2 nm was estimated[7]. Thus the particle size appears to depend on the type of surfactant molecules. It was also found that the content of carbon in the particles depends on the type of surfactant molecules.

3. REDUCTION OF METAL IONS BY KBH$_4$

Metal ions can be reduced to the metallic state in aqueous solution by use of KBH$_4$ or NaBH$_4$. van Wonterghem et al. have shown that boron atoms from KBH$_4$ enter into the alloys

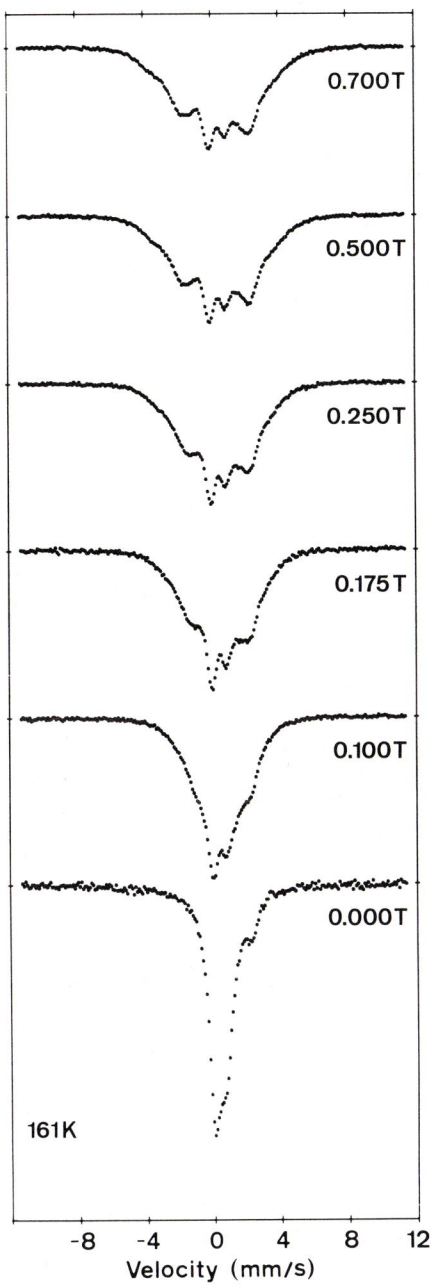

FIGURE 3
Mössbauer spectra of amorphous particles of Fe$_{100-x}$C$_x$ at 161 K in various applied magnetic fields.

and may stabilize an amorphous structure[2]. The method has been used to prepare amorphous alloys of Fe, Co, and Ni with various amounts of boron, e.g. $Fe_{62}B_{38}$[2], $Fe_{44}Co_{19}B_{37}$[2], $Fe_{37}Ni_{28}B_{34}$[2], and $Fe_{52}Ni_{37}B_{11}$[8]. Dragieva et al.[9,10] have reported a similar procedure using $NaBH_4$.

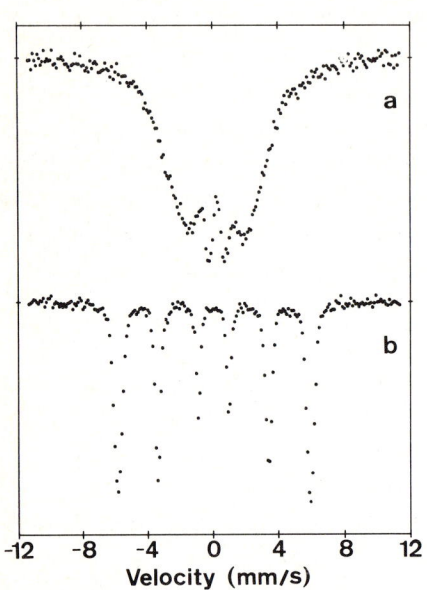

FIGURE 4
Mössbauer spectra of ultrafine particles of $Fe_{44}Co_{19}B_{37}$ obtained at room temperature. a: before annealing; b: after annealing at 725 K.

Figure 4 shows room-temperature Mössbauer spectra of $Fe_{44}Co_{19}B_{37}$ particles prepared by reduction by KBH_4. The spectrum obtained before annealing exhibits very broad lines, indicating a distribution in magnetic hyperfine fields. The isomer shift relative to α-Fe is 0.19 mm s^{-1}, and the quadrupole shift is negligible. The average magnetic hyperfine field is about 18 T, which is considerably smaller than those of crystalline Fe-Co alloys. The magnetic hyperfine field is also smaller than those of amorphous $(Fe_{1-x}Co_x)_{80}B_{20}$ alloys produced by the liquid quench technique[11], but is similar to those of amorphous Fe-B alloys with higher boron content[12]. The spectrum obtained after annealing at 725 K has relatively narrow lines, and the average magnetic hyperfine field is about 36.5 T, whereas the isomer shift has decreased to about 0.03 mm s^{-1}. These parameters are close to those of a crystalline Fe-Co alloy with an Fe:Co ratio of 7:3[13]. Thus the Mössbauer results indicate that amorphous Fe-Co-B particles are formed during the preparation and that these particles crystallize during annealing at 725 K.

Further evidence for this conclusion is obtained from the X-ray diffraction studies. Before annealing, the diffractograph showed only a very broad line, indicating the absence of crystalline phases. After annealing, a sharp diffraction line is observed, corresponding to a well-crystallized Fe-Co alloy[2].

The ultrafine amorphous particles prepared by reduction by KBH_4 had particle sizes in the range 10-100 nm.

4. CONCLUSIONS

The chemical methods for preparation of ultrafine particles of amorphous alloys seem to be applicable for production of alloys with a wide range of compositions and particle sizes. Particles prepared by these methods may have interesting applications in, for example, ferrofluids, magnetic devices, and catalysis.

ACKNOWLEDGMENT

Financial support from the Danish Technical Research Council is gratefully acknowledged.

REFERENCES

1) J. van Wonterghem, S. Mørup, S.W. Charles, S. Wells, and J. Villadsen, Phys. Rev. Lett. 55 (1985) 410.

2) J. van Wonterghem, S. Mørup, C.J.W. Koch, S.W. Charles, and S. Wells, Nature 322 (1966) 622.

3) J. van Wonterghem, S. Mørup, S.W. Charles, S. Wells, and J. Villadsen, Hyperfine Interactions 27 (1983) 333.

4) J. van Wonterghem, S. Mørup, S.W. Charles, and S. Wells, J. Colloid. Interface Sci. (in press).

5) S.W. Charles and J. Popplewell, in: Ferromagnetic Materials, Vol. 2, ed. E.P. Wohlfarth (North-Holland, Amsterdam, 1980) p. 509.

6) S. Bjarman and R. Wäppling, J. Magn. Magn. Mater. 40 (1983) 219.

7) S. Mørup, B.R. Christensen, J. van Wonterghem, M.B. Madsen, S.W. Charles, and S. Wells, J. Magn. Magn. Mater. (in press).

8) S. Mørup, J. van Wonterghem, A. Meagher, and C.J.W. Koch, IEEE Trans. Mag. (in press).

9) I.O. Dragieva, M. Slavcheva, and D.T. Buchkov, J. Less Common Met. 177 (1986) 311.

10) D. Buchkov, S. Nokolov, I. Dragieva, and M. Slavcheva, J. Magn. Magn. Mat. 62 (1986) 87.

11) S. Dey, P. Deppe, M. Rosenberg, F.E. Luborsky, and J.L. Walter, J. Appl. Phys. 52 (1981) 1805.

12) T. Nakajima, I. Nagami, and H. Ino, J. Mater. Sci. Lett. 5 (1986) 60.

13) C.E. Johnson, M.S. Ridout, and T.E. Cranshaw, Proc. Phys. Soc. 81 (1963) 1079.

LABORATORY SCALE PRODUCTION OF SOFT MAGNETIC WIDE AMORPHOUS TAPES : OPTIMUM PRODUCTION CONDITIONS FOR IMPROVED PROPERTIES.

A.R. YAVARI and P. DESRE

LTPCM, CNRS UA 29, Institut National Polytechnique de Grenoble, BP 75, D.U., 38402 Saint Martin d'Hères Cédex, France.

We discuss the importance of the introduction of the planar-flow casting technique in the technology of fabrication of metallic glasses. After discussing major materials considerations such as crucible and substrate compatibility with the liquid alloy, we show how in laboratory scale production, the glassy tapes are prepared under transient conditions which result in non-reproducable observations on as-quenched tapes. After annealing however magnetic and other properties of glassy tapes converge to reproducable values both along the length of each tape and for different tapes provided that they are fully amorphous.

1. INTRODUCTION

While metallic glasses were prepared by electrochemical or vapor-deposition before 1940's, development of metallic glasses by quenching the liquid alloy began in universities in the U.S.A. in the 1960's. The first metallic glasses prepared by liquid-quenching were obtained by the Duwez group at Cal-Tech[1]. Metal-metalloid deep eutectics provided the first easy glass formers. Early work on the thermodynamic and structural characteristics of metal-metalloid glasses was undertaken by David Turnbull of Harvard University and his students namely H.S. Chen, D.E. Polk and G.S. Cargill III. Most of the early metal-metal glasses were reported by the Giessen group at Northeastern University in the mid-1960's and later by Buschow (for a review see Wang[2]).

From a technological point of view, the most important contribution was made by M.C. Narasimhan of Allied Chemical[3]. Prior to his work, most metallic glasses were prepared by melt-spinning which can produce narrow ribbons of fairly irregular shape. In the planar-flow-casting (pfc) technique developed by Narasimhan, the close proximity of the molten metal crucible to the surface of a rotating quenching roller allows fabrication of regular shaped wide amorphous or microcrystalline tapes. This process which is the most widely used for industrial production of amorphous tapes is also useful on a laboratory scale. In this paper we discuss some pertinent aspects of such laboratory scale production of soft-magnetic Fe-B type amorphous tapes.

2. MELT-SPINNING AND PLANAR-FLOW-CASTING (pfc) :

In the melt-spinning process the molten metal is expelled from a small hole of a nozzle onto a moving substrate at 1 to some 10 mm distance. The flow rate out of the crucible depends on the ejection pressure, the height of the liquid column above the nozzle, the nozzle hole size, alloy viscosity etc and for a given set of such conditions the form of the solidifying ribbon depends on the total liquid jet velocity when it reaches the substrate after free-falling and on the substrate speed and these parameters can be related by the Bernoulli equations. However as the liquid jet emerges from the crucible with its various velocity components, it is cooled and deformed as its experiences the wind generated by the substrate motion and in particular if heterogeneities and oxidation are also present in the jet and the puddle of melt that forms on the substrate, velocity and viscosity gradients result in shape instabilities that grow and destabilise the jet and the ribbon is discontinued or perturbed. **Figure 1** shows successive pictures of a melt-spinning experiment with a liquid FeSiB alloy arriving on a rotating copper roller. One can see the onset and

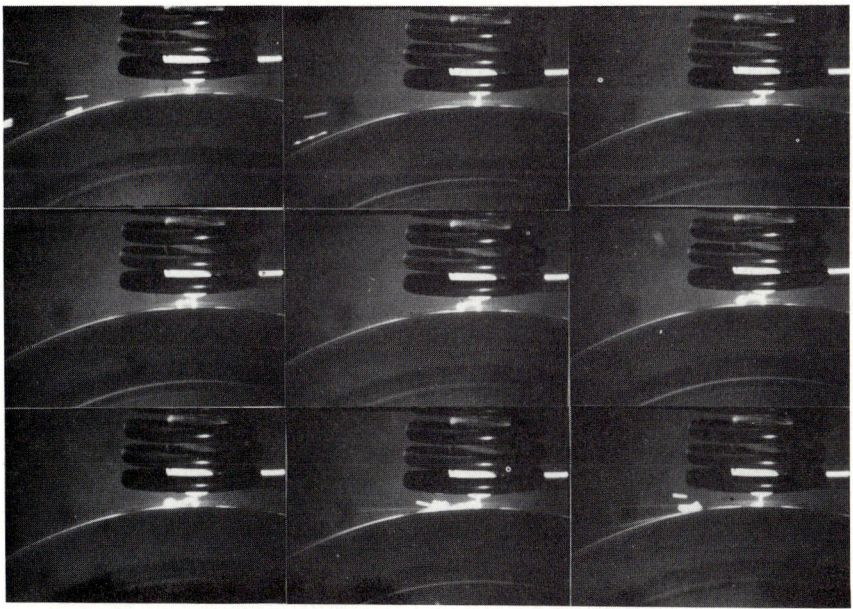

Figure 1 : Melt-spinning : successives images showing growth of a shape instability in the melt on the substrate.
(from left to right, top to bottom).

growth of a shape instability which eventually results in the breaking up of the jet and consequently the emerging solidified ribbon. One of the most important parameters in liquid quenching is the liquid viscosity. In fact in easy glass forming alloys, the liquid viscosity increases rapidly with decreasing temperature[4]. This influences the kinetics of nucleation and growth of crystalline phases[4] but also affets the kinetics of growth of instabilities in liquid jets as deformation rates in the melt depend directly on the viscosity. The jet is therefore more easily destabilised while its temperature is still high and its viscosity still low.

In planar-flow-casting (pfc), an appropriately shaped crucible nozzle is brought so close to the substrate as to eliminate the free-falling jet zone such that the liquid now flows directly from the nozzle onto the substrate. Here the liquid in which shape instabilities can grow rapidly is mechanically supported by the crucible walls brought into the zone where velocity and viscosity gradients are very high. **Figure 2** shows a laboratory scale pfc arrangement with a molten FeSiB alloy in a quartz crucible placed in close proximity of a rotating copper roller. The nozzle hole in melt-spinning is replaced by a slit of solidified width less than 1 mm. While the width of the solidified ribbon in melt-spinning depends on the jet velocity as it arrives on the substrate, pfc usually allows formation of regular shaped ribbons of width exactly equal to the nozzle slit length. The slit is places at a distance well below 1 mm from the substrate[3]. This distance is often of the order of 100 μm. If the liquid solidifies on the substrate before it leaves the nozzle behind, solid-solid rubbing between the alloy and the nozzle will damage the latter. This is avoided by controlling the liquid flow rate and its temperature in relation to the ejection pressure, substrate velocity and nozzle dimensions and its distance from the substrate. On the other hand if these conditions are such that the alloy is still completely liquid (with a low viscosity) beyond the

Figure 2 : Side views of crucible, substrate and incoming ribbon in melt-spinning (left) and planar flow casting (right).

nozzle, it will again be subjected to shape instabilities and a perfect tape is not obtained. Optimum operational conditions must be found for each combination of alloy, nozzle, substrate and ambiant gas.

3. ON THE CHOICE OF CRUCIBLE, ALLOY and SUBSTRATE COMPOSITION.

In pfc the crucible material and the liquid alloy should be such that the former is not wetted by the latter[3]. On the other hand it is best to maximise substrate wetting by the liquid alloy. If the alloy wets the crucible, instead of just flowing in between the nozzle and the substrate, the melt will climb onto and solidify on the nozzle's outer walls. Fortunately for laboratory experiments, quartz which has the added advantage of being transparent and not too expensive can be used as crucible material for Fe-B type melts. This condition does not hold for all glass-forming melts and for example, it is difficult to fabricate tapes of titanium-rich alloys by pfc using quartz nozzles. It is the same for other melts rich in elements that can reduce the quartz, particularly if they have high melting temperatures. The wetting of the substrate by the liquid alloy is less critical for the macroscopic shape of the tape but plays an important role on the microstructure of the ribbon side in contact with the substrate. Better wetting results in smaller sized of gas bubbles and increased metal-metal contact area which in turn increases the quench-rate. While wetting quality can usually be tested by measuring contact angles between a liquid droplet on a solid substrate, in the case of Fe-B type droplets on copper alloy substrates such static measurements are excluded as the alloys often melt at temperatures above those of the substrate. It can be shown however that wetting times are much shorter than solidification times when Fe-B type droplets are introduced on cold substrates[5]. One can therefore obtain contact angles under dynamic conditions before solidification[5] even when the substrate melts before the alloy. **Figure 3** shows two such observations for droplets of a FeSiBC alloy (Allied Chemical 2605-SC composition) on a stainless steel and a copper substrate under a vaccum of 10^{-5} torrs. Although solidification under high vaccum is unrelated to conditions of industrial production of wide tapes by pfc, similar observations can also be made for more practical conditions. The observations of fig.3 nevertheless indicate the importance of wetting conditions measured by the contact angle method because the accompanying SEM micrographs of the droplets'contact surfaces with the two substrate (also shown in fig.3) show clearly an improved microstructure with increased metal-metal contact in the case of good wetting. Fortunately for most Fe-B type alloys, wetting conditions for quenching on usual copper rollers in air, argon or helium gas are reasonably good (data not shown here) and further information is only necessary for optimisation of industrial production. It should be noted that this situation does not hold for all melts and for example platinum tapes usually cannot be formed on copper surfaces by pfc.

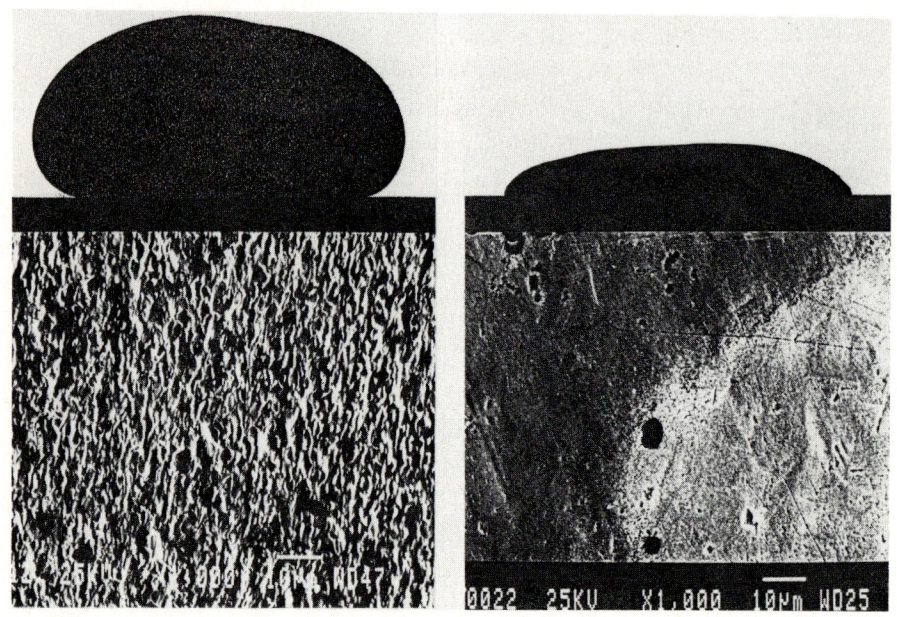

Figure 3 : Droplet form and microstructure of contact surface for FeSiBC droplets solidified on copper (left) and stainless steel (right) substrates under vacuum .

The alloy composition also limits the experimental conditions directly when a fully amorphous tapes is required. In particular when the composition is such that fully amorphous tapes are only obtained at very high quench rates, very thin ribbons must be prepared and when this critical thickness is below about 17 µm, tapes prepared by pfc usually contain small holes. In particular this critical thickness decreases with increasing Fe-content in or near the hypoeutectic regime in Fe-B type alloys. On the other hand these hypoeutectic alloys are interesting because of the reduction of thermal embrittlement with increasing Fe-content[6]. These are among the reasons why full optimisation of the production condition depend sensitively on the alloy composition in the FeSiB system. Near eutectic alloys are the most interesting because their soft-magnetic properties are among the best[7], their embrittlement tendency is reduced[6] and their glass-formation tendency enhanced. Figure 4 shows a FeSiB ternary phase diagram isotherm calculated for T = 1200 K[8]. It is seen that

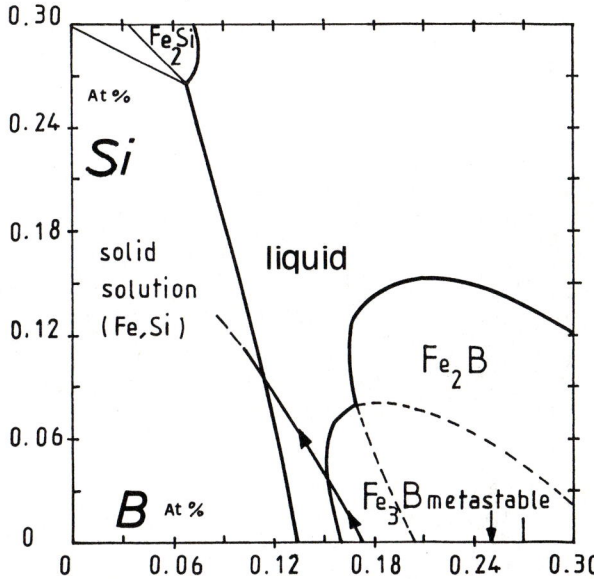

Figure 4 :
FeSiB ternary phase diagram isotherm calculated for 1200 K.

certain compositions are still liquid at these temperatures. The eutectic line is shown, with arrows. Apart from their easier glass-formation, low eutectic compositions also result in less wear and degradation of the crucible and the substrate due to lower pfc operational temperatures.

4. THE TRANSIENT NATURE OF CONDITIONS IN LABORATORY SCALE PLANAR-FLOW-CASTING.

In industrial casting of thin tapes, tape lengths from several hundred meters to tens of kms are produced. These tapes which require about 2 kg per km of length per cm of width are usually produced under steady-state conditions where the alloy and the substrate temperature and surface state are held constant although crucible wear may occur. The properties of such tapes vary little along their length. In laboratory scale pfc most researchers do not need to obtain more than several tens of meters of tapes of 1 to a few centimeters in width. For transition metal base amorphous alloys (~ 30 μm thick), this requires about 2 g per m of length per cm of width amounting to a melt weight of about 50 g. In any case, for a variety of reasons, accessories and added precautionary measures are required for production of lengths over about 50 m. Optimum quench rates require that the tapes adhere to the substrate for a minimum length of 20 to 50 cm especially for Fe-B type glassy tapes which tends to be brittle otherwise. On the other hand, the tape must come off the substrate before a full turn of the roller otherwise it will collide with the incoming melt under the nozzle. This puts a lower limit on the rotating substrate diameter which should be at least some 20 cm. For a substrate of

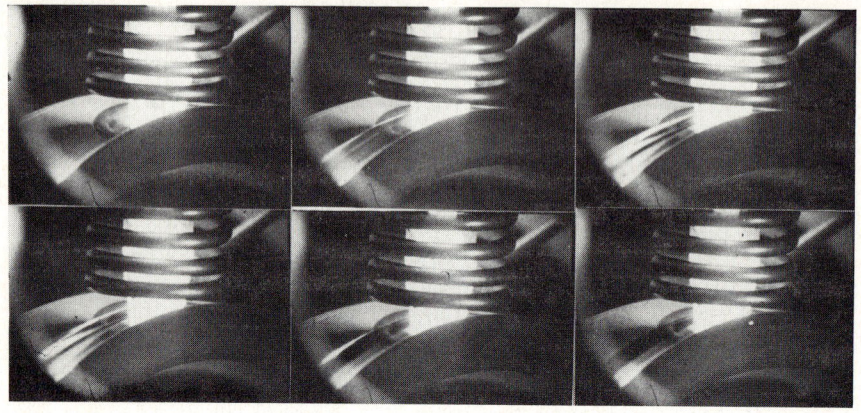

Figure 5 : Planar flow casting : pictures taken during casting of the first five meters of an FeSiB tape showing increasing adherence length for successive images (left to right, top to bottom).

30 cm in diameter, each point on the strip of the substrate wetted by the melt passes under the nozzle once for every meter of the cast tape and since the substrate is initially near room temperature, this results in progressive heating of the substrate until a steady state regime is established. Interestingly, this increase in the substrate surface temperature improves its wetting by the liquid alloy and increases the adherence length which is nearly zero at first and thus results in improved quench-rates for successively cast length segments. Increased quench-rates result in less relaxed amorphous states and high enthalpies ΔH_r of relaxation as measured by calorimetry. Bigot et al[9] measured ΔH_r for several segments along the length of a FeSiB amorphous tape prepared by pfc and found that for a copper roller, ΔH_r keeps increasing up to 50 m of tape length although the sharpest increases occur in the first 10 m or so. It turns out that these first 10 cm are often brittle as-quenched in the case of FeSiB. **Figure 5** shows successive pictures of a laboratory scale pfc experiment with an FeSiB tape being cast on a copper roller. The frame corresponds approximately to one shot per turn of the roller with a circumference of about 1 m. It is seen that the adherence length increases with time and the tape flys off the substrate increasing down stream. These observations are consistent with those of Bigot et al[9] although details often differ slightly for different machines.

The quench rate increases with the adherence length but it also increases with the substrate velocity. **Figure 6** shows SEM micrographs of the substrate-sides of two FeSiB tapes of equal thickness but cast respectively at 30 m/s and

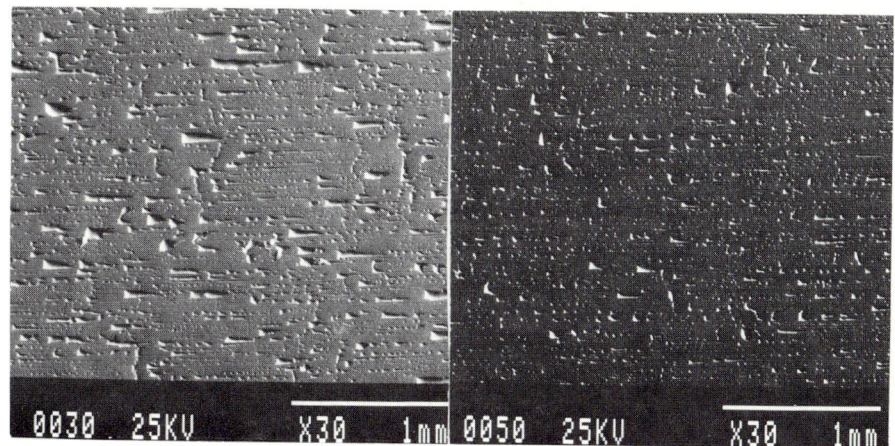

Figure 6 : SEM micrographs showing FeSiB tapes substrate side for casting at 30 m/s (left) and 50 m/s (right).

50 m/s substrate velocity. It is seen that the gas bubbles on the tape cast at a higher speed have smaller sizes thus allowing increased metal-metal contact. This increases the quench rate even on the free side of the tapes as evidenced by the X-ray patterns of **Figure 7** taken from the free sides of each tape. While the more slowly cast tape contains a significant crystalline fraction, this crystallinity is removed by increasing casting rates.

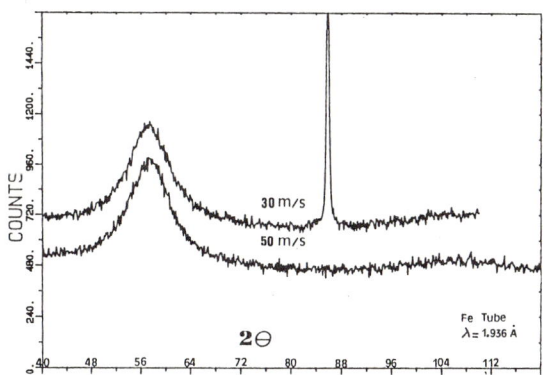

Figure 7 : X-ray diffraction patterns of free side of tapes of Figure 6.

Since the quench rate varies both along the tape length and from tape to tape, the question is raised as to whether it is meaningful to compare magnetic properties of different tapes of the same composition. **Figure 8** shows core loss measurements on an FeSiB 2605-S2 tape supplied by Allied Chemical and a tape of

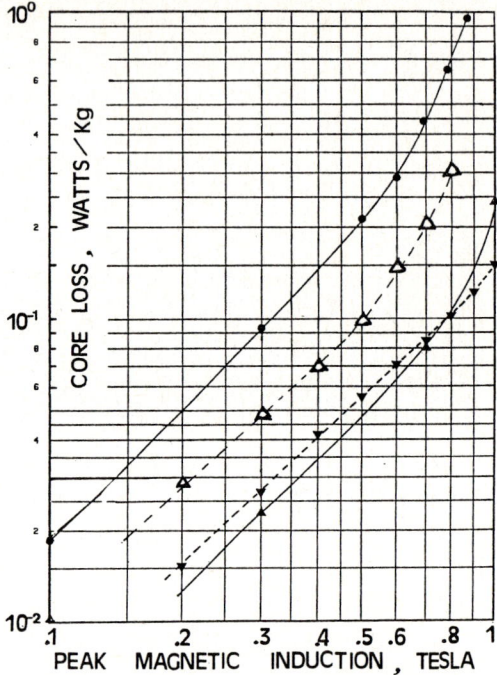

Figure 8 :
Core loss measurements at 60 Hz of Allied Chemical (△ ▲) and our more rapidly quenched (● ▼) FeSiB(S_2) tapes before (△ ●) and after (▲ ▼) 380°C annealing.

similar composition which we have prepared at a much higher casting speed[10]. It is seen that while the latter tape exhibits much higher losses in its as-quenched state, the losses are comparables for the two tapes after a 2 hour anneal at 653 K (380°C). Thus although most properties including magnetic properties vary for different tapes and tape segments of the same composition in their as-quenched states, the magnetic properties converge towards well defined values after relaxation annealing thus allowing comparison between properties of different amorphous compositions, thicknesses and thermomechanical treatments.

REFERENCES.
1) P. Duwez, R.H. Willens and W. Klement, J. appl. Phys. 31 (1960) 1136.
2) R. Wang, Bull. Alloy Phase Diagrams 2 (1981) 269.
3) M.C. Narasimhan, U.S. Patent n°4142571 (1979).
4) A.R Yavari, "Amorphous Metals and Non-Equilibrium Processing", Ed. Von Allman, Les Editions de Physique, (1984) 31-43.
5) A.R. Yavari, to be published.
6) A.R. Yavari, Proceedings, Sixth International Conference on Rapidly Quenched Metals, Montréal, August 1987.
7) N. Decristofaro, A. Freilich and G. Fish, J. Mat. Sci. 17 (1982) 2365.
8) P. Desré et al, to be published.
9) J. Bigot, M. Harmelin and P. Cremer, Proceedings LAM6, Z. Metallkde, in press, 1987.
10) A.R. Yavari, M. Barrue, M. Harmelin and M. Perron, submitted to the J. Mag. Mag. Mat. 1987.

SYNTHESIS OF AMORPHOUS MATERIALS BY MECHANICAL ALLOYING

W. KRAUSS*, C. POLITIS and J.R. THOMPSON**

Kernforschungszentrum Karlsruhe, INFP, Postfach 3640, D-7500 Karlsruhe, FRG
*Edmund Bühler, Postfach 1126, D-7454 Bodelshausen, FRG
**University of Tennessee, Dept. of Phys., Knoxville, TN 37996-1200, USA

Amorphous powders of Ti-Cu-Pd, Zr-Rh, Hf-Cu, and Hf-Ni were prepared by mechanical alloying (MA) in a high energy ball mill. The characterization of the amorphous powders and their formation was done by X-rays, raster-electron-microscopy and differential-thermal-analysis (DTA). The mechanically alloyed amorphous powders were compared with liquid quenched materials. We found that the MA-powders had partly significantly higher crystallization temperatures than the LQ foils. Furthermore, MA led to amorphous alloys for compositions there LQ-methods like splat-cooling or melt spinning failed.

1. INTRODUCTION

Until recently the preparation of amorphous materials for research and industrial applications are done by using rapid solidification techniques. In contrast to rapid solidification the newly developed process of mechanical alloying [1,2] requires no molten alloys for the synthesis of amorphous materials. We performed the mechanical alloying in high energy ball mills starting with elemental powder mixtures of appropriate compositions. During the intense grinding the powder particles are cold-welded, thinned and fractured again and again until a mixture on a very fine scale is reached. In many cases the mixing continues to an atomic level and alloys are formed. Depending on the energy state of the structures amorphous, microcrystalline, crystalline or mixed phases will be obtained. To demonstrate that the materials prepared by MA have the same structure like rapidly solidified materials we also performed liquid quenching by splat-cooling or melt spinning. Stimulated by earlier examinations done in the Nb-Ge and Nb-Ge-Al systems [3] we expected larger homogeneity ranges for amorphous materials prepared by MA than by LQ.

2. EXPERIMENTAL

The formation of amorphous materials was done by two methods, liquid quenching and mechanical alloying. The mechanical alloying, a form of repeated deformation processing, was carried out by a ball milling technique. The containers and balls, used for the alloying process, were made of either hardened steel or Co-bonded tungsten carbide. The materials to be alloyed, generally prepared from 99.9 at.% elemental powders, were processed in an Argon atmosphere and milled at ambient temperature for several hours.

The master alloys used for the liquid quenching experiments were prepared by repeated arc-melting in static Argon atmosphere, gettered with molten Zr. The liquid quenching was done with a commercially available two-piston splat-cooling and in a single-roller melt-spinning equipment [4] in atmosphere of 99.9999% Ar gas at a pressure of 500 mbar. The produced amorphous foils had a thickness between 20 and 50 μm. All materials prepared by LQ or MA were characterized by X-ray diffraction using Ni-filtered Cu K_α-radiation. The DTA-investigations were performed in static He-gas atmosphere with heating rates of 20 K/min using a Netzsch STA409 system.

3. EXPERIMENTAL RESULTS AND DISCUSSION

a. Amorphous alloys of Ti-Cu, Ti-Pd, and Ti-Cu-Pd

By mechanical alloying amorphous powders can be prepared in both Ti-Cu and Ti-Pd binary systems. For the Ti-Cu system, well-known as glass former by rapid solidification, we found a wide amorphous region, extending in composition from about 10 to 90 at.% Ti. The X-ray analysis, done on powders in reflection, yield broad and smooth diffractograms characteristic for the amorphous state. In Fig. 1 the X-ray diffractograms of both LQ and MA amorphous $Ti_{57}Cu_{43}$ are shown. By LQ amorphous alloys could be prepared in the range $Ti_{1-x}Cu_x$ $30 < x < 70$. The performed DTA measurements yield the same exothermic crystallization behaviour for MA and LQ amorphous $Ti_{1-x}Cu_x$.

In contrast with the above results, fully amorphous Ti-Pd powders only can be produced by mechanical alloying. The amorphization of a $Ti_{50}Pd_{50}$ material during milling is illustrated in Fig. 2. For these measurements a part of the total amount of powder was removed from the vial in an Ar-glove-box to prevent contamination of the milled powder by air. In the first diagram, indicated by 0 h, the spectra of elemental Ti and Pd are superimposed. The next curves, taken after a milling time of 4, 8, 12 and 17 h, show the transition from the crystalline to the amorphous state. During the first hours of processing, the intensity of the Ti lines decrease rapidly, while the Pd lines tend to persist longer, as seen by comparison of the spectra after 0 and 4 h milling. The strongest line at $2\Theta=40.1°$ is a superposition of the Pd(111) and Ti(101) reflection peaks. The first maximum of the amorphous phase is centered closely on this angle. After 17 h all crystalline powder is alloyed and converted to the amorphous phase, as seen in Fig. 2. For the fully amorphous alloy an effective particle size of about 2 nm was observed [5].

In order to contrast further the processes of mechanical alloying and liquid quenching, a series of ternary Ti-Pd-Cu alloys have been investigated. By LQ only in a small region near to the $Ti_{50}Cu_{50}$ composition amorphous ternary alloys could be synthesized. MA leads to the amorphous state in a wide region of the Ti-rich Ti-Pd-Cu phase diagram. The differences and the advantage of MA can be seen in Fig. 3. These results give primary support, that the mechanisms whereby amorphization occurs are different for liquid quenching and mechanical alloying. The observed formation of amorphous, crystalline and mixed semi-amorphous alloys in the Ti-Cu-Pd system indicate that the mechanical alloying process cannot be described by a liquid-quenching model in the dimensions of the contact area of 2 welded powder particles [5].

b. Amorphous Zr- and Hf-alloys

The amorphization of Zr-Rh, Hf-Cu and Hf-Ni powder mixtures during mechanical alloying and their crystallization behaviour was examined in respect of the fabrication of amorphous bodies. Amorphous alloys were prepared in the Zr-Rh system near the composition $Zr_{75}Rh_{25}$ by liquid quenching and mechanical alloying. Fig. 4 shows the DTA-curves of MA powder and a LQ splat. The crystallization temperature (T_c) is about 140 K higher for MA than for LQ materials. This difference in T_c was often observed between MA and LQ materials but not with such a high value. This difference, perhaps partially caused by small difference of the alloy compositions, indicates that a contamination of the amorphous Zr-Rh powder by O_2, which lowers T_c [1], can be excluded.

The X-ray diagrams of a mechanically alloyed $Hf_{60}Cu_{40}$ powder mixture for a milling time of 3, 7 and 12 h is plotted in Fig. 5. The Bragg-reflection peaks of pure Cu already disappeared after 3 h milling and only the broadened Hf-lines can be identified. After 12 h milling the X-ray diagram shows the curve typically for an amorphous alloy. For Hf-Ni alloys a little bit longer processing times (e.g. $Hf_{50}Ni_{50}$, 18 h) are necessary for reaching the amorphous state.

From this amorphous powders prepared by mechanical alloying nearly fully dense bodies were formed by cold-pressing, hot-pressing and hot-isostatic-pressing. Fig. 6 gives the micrographs of both a cold-worked and a HIP amorphous body of a $Hf_{35}Ni_{65}$ alloy. Up to now the best results by compacting amorphous alloys were obtained by hot-isostatic pressing below the crystallization temperature [6]. The showsn HIP-sample has a density of about 93%.

4. CONCLUSION

Amorphous alloys can be prepared by the newly developed process of mechanical alloying. This technique is more powerful in respect to the formation of amorphous alloys than the rapid solidification by splat-cooling or melt-spinning with cooling rates of some 10^6 K/sec. This is clearly shown by the results obtained with Ti-Pd alloys. Furthermore, MA-powders can be compacted to dense, amorphous bodies.

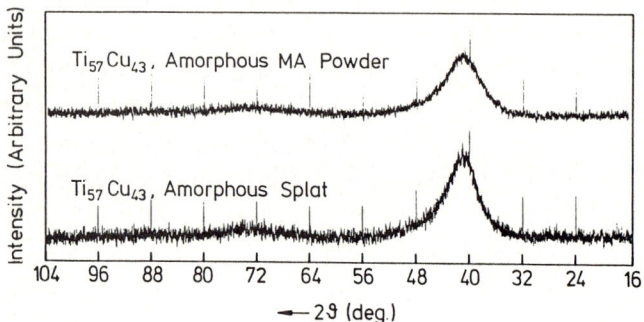

FIGURE 1
X-ray spectra of both mechanically alloyed and rapidly solidified $Ti_{57}Cu_{43}$.

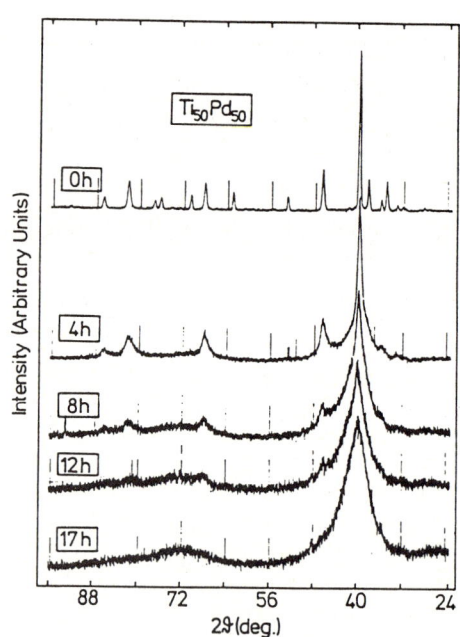

FIGURE 2
Formation of an amorphous $Ti_{50}Pd_{50}$ alloy by mechanical alloying in dependence of the milling time.

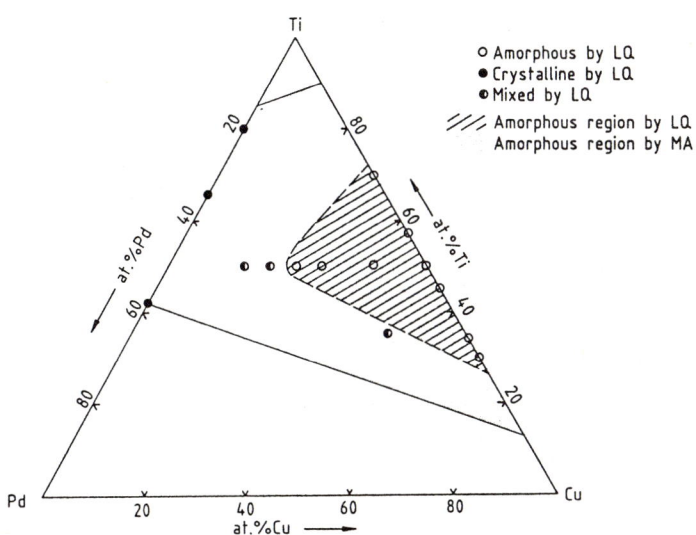

FIGURE 3
Ternary Ti-Cu-Pd diagram with the amorphous phase regions for MA and LQ.

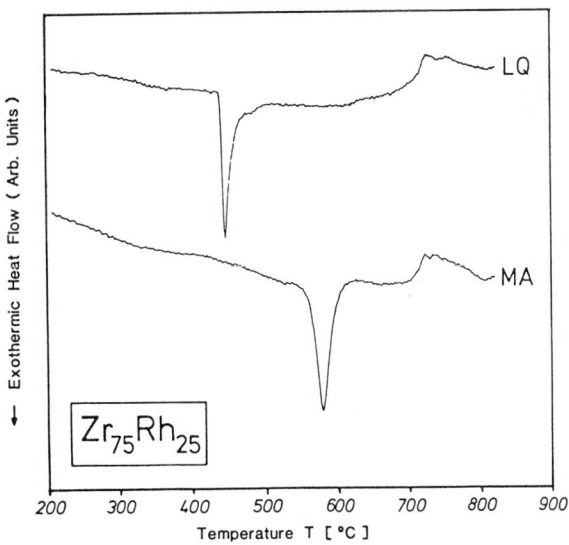

FIGURE 4
DTA-curves of mechanically alloyed and liquid quenched $Zr_{75}Rh_{25}$.

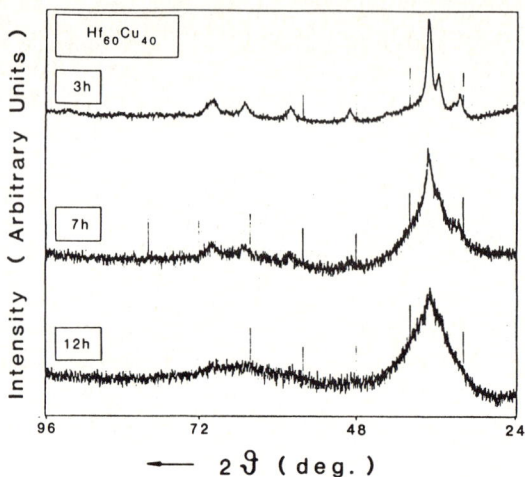

FIGURE 5
X-ray diagrams of a $Hf_{60}Cu_{40}$ powder mixture after 3, 7 and 12 h milling.

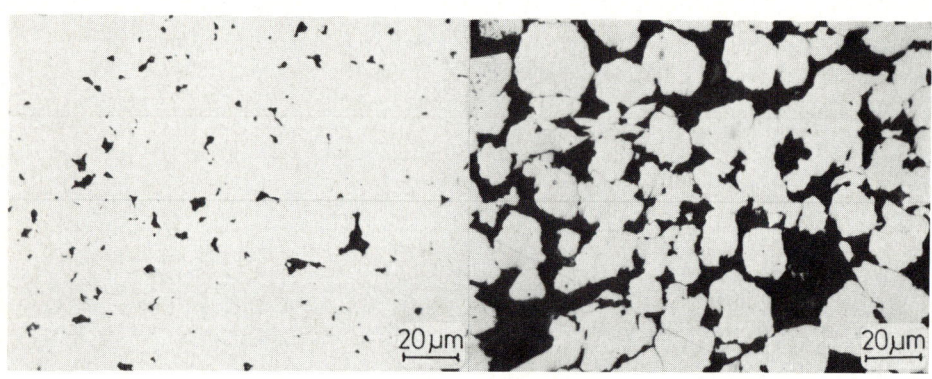

FIGURE 6
Micrographs of cold (left) and hot-isostatic-pressed (right) amorphous $Hf_{35}Ni_{65}$ samples with a pressure of about 2 and 1 kbar, respectively.

REFERENCES

1) C.C. Koch, O.B. Cavin, C.G. McKarmay and J.O. Scarbrough, Appl. Phys. Lett., 43 (1983) 1017.
2) C.Politis and W.L. Johnson, J. Appl. Phys., 60 (1986) 1147.
3) C. Politis, Physica 135B (1985) 286.
4) E. Bühler, Postfach 1126, D-7454 Bodelshausen, FRG.
5) J.R. Thompson and C. Politis, Europhys. Lett. 3 (1987) 199.
6) W.Krauss, H.L. Luo, C. Politis and P. Weimar, Verhandlungen der DPG, Münster, M-19.8 (1987).

WHAT IS WRONG WITH MAGNETIC DOMAINS IN AMORPHOUS THIN FILMS ?

Irena B. PUCHALSKA

Laboratoire de Magnétisme et d'Optique des Solides, C.N.R.S., 92195 Meudon France.

Typical magnetic domains observed in thin and thick, soft and hard amorphous magnetic layers are discussed. Thse structures have been investigated both for fundamental reasons and in view for use in the information storage. The question is raised - have the magnetic domains in amorphous films served this purpose ?

1. INTRODUCTION

The occurence of ferromagnetism was predicted by Gubanov[1] in 1960 and the first pictures of domain structure in thin amorphous films were observed by Mader and Novick[2] in 1965. They showed that 70Co - 30Au films 500 Å thick display ripple and cross-tie walls. In the same year Barglay and Turnbull[3] reported that amorphous Co-P alloys are also ferromagnetic.

The important activity in research of magnetic properties and magnetic domains in amorphous films started really in 1970-ties when Chaudari, Gambino and Cuomo[4] discovered cylindrical domains (bubbles) in Gd-Co amorphous films. This caused a stir in the world of magnetism and gave rise to much research in this field. The impetus for such studies was the original observation that some non-crystalline films exhibit a uniaxial magnetic anisotropy and support stripe domains and bubbles. At that time magnetic bubbles were generally believed to be the most suited structure for magnetic memory devices. The idea to replace epitaxial garnets by amorphous films was very exciting due to the relatively cheap technology and wide choice of alloy composition. The absence of grain boundaries in these materials results in high magnetic susceptibility and low coercivity. On the other hand it has been found that some amorphous compositions display high coercivity ; these materials could be a good candidats for magnetic and magnetooptical recording.

A measur of interest in the magnetic properties of amorphous films is given by a number of sessions and invited and contributed papers presented in 1970-ties and 1980-ties at international conferences on magnetism. New materials had appeared and many promising results were published.

The properties and type of domain and domain walls depend on magnetisation, thickness of the specimen, anisotropy and stresses. Many of these parameters can be induced or changed by deposition process. In consequence it is possible to dispose of a large variety of amorphous thin films with different properties[5]. By choice of composition and type of deposition technique (sputtering, evaporation, chemical deposition) it is possible to obtain the soft materials which can be used for memory devices, flux guidance (circuit element and magnetic heads) and hard materials for different types of recording. In all these applications the information carriers are magnetic domains, magnetic walls or singularities (lines) in the walls.

This paper will deal with some typical domains observed in soft and hard amorphous films.

2. THIN FILMS

Domains which appear in amorphous films are in general similar to those observed in the crystalline layers. Fig. 1 shows Neel walls, cross-tie and magnetisation ripple in the amorphous Fe B film. Ripples observed in thin polycrystalline films were interpreted[6] to be due to local anisotropy supperposed with the magnetic crystalline anisotropy ; ripple wavelengh is written by $\lambda = \dfrac{A^{1/2}}{K_u H/H_k(\alpha)^{1/2}}$ where A is the exchange constant, α angle between easy axis and field H, Ku anisotropy constant. Initially the observation of ripple in the amorphous films was surprising . After Kronmüller [7] this structure is due to some sort of spin dislocations and strain field coupled to magnetic properties through the magnetoelastic and elastic coefficients. Magnetoelastic coupling energy is expressed by $\varphi m = -\dfrac{3}{2} \lambda s \Sigma_{ij} \delta_{ji} \gamma_i \gamma_j$, where λ_s is isotropic magnetostriction coefficient, and δ and γ elastic coefficients. Ripples play an important role in many magnetic properties like magnetisation process, histeressis, susceptibility and their existence is unfavorable for applications.

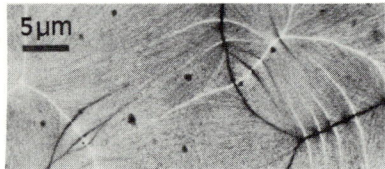

FIG. 1 - Domain structure in amorphous FeB sputtered film. T = 760 Å. Courtesy of S. Schwarzl, F. Stobiecki, H. Hoffmann.

3. MAGNETIC DOMAINS IN SOFT MATERIALS WITH PERPENDICULAR ANISOTROPY

Materials which support bubbles are characterised by an important quality factor $Q = \dfrac{K_\perp}{2\pi M_s^2} > 1$ where $K_\perp$ is perpendicular anisotropy constant and Ms saturation magnetisation.

There have been many discussions concerning the origin of perpendicular anisotropy in atomically disorded but magnetically ordered films. For instance in GdCo alloys the extent of atomic order is less than 25 Å therefore beyond this distance and in the range of the thickness of typical film - the long range order does not exist[4]. In consequence perpendicular anisotropy in amorphous films does not originate from magnetocrystalline anisotropy but from stress induced anisotropy, pair ordering and shape anisotropy (columns).

Discovery of bubbles in amorphous Gd-Co and Gd-Fe films[4] were very promising for applications. It was shown that amorphous rare earth transition metals (R-T) films may provide very small bubbles (e.g.[8,9] and Fig. 2) which garnet cannot provide. It was also shown that sputtered amorphous films such a GdCo, GdFe and also Gd Co Mo (Fig.3) have a number of avantages such as wide range of magnetic properties and an inexpensive technology. These compositions were widely investigated (e.g.[10]) with the view of their use for bubbles devices.

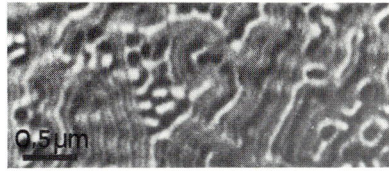

FIG. 2 - Stripe domains and bubbles in 20 Gd 80 Co film. T = 500 Å (Fot. S. Hamzaoui)

FIG. 3 - Stripe domains in amorphous Gd Co M film. (Mo 14 at %). Courtesy of T. Katayama

Some other alloys like chemically deposited CoP and sputtered Ho-Co show perpendicular anisotropy. Cargil, Gambino and Cuomo[11] observed the stripe domains in 83Co 17P layers less than 6 um thick. In similar composition but in thicker films (T = 20 - 30 μm) Puchalska and Sadoc[12] found very mobile circular (bubble-like) domains (Fig. 4) which originated from zig-zag bands. Some important investigations on domains structure and magnetic properties in CoP were reported also by Dietz and Hünseler[13] and by Sanches-Trujillo, M. Riveiro, G. Rivers and Hernando[14,15,16].

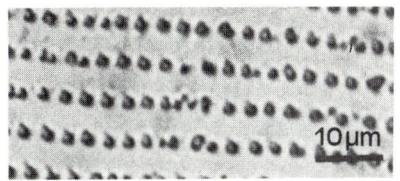
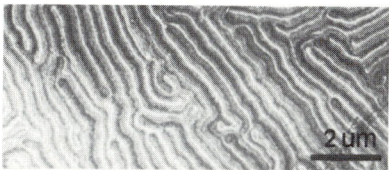

FIG. 4 - Circular domains in 83 Co 17 P film. T = 23,5 μm

FIG. 5 - Domain structure in 20 Ho 80 Co film. T = 1500 Å. Courtesy of T. Suzuki

Another interesting amorphous material which displays perpendicular anisotropy and stripe domains is the HoCo alloy[17,18] (Fig. 5). In the HoCo films bubbles appear only sporadically since the quality factor is not very important. For this reason the domain wall consists of many singularities - Neel Lines (often called "Bloch Lines") which are clearly visible in Fig. 6 and 7. After Suzuki[17] the distance between the two lines is from 100 to a few hundred Å which is in agreement with Hubert's calculations[19].

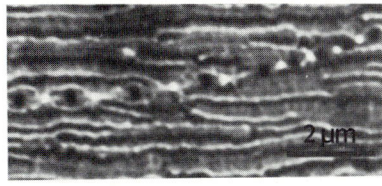
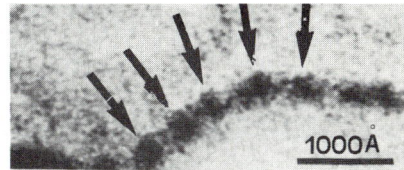

FIG. 6 - Stripes and singularities in Bloch walls in 20 Ho 80 Co film. T = 1500 Å

FIG. 7 - Singularities in Bloch wall in 19 Ho 81 Co. T = 1200 Å. Courtesy of T. Suzuki

3. DOMAINS IN THE SOFT FILMS WITH AN IN-PLANE ANISOTROPY

In thin magnetic samples with in-plane anisotropy zig-zag domain boundaries occur as transition boundaries between two regions of head - on head magnetisation. The zig-zag band can be nucleated in controlled way[20] by applying a local magnetic in-plane field opposite to direction of magnetisation saturation of the sample. The zig-zag band can be split into separated lozenge-type domains (figs. 8 and 9) in a higher in-plane field. It was shown[20,21] that in Ni Co P films these two structures can easily propagate under the influence of transverse field H_t.

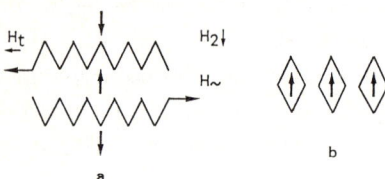 

FIG. 8 - Schematic representation of zig zag band and lozenge domains

FIG. 9 - Kerr effect of zig zag band

In a transverse field H_t superposed on a local ac field (Fig. 8) applied along the easy axis, the zig-zag walls move in a characteristic way : the walls with positive charges run parallel and in the same sense that H_t while the opposite zig-zags run in the opposite direction (Fig. 8). There exists two phases of the motion. In the phase one the velocity of zig-zags increases and then decreases to zero (Fig. 10). In some value of transverse field the walls are no longer mobile ("dead gap"). When the field reaches the second threshold value the zig-zag motion reapears in the same direction as before (phase two).

This unusual transverse right/left motion in two phases was found[20] to be due to effective bias (transverse) field, variable coercivity of the walls and transformation of the asymmetric Bloch walls into asymmetric Neel walls in transverse field (see Fig. 11).

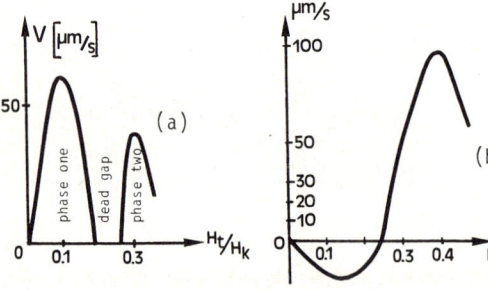

FIG. 10 - Velocity of zig zags (a) and lozenge - type domains (b) versus transverse field Ht

Small lozenge-type domains may propagate in the hard direction in a two wires simple system. The motion of this domain has been discussed in detail previously[21] and is shown in Fig.12. In the case of separated domains there exist two phases of the motion, however in the second phase domains propagate in opposite direction (see Fig. 10b). This difference between the zig-zags and isolated domains is not clear.

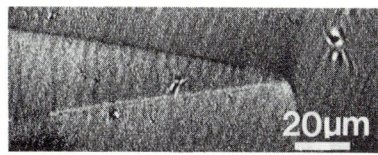

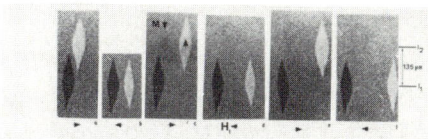

FIG. 11 - Neel segments of zig zag wall

FIG. 12 - Kerr effect of lozenge type domain displacement a hard direction

4. DOMAINS IN HARD MATERIALS

Thin films with an important coercivity (Hc ~ 1000 Oe) are used for magnetic and magnetooptical recording. There exist two types of magnetic recording : perpendicular and longitudinal recording.

Currently the sputtered CoCr films are considered as the most appropriate materials for perpendicular recording. Such films are composed of hexagonal CoCr columns, each of which separated from its neighbouring columns by Cr-rich non-magnetic layers. Such an "architecture" of magnetic films gives rise to an important perpendicular anisotropy. The constant of perpendicular anisotropy K strongly depends on the column-diameter and distance between the columns. It follows that depositing the films by oblique incidence leads to better developped columns. The second positive aspect of the oblique incidence technique is the induction of a small in-plane anisotropy in the sample which causes the domains to be well aligned.

It was been suggested [22] that the obliquely deposited CoAu amorphous films might be a suitable candidates for a magnetic recording medium. Several compositions of amorphous CoAu system display a columnar microstructure and stripe domains. For instance the 72,2Co 27,8Au film deposited at $\beta = 60°$ angle of incidence has quite important quality factor $Q = 0,696$ and displays regular stripe domains (Fig. 13).

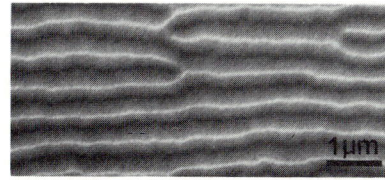

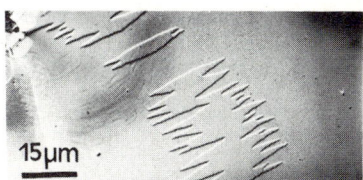

FIG. 13 - Stripe domains in 72,2 Co 27,8 Au film deposited at $\beta = 60°$ T = 2400 Å

FIG. 14 - Zig zags in thin Fe Gd film deposited at $\beta = 65°$ angle of incidence

Materials used for the longitudinal recording are the polycrystalline films of e.g. NiCo, with the in-plane anisotropy and high coercivity. In such films the transition between two domains are zig-zag walls. This structure is not optimal for high density recording ; a sharp transition would be more advantageous. Unfortunately the straight walls do not exist in the high coercivity thin films. The crucial problem for the longitudinal recording is the noise which can be due to the combination of the following : zig-zag fluctuation, the zig-zag structure itself which is quite complex and the ripples. However it is known[23,24] that some hard amorphous thin films (e.g. CoP or FeGd) deposited at important angle of incidence do not display ripple structure (see Fig. 14).

5. BLOCH LINE MEMORY

At the present time the tendency is to use the singularities in the Bloch wall as the information carriers (so called Bloch Lines Memories). This last approach is hoped to lead to devices of 1.6 G bit/cm^2 [25]. The required materials for the Bloch Lines Memories are magnetic layers with important perpendicular anisotropy. Amorphous magnetic layers like 20Gd 80Co or perhaps 19Ho 81Co might serve this purpose but for the time being the bubble epitaxial garnets (e.g. (Y Sm Lu Ca) (Fe Ge)$_5$ O$_{12}$) are used for investigations and for the first prototype.

6. CONCLUSION

It was shown that amorphous films may provide a large variety of magnetic domains. GdCo and GdCoMo films display small bubbles that the garnet cannot provide. But for some reason these films were never used for bubble devices.

We have discussed the domains in soft materials : bubbles, bubble - like domains, zig-zag walls, lozenge-type domains, singularities in the walls, cross-tie, Neel lines, and in hard materials : stripes and flat domains. All these structures should be a good candidates either for magnetic recording or for memory devices. However amorphous films are hardly ever used for these applications.

What is wrong with magnetic domains in amorphous thin films ?

REFERENCES

1) A.I. Gubanov, Fiz Tverd. Tela 2 (1960) 502

2) S. Mader and A.S. Novick, Appl. Phys. Letters 7 (1965) 57

3) B.G. Barglay and D. Turnbull, Bull. Am. Phys. Soc. 10 (1965) 1101

4) P. Chaudari, R.J. Gambino and J.J. Cuomo, IBM J, Res. Develop. 17 (1973) 68

5) P. Grundy, J. Mag. Mag. Materials 21 (1980) 1

6) H. Hoffman, J. Appl. Phys. 35 (1964) 1790

7) H. Kronmüller, IEEE Trans. on Mag. Vol. Mag. 15,5 (1979) 1218

8) S. Herd and P. Chaudari, Phys. Stat. Sol. 18 (1973) 603

9) S. Hamzaoui, Thesis University Paris Sud France (1985)

10) T. Katayama, Microelectronics Journal 12, 5 (1981) 23

11) G.S. Cargil III, R.J. Gambino, J.J. Cuomo, IEEE Trans. Mag. Mag. 10 (1974) 803

12) I.B. Puchalska and F. Sadoc, J. Appl. Phys. 47 (1976) 333

13) G. Dietz and A. Hünsemer, J. Mag. Mag. Materials 6 (1977) 68

14) J.M. Riveiro, M.C. Sanches-Trujillo, IEEE on Mag. Mag. 16, 6 (1980) 1426

15) R. Riveiro, G. Riveiro, M.C. Sanches-Trujillo, J. Mag. Mag. Materials, 31-34 (1983) 1551.

16) J.M. Riveiro, A. Hernando, Phys. Rev. B 32, 8 (1985) 5102.

17) Takao, Suzuki, Jap. J. Appl. Phys. 20 (1981) 2079.

18) S.S. Nandra and P.J. Grundy, Phys. Stat. Sol (a) 41 (1977) 65.

19) A. Hubert, AIP Conf. Proc. No 18 (1973) 178.

20) M. Labrune, S. Hamzaoui and I.B. Puchalska, J. Mag. mag. Materials 60 (1936) 243.

21) M. Labrune, I.B. Puchalska and A. Hubert, J. Mag. Mag. Materials 61 (1986) 321
    M. Labrune and I.B. Puchalska, J. Appl. Phys. 61 (1987) 4213.

22) A. Ovadia, I.B. Puchalska and J.P. Jakubovics, J. Mag. Mag. Materials, 54-57 (1986) 851.

23) I.B. Puchalska, J. Appl. Phys. 50 (1979) 2242.

24) M. Labrune, S. Hamzaoui and I.B. Puchalska J. Mag. Mag. Materials 27 (1982) 323.

25) S. Konishi, IEEE Trans. Mag. Mag. 19 (1983) 1838.

EFFECT OF BOTH THE CONVENTIONAL HEAT TREATMENT AND THE MAGNETIC ANNEALING ON MAGNETOELASTIC COUPLING COEFFICIENT OF AMORPHOUS METAL RIBBONS

Luciano LANOTTE and Flavio PORRECA

Dipartimento di Fisica Nucleare, Struttura della Materia e Fisica Applicata, Piazzale Tecchio 80, 80125 Napoli, Italy+

A new method for calculating the maximum value of the magnetomechanical factor, K_{max}, as a function of the Young modulus in the demagnetized state, E_o, the longitudinal saturation magnetostriction, λ_L, and the magnetic anisotropy energy, k, is applied in ribbon-shaped samples of $Fe_{67} Co_{18} B_{14} Si_1$ Metglas (100x2x0.05 mm^3).
The K_{max} values obtained after conventional heat treatment (CHT) differ from thoose measured after magnetic annealing (MA) in a transversal field, because in the second case a magnetic anisotropy k_{DSRO} is induced by directional short range order (DSRO). Therefore it is also possible to evaluate k_{DSRO} by means of K_{max}, Eo and λ_L measurements.
For more extensive tractation of both the theory and the experimental results concerning this subject see ref. 2.

1. EXPERIMENTAL

In the CHT the temperature of the samples has been rised by steps of 15°C/min. The specimens ha-e been kept in He inert atmosphere and heated for 60 mins to prefixed temperature T=150, 250, 300 and 350 °C. At the last temperature the treatment were performed for a time "t" equal to 2 and 3 hours too. A cooling rate of 20°C/min has been operated. All temperatures were measured with accuracy of ± 0.2 °C.

Also in the MA experiments the sample heating went on for 60 mins, while the increments of temperature were of 10°C/min and the cooling was performed more quickly (40°C/min) in order to congeale the induced anisotropy. Also in this case other treatments at 350°C for 2 and 3 hours were made.
The technique of magnetoelastic waves resonance was used in order to measure the Young modulus[1].

+ Gruppo Nazionale Struttura della Materia del C.N.R. e Consorzio Interuniversitario Struttura della Materia del M.P.I.

We used the classical fluxmetric technique for determining the magnetization curves of the treated samples, both free and under tensile stress, for the evaluation of magnetic anisotropy energy and magnetostriction.

2. THEORETICAL.

Since linear magnetization M_L (H) was observed, the hypothesis of uniaxial anisotropy was used and the Young modulus at constant H was obtained:[2]

$$1/E_H = 1/E_M + \lambda_L^2 H^2 /(4 M_S H_a) \qquad (1)$$

where E_M and E_H are the linear elasticity modului at constant M and H, respectively, M_S is the saturation magnetization and H_a is the anisotropy field. Taking into account that $k = M_S H_a /2 = -\lambda_L \sigma_i'$ where σ_i' is the residual internal stress (average) from (1) it results:

$$K_{max} = \frac{\Delta E^{\frac{1}{2}}}{\Delta E_M^{\frac{1}{2}}} = \left(\frac{-E_M M_S^2 H^2}{2 \lambda_L \sigma_i'^3}\right)^{\frac{1}{2}} = \left(\frac{-2 \lambda_L E_H}{\sigma_i'}\right)^{\frac{1}{2}} = \left(\frac{2\lambda_L^2 E_H}{k}\right)^{\frac{1}{2}} \qquad (2)$$

After MA the magnetic anisotropy is due to both magnetostrictive contribute, k_{MS}, and to DSRO, but one can describe directional short range ordering effects by means of an "apparent" internal stress σ_i^+ : $k_{MS} + k_{DSRO} = -\lambda_L \sigma_i^+$

Therefore, with obvious means of the simbols, we assert:

$$k_{CHT} = -(\lambda_L \sigma_i')_{CHT} \quad ; \quad k_{MA} = k_{CHT} + k_{DSRO} = -(\lambda_L \sigma_i^+)_{MA}$$

namely

$$k'_{DSRO} = (\lambda_L |\sigma_i^+|)_{MA} - (\lambda_L |\sigma_i'|)_{CHT} \qquad (3)$$

On the other hand from relation (2)

$$k = 2 \lambda_L^2 E_o / K_{max}^2 \qquad (E_H = E_o)$$

so that another possible evaluation of k_{DSRO} can be obtained by the equation:

$$k''_{DSRO} = 2\left[\left(\lambda_L^2 E_o / K_{max}^2\right)_{MA} - \left(\lambda_L^2 E_o / K_{max}^2\right)_{CHT}\right] \qquad (4)$$

3. RESULTS AND DISCUSSION

It has been made the comparison between the experimental value of K_{max}, namely $K_e = ((E_S - E_o)/E_S)^{\frac{1}{2}}$, obtained from direct determination of the Young modulus, and the semitheoretical ones $K_{max_{MA}}$ and $K_{max_{CHT}}$ calculeted from relation (2) by measurements of $E_o = E_H$, λ_L, σ_i' and k (fig.1) (we retain

$$\lambda_L = \lim_{\Delta\sigma \to 0} \sum_0^{M_S} \Delta M / \Delta\sigma \quad ; \quad \sigma_i = 1/\lambda_L \sum_0^{M_S} H \cdot \Delta M$$

The low differences (4%), between K_e and K_{max} values, support the validity of the method, for the evaluation of the directional ordering anisotropy (relation (4)) explained at the hand of the preceeding section.

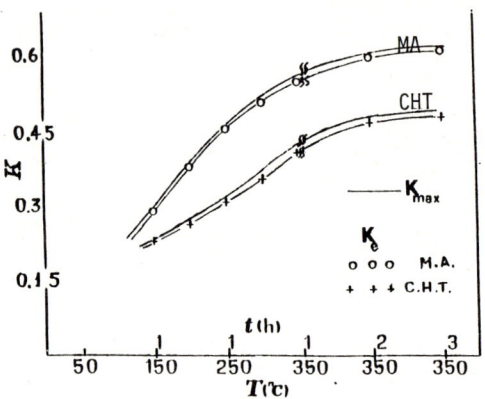

FIGURE 1

The experimental "K_e" and semitheoretical "K_{max}" behaviour of magnetoelastic coupling factor after MA and CHT. T= treatment temperature; t= treatment time.

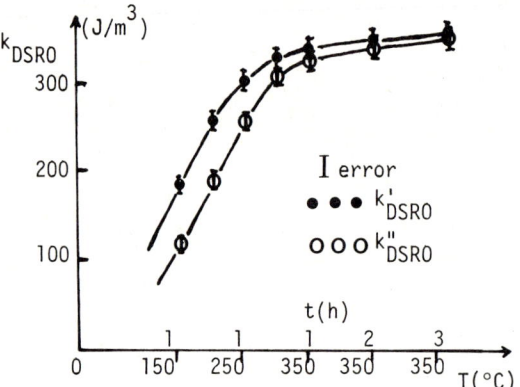

FIGURE 2

Directional short range ordering induced by MA. T= treatment temperature; t= treatment time

In fig. 2 we reported the curves of k_{DSRO} measured by means of the classical evaluation k'_{DSRO} and also by means of the expression deduced by us k''_{DSRO}.

When, established the temperature at 350°C, the treatment time is increased over 1 h, the curves of k'_{DSRO} and k''_{DSRO} become more and more coincident (in the limit of the experimental errors). Therefore the directional short range ordering gives rise to an induced anisotropy that can be evaluated in an original manner by means of the relation

$$k''_{DSRO} = 2\left[(\lambda_L^2 E_o / K_e^2)_{MA} - (\lambda_L^2 E_o / K_e^2)_{CHT}\right] \qquad (5)$$

when the MA effects have reached the saturation.

REFERENCES

1) L. Lanotte, C. Luponio and F. Porreca, J. Appl. Phys. 54 (1983) 375.
2) L. Lanotte, C. Luponio and F. Porreca, Il Nuovo Cimento D (1987) in course of publication.

STUDY OF THE PREPARATION DEPENDENCE OF THE QUALITY OF DIFFERENT SOFT MAGNETIC AMORPHOUS RIBBONS

Herbert SASSIK and Rudolf WEZULEK
Institute for Experimental Physics, Technical University Vienna, A-1040 Karlsplatz 13, Vienna, Austria

In the last years the interest in rapid solidification has been increasing and was subject of extensive studies. The primary consideration is to obtain a high cooling rate. Its control is essential to maximise the advantages of this process. The cooling rate can be measured directly by pyrometer[1], photocaloric instrumention[2] or indirectly by examination of different properties, which depend on cooling rate (microstructure, grain size, dendrite structure[3]). Surface quality of amorphous ribbons[4] and hysteresis loop and magnetostriction[5] are also usefully properties reflecting changes of the cooling process. The planar flow casting should be stable and support us with reproducable homogenous, well definable wide ribbons.

Amorphous ribbons of Co-Fe-B, Co-Si-B, Ni-Si-B were obtained by planar flow casting under atmospheres of He, Ar or vacuum on a CuZr-wheel ($\emptyset$ = 200 mm, $20 \leqslant v \leqslant 30$ m/s) in slotted quartz nozzles (10 x 0,5 mm^2) with an injection pressure 100 - 300 mbar. An accurate adjustment of the nozzle kept the slot properly positioned with respect to the wheel. A lift-off and a gaswind rejection to protect the melt puddle were used. The wheel was balanced to a few milligrams and highly polished with a height change of some micrometer. Selected results of optical micrographs are presented in fig. 1 for $Ni_{60}B_{40}$. The surface quality is improved drastically by reducing the surrounding gas pressure and by application of He gas. The free surface becomes smooth very early (1000 to 500 mbar) because the gas pockets are not forced under the ribbons, followed by disappearing gas pockets on the wheel side (500 to 10^{-3} mbar). At higher vacuum, both sides of the ribbons are equal, with constant cross section, smooth and shiny without any serrated edges. The change of thickness is only a few µm within one circumference of the wheel (620 mm), but the ribbons were only partially amorphous especially on the free side.

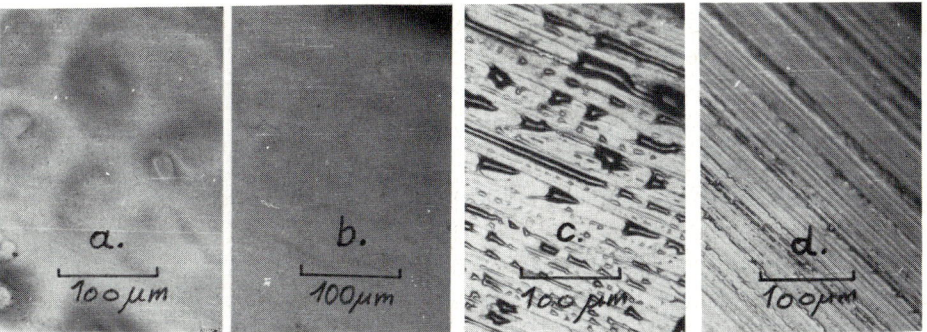

Figure 1: a) on air, free side; b) 500 bmar He, free side;
c) 500 bmar He, wheel side; d) 10^{-1} mbar He, wheel side

By measuring the stress dependent hysteresis loop (Details 5, $H_{max} \leq$ 5 Oe $\simeq$ 400 A/m, 10 Hz, $\sigma \leq$ 400 mPA) H_c is obtained. The magnetostriction constant λ_s was detected by a small angle magnetization rotation method[6]. Selected results are presented for $Co_{80-x}Fe_xB_{20}$ ($0 \leq x \leq 6$). At $x = 6$ λ_s and at $x = 5$ H_c are a minimum respectively. The effect of reduced He-atmosphere is demonstrated by the following results: casting on air led to ribbons[7] of $Co_{80}B_{20}$ and $Fe_{80}B_{20}$ with a pronounced scattering of H_c along the ribbon length ($\pm$ 0,015 Oe and $\pm$ 0,005 Oe). Fig.2 demonstrates for ribbons of this work H_c along the ribbon axis. H_c is reduced from 0.05 to 0.015 Oe by substituting Co with Fe; ΔH_c changes from $\pm$ 0.005 to $\pm$ 0.001. The short distance scattering is much smaller. For $x = 6$ however the ribbon shows a very pronounced scatter due to changes in the nozzle-wheel distance, what is content of further examinations[8]. In every case the stress dependence of H_c did not reveal any stresses along the ribbons. λ_s measurements were performed, $-5,6 \cdot 10^{-6} \leq \lambda_s \leq + 0,15 \cdot 10^{-6}$ for $0 \leq x \leq 6$, independent of external stress, because in principle λ_s is an interatomar property in contrast to H_c[8]. Along the length λ_s is absolutely constant. Both measurements cannot be falsified by a wrong detection of the cross section, because this cancels out. Only effects of thickness due to domain structure ordering might be existent[9].

Summarizing it was possible to find an experimental realisation for optimal process parameters, which allow to produce soft magnetic amorphous ribbons, which are unstressed, homogeneous and reproducable within a few percent without annealing.

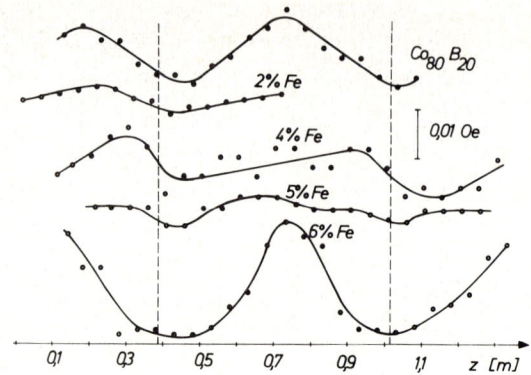

Figure 2: Change of H_c along the ribbon length, $0.05 \geq H_c \geq 0.015$ for $0 \leq x \leq 6$

ACKNOWLEDGEMENT

This work was supported by Austrian Science Fonds P 5020. We gratefully acknowledge the fruitful discussion with R. Grössinger.

REFERENCES

1) E. Vogt, G. Frommeyer, Proc. Rapidly Solidified Materials San Diego, USA, eds. P.W. Lee, R.S. Carbonara (ASM, Fe. 1986) 291

2) A.G. Gillen, B. Cantor, Acta Metall 33 10 (1985) 1813

3) J. Wittig, G. Frommeyer, E. Vogt in 1) 273

4) S. Huang, H. Fiedler, Mat. Science and Eng. 51 (1981) 39

5) R. Grössinger, H. Sassik, A. Lovas JMMM 41 (1984) 107 and Proc. MRS Conf. Strassbourg, Ed.M.v. Allmen (Ed. de Physique 1984)

6) K. Narita, J. Yamasaki, H. Fukunaga IEEE Mag. 16 (1980) 435

7) R. Grössinger, F. Haslinger, H. Sassik, DPG Spring Meeting Freudenstadt FRG (1986) AM 95

8) R. Grössinger, A. Pönninger, G. Herzer, this Conference

9) A. Veider, G. Badurek, R. Grössinger, H. Kronmüller JMMM 60 (1986) 182

MAGNETIC PROPERTIES OF AMORPHOUS $Fe_{80}Sb_{20}$ THIN FILMS*

Iwona GOSCIANSKA[a], H. RATAJCZAK[b], M. KONC[c], and P. SPISAK[c]

[a]Institute of Physics, A. Mickiewicz University, ul. Matejki 48/49, 60-769 Poznan, Poland
[b]Institute of Molecular Physics, Polish Academy of Sciences, ul. Smoluchowskiego 17/19, 60-179 Poznan, Poland.
[c]Department of Experimental Physics, Faculty of Sciences, P.J.Safarik University, Nam. Februaroveho vitazstva 9, 04154 Kosice, Czechoslovakia.

Hall effect and influence of annealing on coercive field in amorphous FeSb thin films have been studied, as well as the temperature dependence of resistivity. Thickness and temperature dependence of M_s and R_s have been found and correlate with the structural transitions.

1. INTRODUCTION

Amorphous alloys of Fe with different elements have been extensively studied. However, the properties of amorphous alloys of FeSb remain rather insufficiently known[1]. Investigations of magnetic properties of these alloys over a wide range of composition and temperature proved very interesting behaviours[1].

In the present paper we have studied some magnetic properties of $Fe_{80}Sb_{20}$ thin films. The chosen composition is Fe-rich enough to ensure pure ferromagnetic interactions in the studied alloy.

2. EXPERIMENTAL

Thin films of FeSb alloy have been deposited by flash evaporation technique in ultra-high vacuum of about 10^{-7} Pa onto glass substrates kept in magnetic field and cooled with liquid nitrogen. As the initial material for evaporation premelted and granulated alloy of $Fe_{80}Sb_{20}$ was used. This method supplied us with good quality amorphous films of different alloys[2]; thus, we expected the same for the FeSb films. The films studied were from 8 to 55 nm thick. Because of small thickness the film composition was not verified.

The resistance of the films was measured by a d.c. four terminal method in a vacuum furnace. The coercive field, H_c, of the films was determined from the in-plane hysteresis loops obtained by the magnetooptic Faraday effect.
From the Hall effect measurements spontaneous Hall coefficient, R_s, and saturation magnetization, M_s, were determined.

---
*This work was supplied by Institute of Physics, Polish Academy of Sciences, in Warsaw

3. RESULTS AND DISCUSSION

An exemplary plot of the temperature dependence of resistivity is shown in Fig.1. The studied $Fe_{80}Sb_{20}$ films exhibit two regions of resistivity decrease: (1) of about 10% in temperature range of about 350 to 580 K, being probably due to structural relaxation, (2) of about 16% in the region from about 580 to 660 K. The latter one with larger slope is rather due to crystallization. However, such a diffused crystallization region indicates growth of existing nuclei and microcrystallites. Thus, the studied as-deposited films contained some amount of microcrystalline phase. Electron diffraction observations of our films confirmed the presence of microcrystallites in amorphous matrix.

The Hall effect and the coercive field were measured for samples with different thicknesses, d. The dependences of both, M_s, and H_c, on thickness (Fig. 2) are rather contrary with those which should be expected for homogeneous amorphous thin films. Perhaps, this effect might be explained as being due to the crystalline phase content in our films.

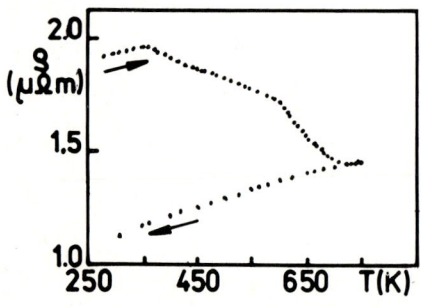

FIGURE 1
Heating temperature dependence of resistivity for sample 44 nm thick

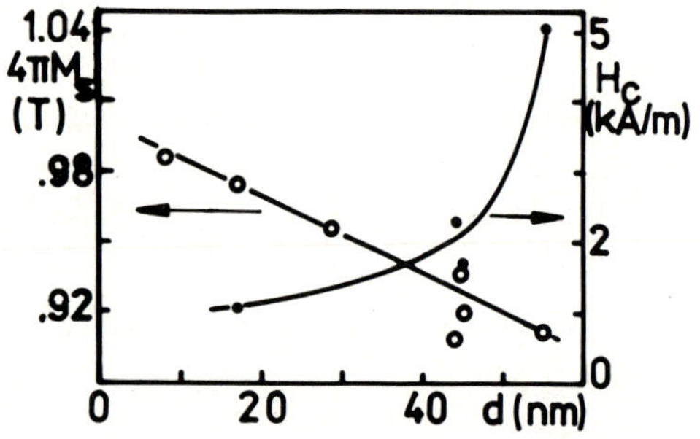

FIGURE 2
Saturation magnetization and coercive field dependence on thickness

$H_{c\parallel}$ and $H_{c\perp}$ were measured parallel and perpendicularly to the magnetic fiel applied during film deposition, respectively. The annealing caused increase in both coercive fields (Tab.1). Similar effect was already observed [2].

The temperature dependences of M_s and R_s (Fig. 3) correlate very well with the $\rho(T)$-curve. Thus, the linear decrease in M_s and the increase in R_s occur in the temperature region of low-temperature structural relaxation. The change in $M_s(T)$ slope and the rapid decrease in $R_s(T)$ at about 400 K reflect surely the crystallization onset by growth of nuclei and microcrystallites. In the temperature range from 500 to 600 K relaxation processes dominate, and the crystallization follows above 600 K. Then, $M_s(T)$ changes its course and may be its sign as well. $R_s(T)$ decreases again in this temperature range.

TABLE 1. Coercive field of sample 44 nm thick on annealing

Annealing conditions		$H_{c\parallel}$ [kA/m]	$H_{c\perp}$ [kA/m]
as-deposited		2.2	1.75
annealed at	473 K, for 0.5 h	2.6	2.7
	523 K, for 0.5 h	2.9	3.0
	523 K, for 1.5 h	3.1	3.1
	573 K, for 0.5 h	4.6	4.9
	573 K, for 1.5 h	5.6	5.7

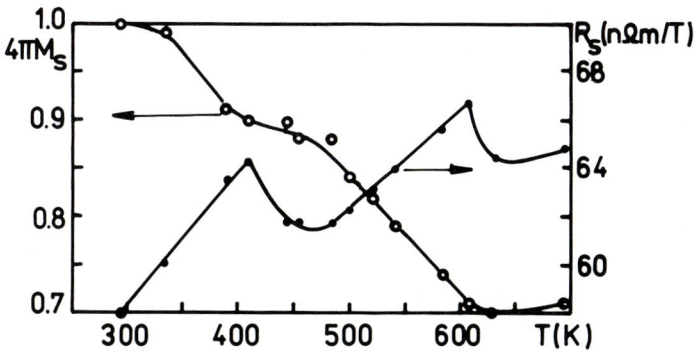

FIGURE 3
Temperature dependence of saturation magnetization and spontaneous Hall coefficient for the sample 44 nm thick

4. CONCLUSIONS

The studies of Hall effect, the temperature dependence of resistivity and the influence of annealing on coercive field in $Fe_{80}Sb_{20}$ thin films deposited on a nitrogen-cooled substrates suggest that the films contain some amount of microcrystalline phase in an amorphous matrix. This affects the changes in temperature dependences of saturation magnetization and spontaneous Hall coefficient in the temperature region where crystallization by growth of nuclei together with structural relaxation occur. Again, their change is observed on crystallization at higher temperatures.

REFERENCES

1) C.L. Chien, Gang Xiao and K.M. Unruh, Phys. Rev. B32 (1985) 5582.

2) H. Ratajczak and I. Goscianska, phys. stat. sol. (a) 96 (1986) 563;
Acta Phys. Pol. A (to be published)

MAGNETIC PROPERTIES AND INDUCED ANISOTROPY IN $(Fe_{.79}Co_{.21})_{75+x}Si_{15-1.4x}B_{10+0.4x}$ METALLIC GLASS RIBBONS

M.L. FDEZ-GUBIEDA, O.V. NIELSEN and P. VOITANIK
Department of Electrophysics, The Technical University of Denmark, DK-Lyngby, Denmark

Amorphous ribbons of composition $(Fe_{.79}Co_{.21})_{75+x}Si_{15-1.4x}B_{10+0.4x}$ have been subjected to three different kinds of annealing treatments: field-, stress-, and stress/field annealing (a tensile stress and an external magnetic field perpendicular to the ribbon axis were applied simultaneously). There is a large difference between the three kinds of induced anisotropies and all of them depend strongly on composition. Saturation polarization and magnetostriction have also been determined at room temperature.

1. INTRODUCTION

Magnetic anisotropy can be induced by applying an external magnetic field during annealing below the Curie temperature. The easy axis of the field induced anisotropy is determined by the magnetization direction during the annealing. This anisotropy is supposed to originate from atomic pair ordering[1,2].

It is known too that magnetic anisotropy can be induced by stress annealing[3,4]. It has been proposed that the stress induced anisotropy K_σ is composed of two contributions[4]: An anelastic recoverable, K_{an}, and a plastic irrecoverable, K_{plast}, i.e. $K_\sigma = K_{an} + K_{plast}$. The origin of this anisotropy is not yet clear[5].

Recently, the effect on the induced magnetic anisotropy of combined stress σ and field H during the annealing process has been studied in Co-rich samples[6].

2. EXPERIMENTAL PROCEDURE

Amorphous ribbons of composition $(Fe_{.79}Co_{.21})_{75+x}Si_{15-1.4x}B_{10+0.4x}$ (x=0,2,4,6,8,10) were prepared by the single roller quenching technique[7].

The easy ribbon axis anisotropy, K<0, and the magnetostriction value, λ_s, were determined using the stress dependence of the "small signal inverse Wiedemann effect"[8]. The hard ribbon axis anisotropy, K>0, was calculated from the saturation field, H^k, as[3]:

$$K = \frac{1}{2}\mu_0 M_s H^k$$

3. EXPERIMENTAL RESULTS AND DISCUSSION

As seen from fig. 1, the saturation polarization and magnetostriction depend strongly on composition. All samples show a high positive λ_s value. The saturation polarization value increases with decreasing metalloid content.

$(Fe_{0.79} Co_{0.21})_{75+x} Si_{15 - 1.4x} B_{10+0.4x}$ *metallic glass ribbons*

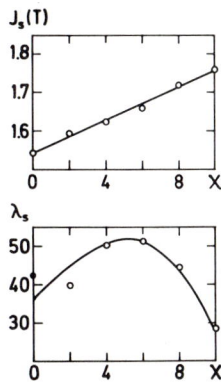

FIGURE 1
Saturation polarization and magnetostriction vs. composition

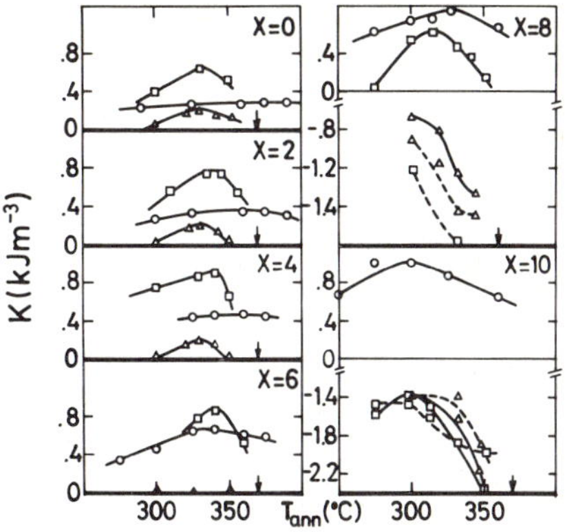

FIGURE 2
Anisotropy induced by (o) field-, (△) stress-, (□) stress/field annealing

Samples were annealed 2h at T_{ann} while a magnetic field (4400 Oe) was applied transverse to the ribbon axis. The maximum of the hard ribbon axis anisotropy induced in this way fig. 2 (o) was increasing with decreasing metalloid content.

Magnetic anisotropy was also induced by stress annealing 1h at T_{ann} with σ= 600 MPa. These samples were first preannealed 1h at T_{pre}(↓fig. 2). This anisotropy fig. 2 (△) showed a pronounced dependence on metalloid content, changing from a hard ribbon axis to an easy ribbon axis anisotropy. The x = 10 and x = 8 alloys were subjected to a stress relaxation. The annealing conditions were: 370°C, 0.5h for x = 10, and their former temperatures, 1h for x = 8. The remaining anisotropy (--- fig. 2) was an easy ribbon axis anisotropy in both cases. While for x = 8 the anelastic component of the anisotropy, K_{an}, was a hard ribbon axis anisotropy, for x = 10 it was an easy ribbon axis anisotropy.

A magnetic field (4400 Oe) was applied perpendicular to the ribbon axis during the stress annealing process in order to study the influence on the induced anisotropy of the magnetization direction during the annealing, fig. 2 (□). Although this field was high enough to keep the magnetization oriented perpendicular to the ribbon axis during the annealing, it was found that for x = 10 the stress/field induced anisotropy was parallel to the ribbon axis. The x = 10 and x = 8 samples were subjected to relaxation (--- fig. 2). The relaxation conditions were the same as used before. In this case, the recoverable component of the anisotropy for x = 10 was also a hard ribbon axis anisotropy.

From fig. 3 it is seen that when the metalloid content is larger than 21% (x<4), the effect of applying a tensile stress during field annealing enhances the effect of the field annealing. For decreasing metalloid concentrations, the

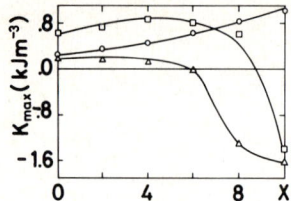

FIGURE 3
Maxium (O)field- (Δ)stress- (□)stress/field induced anisotropy vs. composition

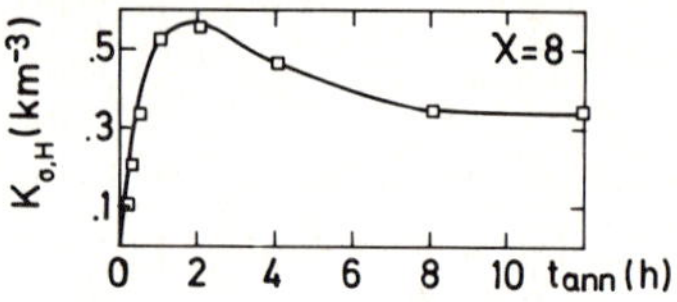

FIGURE 4
Stress/field induced anisotropy vs. annealing time

enhancement is decreasing, and when the metalloid content is less than 18% (x>7), the application of a tensile stress gives a contribution of opposite sign.

Finally, isothermal annealing treatments at 300°C with an applied stress 600 MPa and a magnetic field 4400 Oe applied perpendicular to the ribbon axis were performed on the x = 8 alloy, fig. 4. Initially, the stress/field induced anisotropy increases with time and reach saturation after 1h. Thereafter, the induced anisotropy decreases as the annealing time increases, arriving at a constant value after 8h.

4. CONCLUSIONS

It has been shown that the stress/field induced anisotropy can not be considered as the simple addition of the anisotropies separately induced by stress- and field annealing. This conclusion is in accordance with the previous observations in Co-rich samples. The main effect of applying a magnetic field perpendicular to the ribbon axis during stress annealing is to shift the stress induced anisotropy towards more positive values. The magnitude of the shift depends strongly on composition.

REFERENCES

1) Neel, L., C.r.hebd.Seanc.Sci., Paris, 237, (1953) 1613.
2) Taniguchi, S., Sci.Rep.Res.Insts Tohoku Univ., A7 (1955) 269.
3) O.V.Nielsen and H.J.V.Nielsen, J.Magn.Magn.Mat. 22 (1980) 21.
4) O.V.Nielsen, L.K.Hansen et al. J.Magn.Magn.Mat. 36 (1983) 73.
5) J.Haimovich, J.Jagielinski and T.Egami, J.Appl.Phys. 57 (1985) 3581.
6) M.Vazquez, E.Ascasibar et al. J.Magn.Magn.Mat. 66 (1987) 8.
7) V.Madurga, E.Ascasibar et al. Anales de Fisica B79 (1983) 82.
8) O.V.Nielsen, J.Magn.Magn.Mat. 24 (1981) 81.

MAGNETIC ANISOTROPY IN MODULATED Nd-Fe MULTILAYERS

L. T. BACZEWSKI*, M. PIECUCH, J. DURAND†, P. DELCROIX, G. MARCHAL

Laboratoire de Physique du Solide (U. A. au C. N. R. S. n° 155), Université de NANCY - I, B. P. 239 - 54506 VANDOEUVRE LES NANCY CEDEX (France)

1. INTRODUCTION

Rare-Earth-transition metal (RE-TM) amorphous thin films have recently attracted considerable attention[1,2,3] not only because of basic physical interest but also as a material technologically useful in perpendicular magnetic recording. On the other hand Nd-Fe based alloys are the best permanent magnets up to date and their main phase $Nd_2Fe_{14}B$ has a layer-like structure. So compositionally modulated Nd-Fe films can help to understand the basic mechanisms responsible for the strong perpendicular anisotropy found in Nd-Fe thin films.

In this paper we present the results of magnetic properties investigation of modulated Nd-Fe multilayers in comparison with amorphous Nd-Fe thin films.

2. EXPERIMENTAL

The samples were prepared by alternate evaporation of Nd and Fe layers in an ultra-high vacuum chamber onto mica substrates at 410 K. Then the films were transferred to kapton polyimide. Quality of the samples regarding crystals dimensions, periodicity and interface sharpness has been checked by electron microscopy and X-ray diffraction. All the films were continuous and crystals of iron and neodymium - 200 Å and 40 Å in width respectively - were found.

Magnetic properties were studied by means of ^{57}Fe Mössbauer spectroscopy and vibrating sample magnetometer (VSM).

3. RESULTS AND DISCUSSION

Samples of the following iron layers thickness were studied : 31 Å, 22 Å, 16 Å and 13 Å while Nd layers thickness was kept constant about 38 Å except for the sample with 16 Å of Fe for which Nd layers thickness was 32 Å. These values would correspond to the compositions for homogeneous alloys : $Nd_{29}Fe_{71}$, $Nd_{37}Fe_{63}$ $Nd_{40}Fe_{60}$, $Nd_{50}Fe_{50}$, respectively. Magnetic hysteresis loops were obtained for each sample at room temperature (RT) and 4.2 K for the direction of applied magnetic field (up to 20 kOe) parallel and perpendicular to the sample plane.

---

* On leave from : Institute of Physics, Polish Academy of Sciences, Warsaw, (Poland)

In fig. 1 the hysteresis loops at 4.2 K of $Nd_{50}Fe_{50}$ sample for both directions of applied field are shown. At H=20 kOe the complete saturation was not reached, but in order to make a comparison with the results for amorphous thin films for which magnetization was measured at the same field (for $H^{\perp}$)[4] we also denote this value as M_S.

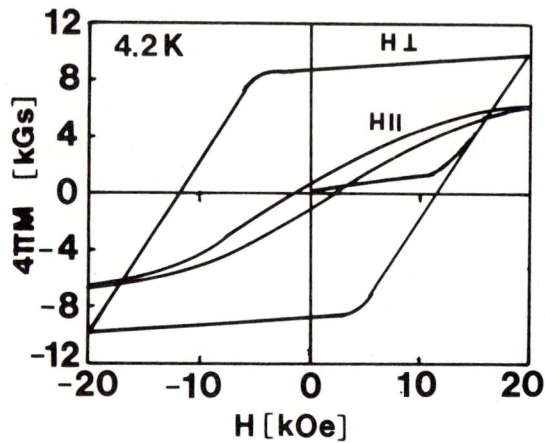

FIGURE 1
Hysteresis loops of $Nd_{50}Fe_{50}$ (13 Å Fe and 38 Å Nd) multilayer sample for the direction of applied magnetic field parallel and perpendicular to the sample plane at 4.2 K

For all investigated samples strong perpendicular magnetic anisotropy was found at cryogenic temperature.

However at room temperature all the samples behave like soft magnetic materials with weak perpendicular anisotropy which decreases with the increasing of Fe content, vanishing for 71 at % of Fe where easy magnetic axis is in plane of the sample.

Temperature dependence of perpendicular anisotropy was confirmed also by Mössbauer spectroscopy and the temperature measurements of saturation magnetization but the detailed discussion will be submitted elsewhere[5].

Saturation magnetization $4\pi M_S$ at H = 20 kOe, T = 4.2 K for the investigated multilayers are presented in fig. 2. The results are compared with those obtained by TAYLOR et al.[4] for amorphous Nd-Fe thin films. The values of $4\pi M_S$ are always higher for multilayers but the difference largely increases with increasing of iron content. Under the assumption that Fe atoms have the magnetic moment of 2.2 μ_B/at like in pure iron, magnetic moment of Nd atoms at 4.2 K was calculated. This assumption is certainly oversimplified regarding the structure of Nd-Fe interface but can be useful as a zero order approximation. For the samples : $Nd_{29}Fe_{71}$, $Nd_{37}Fe_{63}$, $Nd_{40}Fe_{60}$ and $Nd_{50}Fe_{50}$ magnetic moments of Nd atoms in μ_B/at are the following : 2.52, 1.3, 1.39, 1.48 respectively. It can be seen that high content of Fe in bcc alpha phase induces a large magnetic moment in Nd. Thinner iron layers have much weaker influence on Nd atoms.

For explaining a bigger difference in $4\pi M_S$ for high Fe content (shown in fig. 2) we assume that ferromagnetic coupling occurs between Nd and Fe layers. For high iron content ($\omega \geq 25$ Å of Fe) we have seen in Mössbauer spectra pure bcc alpha iron and for smaller Fe layers width - amorphous-like structure. It seems that for bcc iron the magnetic coupling between Nd and Fe is much stronger than for iron in amorphous structure.

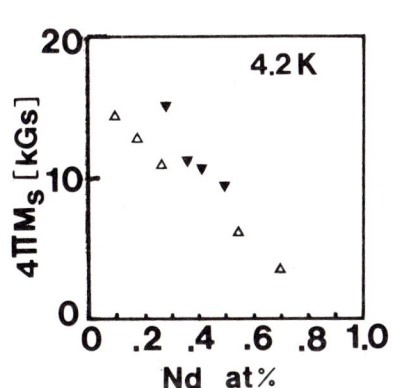 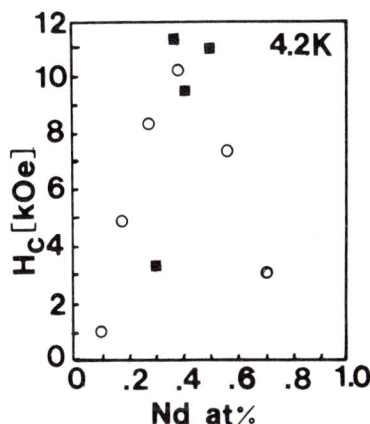

FIGURE 2
Saturation magnetization $4\pi M_S$ at 4.2 K for Nd-Fe multilayers as a function of Nd content (▼). The data for amorphous thin films (△) are from ref. 4

FIGURE 3
Coercivity H_c of Nd-Fe modulated multilayers at 4.2 K (■) as a function of composition. The data for amorphous thin films (o) are taken from ref. 4

The coercive force at 4.2 K for Nd-Fe modulated multilayers as a function of Nd content in comparison to the data for amorphous thin films obtained by e-beam evaporation are shown in fig. 3.

It could be seen that for both systems a significant maximum of H_c exists near the composition of 40 at % of Nd. Similar behaviour can be seen for amorphous Nd-Fe thin films[1,4] and melt-spun amorphous Nd-Fe alloys[6] where the maximal values were found also for the compositions between 40 and 50 at % of Nd.

From the shape of M (H) curves (abrupt increase of M near $H = H_c$) for Nd-Fe multilayers it seems that the domain wall pinning process can be regarded as coercivity mechanism.

†We would like to dedicate this paper to our colleague and friend Prof. J. DURAND who died suddenly during the final redaction of this text.

REFERENCES

1) T. Suzuki, J. Magn. Mag. Mater, 50 (1985) 265

2) T. Morishita, Y. Togami, K. Tsushima, J. Phys. Soc. Japan 54 (1985) 37

3) N. Sato, J. Appl. Phys. 59 (1986) 1942

4) R. C. Taylor, T. R. Mc Guire, J. M. Coey, A. Gangulee, J. Appl. Phys. 49 (1978) 2885

5) L. T. Baczewski, M. Piecuch et al.-to be published

6) J. J. Croat, J. Appl. Phys. 53 (1982) 3161

THE MÖSSBAUER STUDY ON MAGNETIC ANISOTROPY OF THE AMORPHOUS ALLOY $Fe_{71}Ni_{10}B_{13}Si_4C_2$*

** XU Zuxiong, **MA Ruzhang and ***HAN Yongzhu
** Beijing University of Iron and Steel Technology, Beijing, P.R. China
***Beijing Institute of Metallurgy, Beijing, P.R. China

Magnetic anisotropies of the amorphous alloy $Fe_{71}Ni_{10}B_{13}Si_4C_2$ annealed in magnetic field are studied using Mössbauer spectroscopy and X-ray diffraction. It is shown that the distribution of moments in the annealed sample are determined by both stress-produced and thermomagnetic treatment-induced magnetic anisotropies.

1. INTRODUCTION

There are reported a lot of Mössbauer studies on magnetic anisotropies and their changes in common annealed amorphous alloys. It is shown that the abnormal changes of the out-plane magnetic anisotropy are ascribed to the effect of the surface crystallization[1] or surface oxidation[2]. The present paper shows an abnormal out-plane magnetic anisotropy of a amorphous ribbon annealed in magnetic field and discusses its correlation with surface crystallization.

2. RESULTS

An amorphous $Fe_{71}Ni_{10}B_{13}Si_4C_2$ alloy ribbon 10 mm wide and 20 μm thick was prepared by melt-spinning thechnology. Suppose the major axis of the ribbon is X, the minor axis is Y and the normal is Z. The sample was treated at 430°C for 1 h in Ar atomosphere with magnetic field exerted along axis X. Transmission Mössabauer spectra at room temperature were measured in two geometries, i.e. in ribbon perpendicular to γ-ray as shown in fig.1 and in ribbon rotated at 30 degree around axis Y. The angular factors R_1 and R_2, which are the area ratio of second line to third line in a Mössbauer spectrum, were obtained and listed in table 1. X-ray diffraction patterns are shown in fig.2.

3. DISCUSSION

3.1. Magnetic Anisotropy

For a magnetic texture the angular factor of the Mössbauer spectrum can represent the average orientation or the orientation preference $\langle \cos^2 4 \rangle$ with respect to the γ-rays, which is abbrevated as P hereafter. As well known, the greater P, the more moments distribute in or near the orientation of the γ-rays. It seems reasonable to suppose that axes X, Y and Z are three main

---
*This work is supported by Ministry of Chinese Metallurgical Industry

Table 1. The magnetic anisotropies of amorphous alloy $Fe_{71}Ni_{10}B_{13}Si_4C_2$

sample	R_1	R_2	Px	Py	Pz
as-quenched	3.33	2.88	0.38	0.53	0.09
annealed at 430°C	2.80	2.35	0.51	0.31	0.18

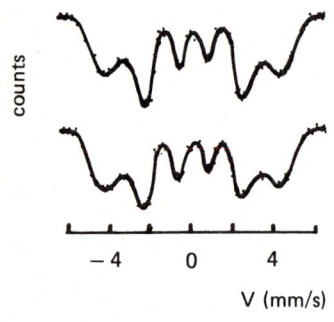

FIGURE 1

The Mössbauer spectra:
a) as-quenched, b) annealed.

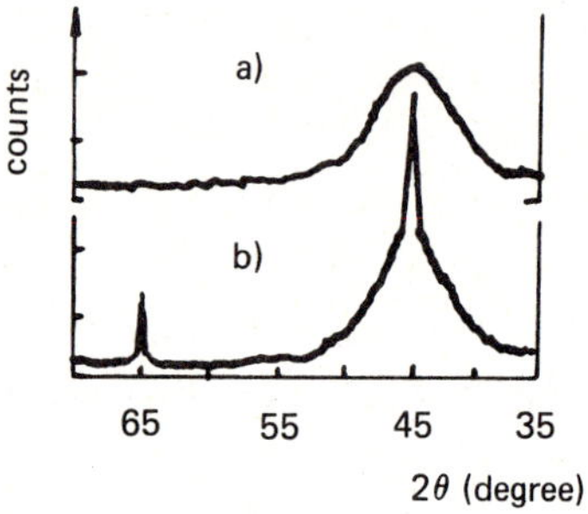

FIGURE 2

The X-ray diffraction pattern in dull side: a) as-quenched, b) annealed.

axes for describing orientation preferences of moments in the as-quenched or annealed ribbons. Therefore, from the measured angular factors we can calculate their three orientation preferences in main axes Px, Py and Pz, i.e. $<\cos^2\alpha>, <\cos^2\beta>, <\cos^2\theta>$, where α, β and θ are the angle between a moment and axes X, Y and Z respectively. The calculation is carried out following the group of equations

$Pz = (4-R_1)/(4+R_1)$,
$Px/4 + 3Pz/4 = (4-R_2)/(4+R_2)$,
$Px + Py + Pz = 1$.

The obtained parameters are also listed in table 1.

Although the magnetic measusrments show that the magnetic properties B_{50} and Hc of the annealed sample are appreciably improved. It is interesting that the changes of its orientation preferences of moments are complicated. The parameters Px y Py show taht moments in the annealed sample preferentially distribute near axis X. Obviously, this change can be attributed to the magnetic anisotropy induced during annealing with longitudinal magnetic field. However, the parameter Pz of the as-quenched sample shows a small component of the zig-zag domain where the moment direction is perpendicular to the ribbon plane.

This component grows after annealing with magnetic field. The later change has not been reported in literature and suggests that there appear some complicated changes in the structure and they produce a effect matchable to the induced magnetic anisotropy.

3.2. Surface crystallization

The transmission Mossbauer spectra in fig. 1 show that there are no apparent sharpline of the crystaline phase and no obvious bulk crystallization. However, the X-ray diffraction shows that there appear sharp lines of the iron-rich crystalline phase in addition to the broad line of the amorphous phase in the annealed sample as shown in fig. 2. Because the Fe-base alloy absorbs Cu-Kα X-ray intensively, the reflection diffraction provides mainly the information about the outer surface layer thick about 1 μm. Therefore, in combination with the transmission Mössbauer study the X-ray diffraction analysis of surface layer shows that both dull and shiny surface layers of the amorphous alloy crystalize partially without bulk crystallization during annealing at 430°C.

According to 1, it seems that both crystallized surface layers exert a compressive stress on the body due to the density difference between the body and surface layer. It is the major cause of the above mentioned abnormal behavior of out-plane magnetic anisotropy. Thereby, the orientation preferences of moments in the sample annealed in a magnetic field depend not only on the induced longitudinal magnetic anisotropy but also on the out-plane magnetic anisotropy which is produced by the compressive stress.

4. CONCLUSION

The surface crystallization occurs during annealing with magnetic field and exerts a considerable influence on the magnetic anisotropy and the orientation preferences of moments in the amorphous alloy. Therefore, the distribution of moments in the sample annealed with magnetic field are determined by both magnetic anisotropies stress-causes and thermomagnetic treatment-induced.

REFERENCES

1) H.N. Ok and A.H. Morrish, Phys. Rev. B23(1981)2257.
2) U. Gonser, M. Ackermann and H. -G. Wagner, J. Mag. Mag. Mat. 34-38(1983)1605
3) Xu Zuxiong, Ma Ruzhang and Ping Jueyun, Scientia Sinica A 28(1985)215

INDUCED ANISOTROPY AND POWER LOSSES OF COBALT - BASED AMORPHOUS ALLOYS

A.A. GLAZER, A.P. POTAPOV, I.E. STARTSEWA and V.V. SHULIKA
Institute of Metal Physics, Ural Division of the USSR Academy of Science, 620219, Sverdlovsk

The induced anisotropy of $Fe_5Co_{80-x}Si_{15}B_x$ amorphous alloys with X = 8, 9, 10, 11, 12 and its influence on power losses after various anneals with or without magnetic field have been studied.

The values of induced magnetic anisotropy in amorphous alloys $Fe_5Co_{80-x}Si_{15}B_x$ was shown to vary significantly with boron content X. Since the anisotropy determines domain structure peculiarities and consequently affect the power losses it is of interest to investigate the influence of boron content on the power losses after various anneals with or without a magnetic field.

The work was carried out on toroidal samples of the aforesaid alloys with X = 8,9,10,11,12. The induced anisotropy constant K_u, saturation induction B_s and Curie point T_c were determined from quasi-static hysteresis loops. The total power losses were calculated from area of the dynamic loops measured by the stroboscopic method over the 20 - 80 kHz frequency range and at flux densities B_m between 0,1 and 0,5 T.

Figure 1 presents values of B_s, K_u, T_c and specific resistivity ρ as functions of X. The induced anisotropy constant K_u is seen to lower as the boron content is raised. Apparently this is so because B_s decreases and T_c lowers to a temperature at which the diffusion is unfavoured.

The frequency dependences of the total P_t and hysteresis P_h power losses at different B_m were determined on all samples in the as-cast state (1) and after successive anneals in vacuum at 300°C in longitudinal (2) or transverse magnetic field (3) and also after quenching from 420°C (4). On some alloys the same parameters were also measured after annealing at 300°C in a zero magnetic field.

The concentration dependences of the total specific losses for 80 kHz and B_m = 0,3 T after the above treatments are depicted in Fig.2a. Annealing in a longitudinal as well as zero magnetic field raises the power losses for the alloys with a boron content below 10 at% but decreases them for alloys with a higher boron content. Transverse magnetic field annealing and quenching from 420°C lead to reduction of the total losses for all alloys. The classical eddy-

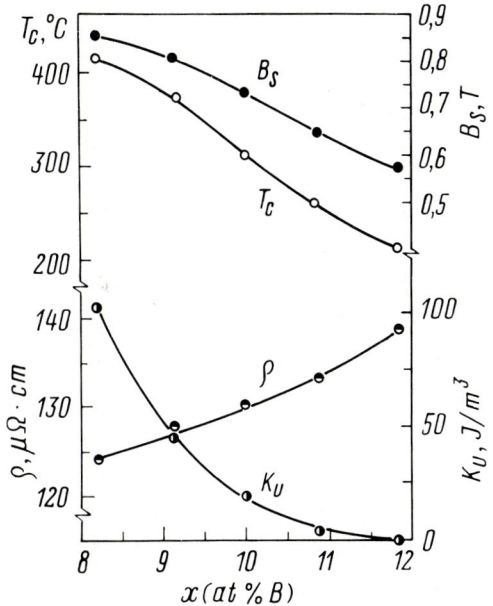

FIGURE 1

Concentration dependences of saturation induction B_s, Curie point T_c, resistivity ρ and induced anisotropy constant K_u.

current losses are displayed by curve 5.

The hysteresis component of power losses for as-cast alloys as well as that for alloys subjected to different annealings is given as function of boron content in Fig.2b. A comparison of Figs.2a and 2b shows that the hysteresis component amounts to a few percent of the total losses. For the alloys with 8,2 and 9,1 at% B where magnetic annealing leads to noticeable induced anisotropy (see Fig.1), the levels of power losses differ largely for the samples that have undergone longitudinal and transverse magnetic annealing. The hysteresis component varies weakly under magnetic annealing in a longitudinal field, but the total losses increase by almost a factor of 2. This means that the anomalous component becomes twice as large too, apparently because of an increase in magnetic domain width. After annealing, the total losses decrease because anomalous component reduces due probably to the domain refinement resulting from the production of a transverse magnetic texture. Zero field annealing of the sample with 9,1 at% B leads to an appreciable increase in the total power losses and in the hysteresis component (latter rises by a factor 8), although the internal stresses and the stresses due to the toroidal samples being tape-wound are

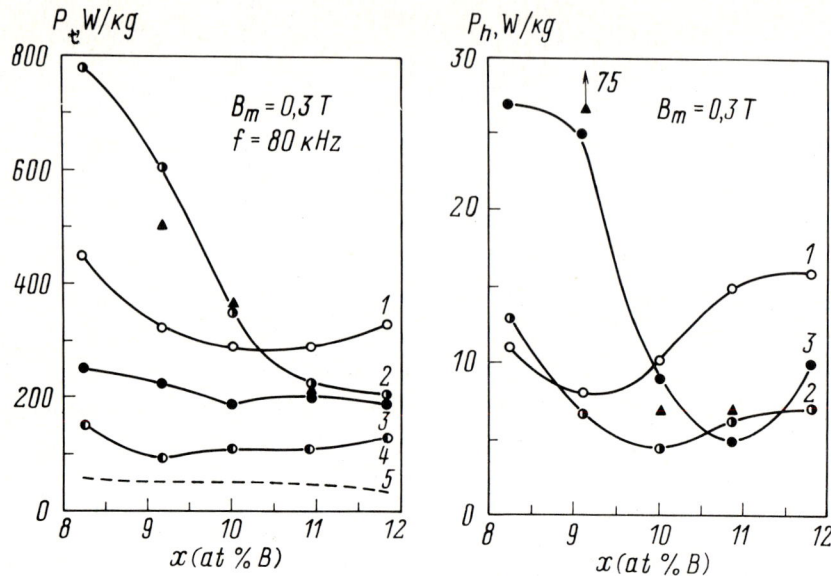

FIGURES 2a AND 2b.

Total P_t and hysteresis P_h losses versus boron content X for samples after above treatment (1)-(4) and also after annealing at 300°C in a zero field (▲).

relieved. This is associated with the domain structure stabilization by the local induced anisotropy.

No induced anisotropy occurs for alloys with 10,9 and 11,8 at% B (see Fig.1) and therefore heat treatment with or without a magnetic field leads to identical power losses, both of the total losses and the hysteresis component. Although the decrease in total losses by about 150 W·kg^{-1} is accompanied by stress relief and structure homogenisation it is mainly accounted for by the anomalous component of losses.

The highest decrease in losses takes place after quenching from 420°C in water. For the alloys with X > 10 the values of losses after quenching decrease relative to those indicated on curves 2 and 3 simply due to higher temperature heating. For the alloys with X < 10 the removal of the local induced anisotropy by quenching and the domain structure destabilization are essential.

The investigation show that for alloys with X≤10 magnetic annealing can largely affect power losses. No induced anisotropy arises for the alloys with X > 10 as the Curie point is low. Therefore annealing in a longitudinal or transverse magnetic field as well as zero field annealing leads to the same results.

NEAR SURFACE MAGNETIZATION PROCESS OF METGLAS 2605 SC AMORPHOUS RIBBONS

M. TEJEDOR, A. FERNANDEZ AND H. RUBIO

Departamento de Física, Facultad de Ciencias, Universidad de Oviedo, 33007 Oviedo, Spain

The magneto-optical transversal Kerr effect is employed for the study of near surface magnetization (< 100 nm) of the amorphous ribbon Metglas 2605 SC. The results are compared with bulk magnetization obtained by induction method. The samples are measured without demagnetizing field (toroidal arrangement). Differences between initial susceptibilities, coercive forces and approach to saturation found in surface and bulk of the sample are presented and discussed.

1. INTRODUCTION

The process of the magnetization of the ferromagnetic materials in an external field has been the objet of many works for very long time and continue nowadays the interest of this theme. In recent years many investigations have been reported about magnetization in the surface of the sample and in the bulk, constituting a new topic known as "surface magnetism". The methods employed for this study are the magneto-optical Kerr effect, Mossbauer spectroscopy and perturbed angular correlation spectroscopy[1]. The interest of these studies is the knowledgment of the distribution of anisotropies present in the samples, and other properties.

It is particularly interesting this study in the case of amorphous magnetic ribbons in which differences between anisotropies in surface and bulk have been found and interpreted in function of a special surface and bulk distribution of tensile and compressive stresses.

In this work we study the well known highly magnetoelastic material Metglas 2605 SC comparing the hysteresis loops obtained by transversal Kerr effect in the surface and hysteresis loops of the bulk obtained by the induction method.

2. EXPERIMENTAL PROCEDURE

For measuring we excited the samples (25 μm thickness and 2.5

cm width) in the toroidal form (8 cm diameter) with a very low
frecuency (30 mHz) magnetic field. The bulk magnetization is obtained from a fluxmeter connected to a secondary around the sample.
The flux is compensated in order to obtain M-H loops. The light
for the transversal magneto-optical Kerr effect employed to obtain
surface magnetization, is produced by a stabilized light source.
It is conducted to the sample by means of an optic fiber with
lens and polarizer at its end. Another optic fiber with condensing
lens and analyzer conducts the reflected light to a sensitive vacuum photodiode, whose output is conected to an X-Y recorder to
obtain a graphic representation of the loops.

This system of measurement employing optic fibers and almost
constant field presents great advantages. Optic fibers allow us
carry easily the light to the sample and also work in a illuminated room if we cover the sample with a dark cloth. A very stable
signal is obtained with this disposition.

The noise is balanced employing a photodiode system with a
great time constant (0.1 s). Also, this system is practically insensitive to mechanical vibrations and the use of very low frecuencies avoid induced currents on the sample, which is an important effect due to the softness of these amorphous materials.

3. RESULTS AND DISCUSSION

Representative major and minor hysteresis loops of the bulk
and surface magnetization obtained along the ribbon axis, are shown
in figure 1.

As it can be seen from hysteresis loops the initial susceptibility is greater for surface magnetization. The major loops show
that the surface magnetization saturation is reached much more
quickly than in the bulk, while the coercive field of the bulk is
smaller than that of the surface, nearly half value.

This means that in the surface magnetization process the wall
displacement prevails while in the bulk is the magnetization rotation the main mechanism. The coercive field is greater in the surface because the process is more ireversible due to the interaction
wall-surface. These results suggest that the anisotropy in the surface is mainly directed along the ribbon axis, while in the bulk
there must be a considerable amount of material with easy axis
in a transversal direction to the ribbon axis. This agrees

with the results of Savage and Spano[2] that stimate a volume fraction of transversal magnetization near 50%.

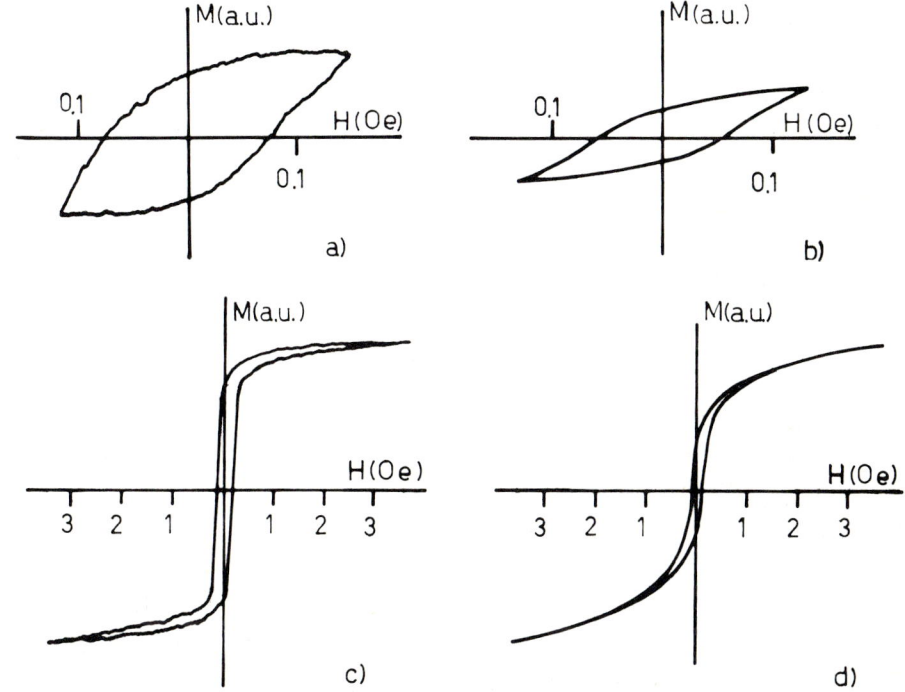

Fig. 1. Major and minor hysteresis loops; a), c): surface.
b), d): bulk.

The origin of these anisotropies can explained assuming a special distribution of internal stresses in bulk and surface of the ribbon, in an analogous way to that adopted by Ok and Morrish[3] to explain the origin of anisotropies in partially cristalized amorphous ribbons.

Further experiments must be maded to propose a model. This is the object for future works.

REFERENCES

1) H. de Waard, J. A. Sawicki, B. D. Sawicka, F. Pleiter and K. Frentz, J. Magn. Magn. Mat. 49 (1985) 50

2) H. T. Savage and M. L. Spano, J. Appl. Phys. 53 (1982) 8092

3) H. N. Ok and A. H. Morrish, Phys. Rev. B 23 (1981) 2257

MEASUREMENT OF SATURATION MAGNETIZATION AND ANISOTROPIES IN AMORPHOUS RIBBONS USING TORQUE MAGNETOMETERS

M. TEJEDOR, B. HERNANDO, J.A. GARCÍA[†] and J. CARRIZO
Departamento de Física, Facultad de Ciencias, Universidad de Oviedo, 33007 Oviedo, Spain
[†]Escuela Superior de la Marina Civil, Gijón, Spain

Saturation magnetization and magnetic anisotropies are important parameters for technological applications of amorphous magnetic ribbons. In this work we try with this kind of samples methods usually employed for the measurements of these magnitudes in magnetic thin films. Some changes have been made in order to consider the special characteristics of these samples: high saturation magnetization and shape anisotropy and low intrinsic anisotropies. The results for various standard samples are presented and discussed.

1. INTRODUCTION

Magnetic amorphous ribbons obtained by quenching have geometrical characteristics similar to those of magnetic films; that is, low ratio between thickness and the width and length of the ribbon. The typical dimensions of these ribbons are about 25μm thickness by various mm width and indefinite length. If we consider a 2x2mm square piece of ribbon its demagnetizing factor in the perpendicular direction is 0.987x4π approximately considering it as an oblate spheroid. This means that we make an error less than 2% if we assume that the demagnetizing factor is 4π as in a magnetic thin film.

Several methods that employ torque magnetometers for the measurement of saturation magnetization and anisotropies in magnetic thin films are based on this particular geometry. In this work we use the main and more useful methods with magnetic ribbons because they are more simple than the standard ones if one disposes of a torque magnetometer.

2. EXPERIMENTAL METHOD

Miyajima et al.[1] developed a method, originally thought for measuring magnetic films, which consists of the measurement of the torque L exerted on the magnetic sample (in our case a magnetic amorphous ribbon) by a variable field H whose direction is 45º to the ribbon plane. If we plot $(L/H)^2$ vs L, we obtain, if the applied field H is strong enough, a linear relation between these magnitudes. The saturation magnetization M_s and the effective anisotropy constant K are obtained without any aproximation from the slope A and the intercept B of the plot in the following manner:

$$M_s = (2B)^{1/2}/V \quad (1) \qquad K = B/(V \cdot A) \quad (2)$$

Where V is the volume of the sample.

For amorphous ribbons the shape anisotropy K_s cannot be assumed as $2\pi M_s^2$ rigourosly because of the geometrical characteristics of ribbons, that are not so drastic as film ones.

Then, if K_u is the intrinsic uniaxial anisotropy of the ribbon, using the Miyajima convention ($K_u < 0$ corresponds to a uniaxial easy axis perpendicular to the plane of the sample):

$$K = (1/2)(N_\perp - N_\parallel)M_s^2 + K_u \quad (3)$$

Where $N_\perp$ and $N_\parallel$ are the demagnetizing factors in perpendicular and longitudinal directions of the ribbon.

A scheme of the used magnetometer[2] is presented in figure 1, and figures 2 and 3 show two examples of amorphous ribbon measurements using the Miyajima method.

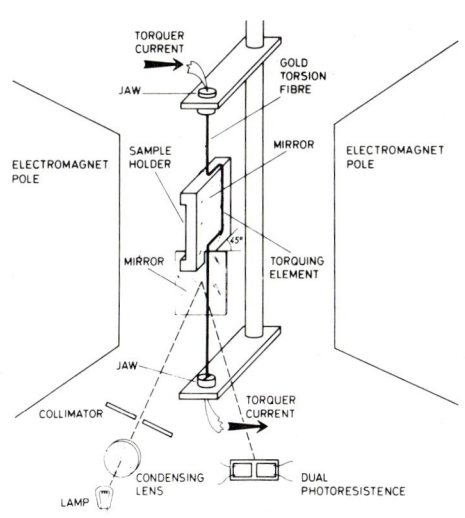

Fig. 1. Scheme of the magnetometer

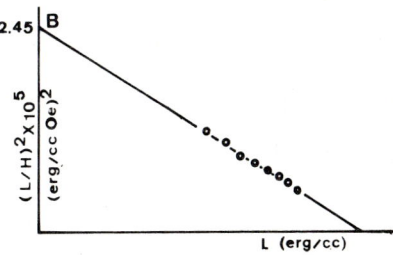

Fig. 2. $(L/H)^2$ vs L, (2826 MB)

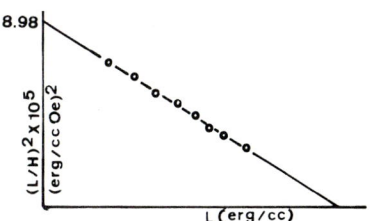

Fig. 3. $(L/H)^2$ vs L, (2605 SC)

Another method developed by Artman[3] is based on the measurement of the slopes of the torque curve $L(\emptyset)$ at the points $\emptyset=0$, and $\emptyset=\pi/2$ For our case it is enough the slope P at $\emptyset=0$ that is related with the anisotropy constant K and saturation magnetization M_s in the following manner[4]:

$$P = 2KV/[1+2K/(M_s H)] \qquad (4)$$

Taking into account (3) and assuming $L_{max}=KV$ can be easily derived:

$$M_s = 2L_{max}P/[HV(2L_{max}-P)] \qquad (5)$$

$$K_u = (L_{max}/V)[1-2(N_\perp-N_\parallel)L_{max}P^2/(H^2V(2L_{max}-P)^2)] \qquad (6)$$

The assumption $L_{max}=KV$ is true only if the magnetic field H is greater than the anisotropy field $H=2K/M_s$. It leads to the condition of validity $P \geq L_{max}$ that reduces the range of applicability of this method.

Figure 4 shows a torque curve of Metglas 2826 as a demonstrative example of this method.

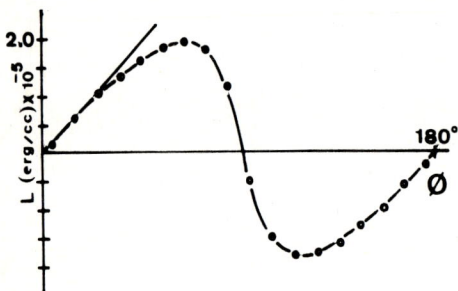

Fig. 4. Torque curve of Metglas 2826.

TABLE I
The results of the measurement for M_s and anisotropy constants for several amorphous ribbons.

SAMPLE	$4\pi M_s$ (kG)	ANISOTROPIES($10^6 \frac{erg}{cm^3}$)		
		K	K_u	K_s
2826	7.5	2.1	-0.2	2.3
2826MB	8.8	3.5	0.2	3.3
2605	18.0	11.0	-2.0	13.0
2605SC	16.8	7.7	-3.3	11.0
2605S2	14.9	8.5	-0.3	8.8
$Fe_{80}B_{20}$	12.7	6.3	-0.2	6.5

3. RESULTS AND DISCUSSION

The results in table I demonstrate that the saturation magnetization obtained by the Miyajima method agrees very well (less than 5% in difference) with the values obtained by other methods (extraction method, vibrating sample magnetometer, etc.).

The only exception is with the results for $Fe_{80}B_{20}$ amorphous ribbon. This disaccordance is probably due to the fact that this amorphous ribbon was very old and partially oxidized.

Nevertheless the values obtained for the anisotropy constants show a clear disagreement with some of the values found in the literature, being greater this disaccordance when lesser is the anisotropy constant.

These results can be explained as due to the errors in the measurements. A simple analysis demonstrates that the error made in the measurement of M_s taken into account in the study of Miyajima is less than 5%. Nevertheless the error made for the measurements of K can exceed largely 100%. This is because the intrinsic anisotropies of the samples are very small in comparison with shape anisotropy.

With the Miyajima method we measure the whole anisotropy which corresponds in this case practically to the shape anisotropy; so, the errors in these measurements are of the order of the value of the intrinsic anisotropy and we cannot get precission in its obtention. K_u usually is about two orders of magnitude smaller than the numbers given in table I.

On the other hand the value of M_s, as we have mentioned above, is measured with precission because it is deduced from the shape anisotropy that is measured with a small error.

The same considerations can be made for the Artman method, because it basically has the same fundament.

REFERENCES

1) H. Miyajima, K. Sato and T. Mizoguchi, J. Appl. Phys. 47 (1976) 4669

2) M. Tejedor, A. Fernandez, B. Hernando and J. Carrizo, Rev. Sci. Instrum. 56 (1985) 2160

3) J.O. Artman, IEEE Trans. Magn. MAG-21 (1985) 1271

4) M. Tejedor, A. Fernandez and B. Hernando, submitted to IEEE Trans. Magn.

MAGNETIC PROPERTIES OF AMORPHOUS Fe-B-Sc ALLOYS

M. GHAFARI

Laboratorium für Angewandte Physik. Universität Duisburg and SFB166, FRG

and

R.K. DAY and J.B. DUNLOP

CSIRO Division of Applied Physics, Sydney, Australia 2070

We report Mössbauer measurements of amorphous Fe-rich Fe-B-Sc alloys as function of B-content. The magnetic hyperfine field distributions ($P(H_i)$) for Fe-rich Fe-Sc alloys consist of a high and a low field tail. In the ternary Fe-rich Fe-B-Sc system the fraction of low field tail decrease rapidly as function of B-content. In addition the presence of B- atoms leads to an increase of Curie temperature and average magnetic hyperfine field ($\overline{H}_i$).

1. INTRODUCTION

Recently a new system of amorphous metal-metal alloys (Fe-Sc) were prepared by the rapid quenching method [1]. In order to obtain more information about magnetic behaviour of these alloys we combined the binary Fe-B and Fe-Sc alloys in a ternary $Fe_{92-x}B_xSc_8$ ($0 < x < 15$) system. The thermal stability, Curie temperature and magnetic hyperfine field increase with increasing B-concentrations. We report here measurements of Mössbauer spectroscopy on $Fe_{92-x}B_xSc_8$ alloys.

2. EXPERIMENTAL

Amorphous ribbons were prepared by single roller melt quenching in helium gas. The ribbons were examined with Cu Kα X-radiation in a Philips diffractometer. Samples outside the range x=0 to x=15 were only partially amorphous. Between x=0 and x=15, broad peaks characteristic of amorphous materials were observed. Mössbauer measurements were performed at 4.2 K using a conventional constant acceleration spectrometer with stationary absorber and moving source (Co^{57} in rhodium).

3. RESULTS AND DISCUSSION

Mössbauer spectra of amorphous $Fe_{92-x}B_xSc_8$ ($0 < x < 15$) are shown in Fig.1. Also included are the $P(H_i)$, which are determined by computer analysis of spectra using the program of LeCaer and Dubois[2]. Good fits to the asymmetrical spectra were obtained by assuming a linear relation between isomer shift and magnetic hyperfine field.

The value of average quadrupole splitting determined in the paramagnetic state were used for the determination of the magnetic hyperfine field distributions. The best fits were obtained with a dipolar field of about $H_d = 1$ T.

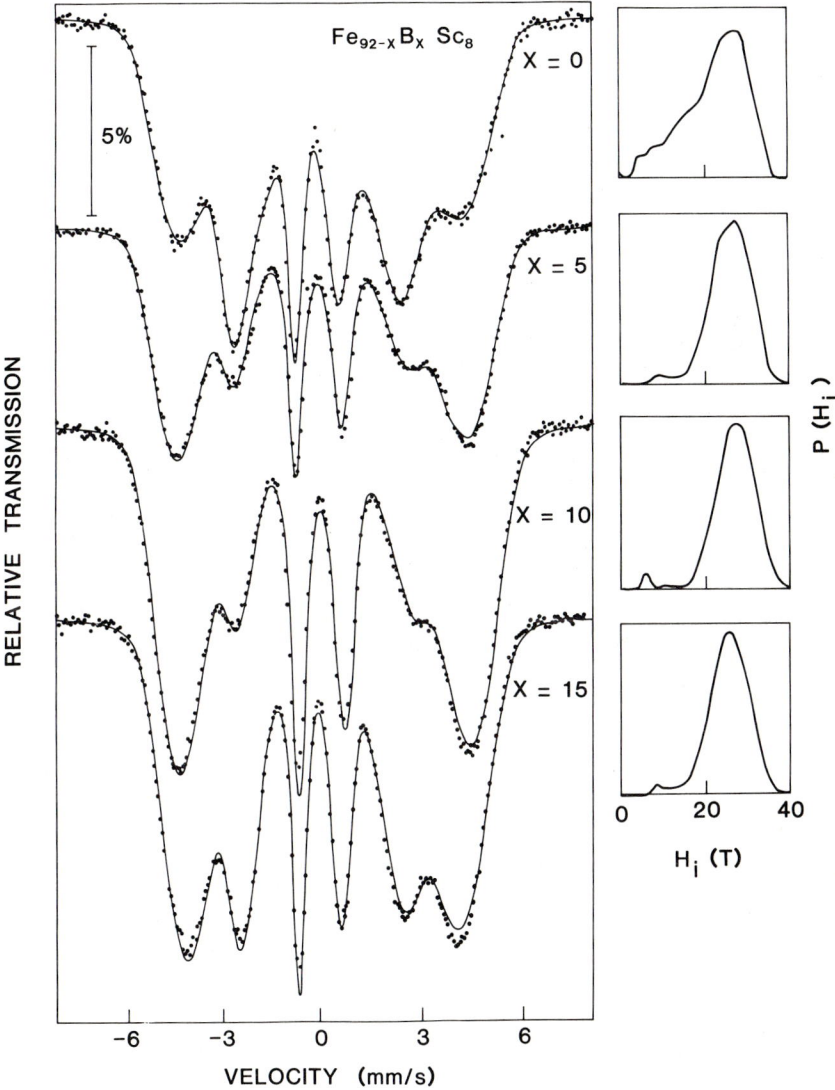

FIGURE 1

Mössbauer spectra and magnetic hyperfine field distributions ($P(H_i)$) of amorphous Fe-B-Sc alloys at 4.2 K

The average hyperfine field and the Curie Temperature (T_c) of amorphous $Fe_{92-x}B_xSc_8$ alloys are shown in table 1.

Table 1
The average magnetic hyperfine field ($\bar{H}_i$) at 4.2 K and Curie temperature of amorphous $Fe_{92-x}B_xSc_8$ alloys ($0 < x < 15$)

$Fe_{92-x}B_xSc_8$	T_c (K)	$\bar{H}_i$ (T)
x= 0	100	22.9
x= 5	298	25.4
x=10	390	25.1
x=15		24.1

$P(H_i)$ shows a high field peak and a low field tail for x=o at T=4.2 K. A considerable fraction of low field tail decrease in ternary Fe-B-Sc system. It has been suggested[3] that the low field components is caused by iron atoms which are close packed. In the ternary system are probably the closed packed Fe atoms destroyed by boron atoms.

We propose the following explanation for the increase of $\bar{H}_i$ and Curie temperature in amorphous Fe-rich Fe-B-Sc system as a function of boron atoms. Due to the large difference in electronegativity between B and Sc in ternary Fe-B-Sc system there is a preferred chemical binding between Sc and B atoms. This means that the influence of Sc atoms on Fe atoms in ternary alloys is less than that in binary (Fe-Sc) alloys. The influence of Sc atoms is also responsible for decreasing of the Curie temperature and $\bar{H}_i$. Therefore one can assume that in ternary alloys the T_c and $\bar{H}_i$ increase compared to binary Fe-Sc alloys.

REFERENCES

1) R.K.Day, J.B.Dunlop, C.P. Foley, M. Ghafari and H.Pask
Sol. St. Commun. 56 (1985) 843.

2) G. LeCaer and J.M. Dubois, J. Phys. E. 12 (1979) 1083.

3) T. Sohmura and F.E. Fujita, J. Phys. F 10 (1980) 743.

STATIC CRITICAL PROPERTIES OF AMORPHOUS FERROMAGNETS FROM MOSSBAUER MEASUREMENTS

Katarzyna BRZÓZKA, Kazimierz JEZUITA and Janusz SZLANTA[+]

Department of Physics, Technical University, Malczewskiego 29, 26-600 Radom, Poland

Mossbauer studies on the static critical behaviour of the amorphous ferromagnet $Fe_{40}Ni_{40}Si_{13}B_7$ are presented. Different methods are used to determine parameters of hyperfine field distributions in the critical region, below T_c. The influence of these methods on the obtained values of T_c and the critical exponent β is shown.

1. INTRODUCTION

The nature of the ferromagnetic-paramagnetic transition in amorphous ferromagnets has fascinated both theorists and experimentalists for more than a decade now. However only in a few Mössbauer works a quantitative analysis of the critical behaviour was carried out to date[1].

The main purpose of the present work is to analyse the critical behaviour of the magnetic hyperfine field H below T_c. Different methods[2-7] can be used to determine parameters of a hyperfine distribution from Mössbauer spectra. But not all are useful in the critical region.

We analyse and compare our results[8] for the amorphous ferromagnet $Fe_{40}Ni_{40}Si_{13}B_7$ obtained by using a lorentzian field distribution P(H) with those obtained here by the polynomial approximation of the field distribution.

2. HYPERFINE FIELD DISTRIBUTION AND CRITICAL BEHAVIOUR

As a first attempt, we have fitted the spectra to six lorentzian peaks with a lorentzian field distribution $P(H)$[6]

$$P(H) = \frac{1}{1 + 4\left(\frac{H - H_{eff}}{\Delta}\right)^2}$$

[+] Work supported by the Institute of Physics Polish Academy of Sciences under Programme CPBP 01.04.II.

The main problem of the analysis in the critical region is the reduction of the number of fitting parameters, without special assumptions about the shape of the field distribution. For this purpose we use the seventh order polynomial approximation of P(H). Here we analyse the temperature dependence of the peak H_{peak} and the mean H_{mean} hyperfine fields as is shown in Figures 1 and 2.

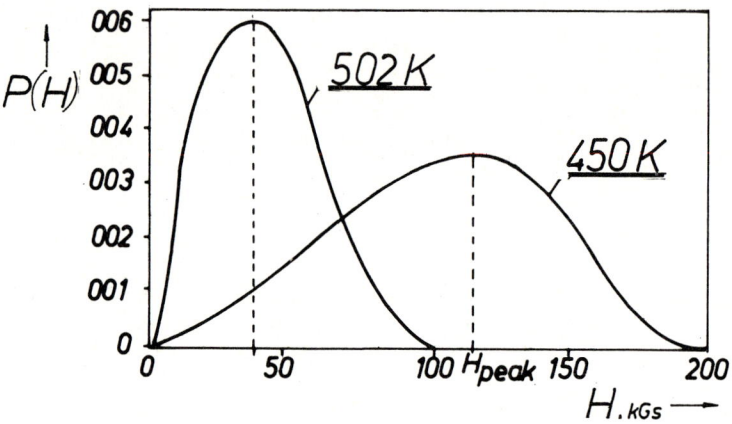

FIGURE 1
Selected hyperfine field distributions for the polynomial approximation method

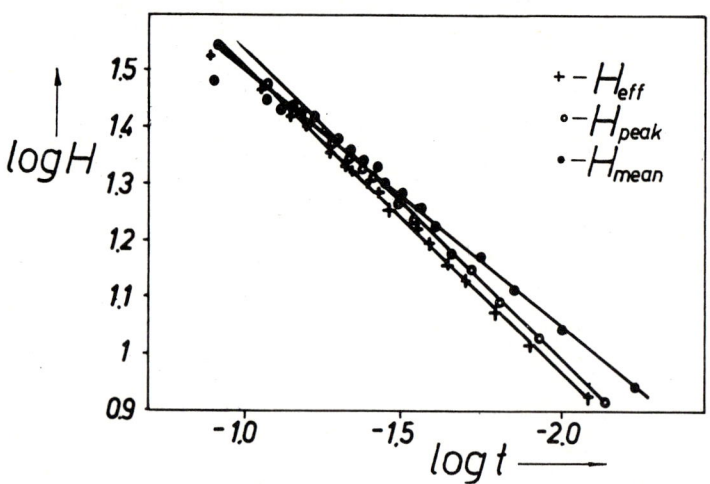

FIGURE 2
Scaling properties of the hyperfine fields

The following values of the Curie temperature T_c and the statical critical exponent have been derived from a fit to the scaling low for: the effective hyperfine field H_{eff}, $T_c=509.6\pm0.3$ K and $\beta=0.55\pm0.03$, the peak hyperfine field H_{peak}, $T_c=509.9\pm0.3$ K and $\beta=0.55\pm0.02$, the mean hyperfine field H_{mean}, $T_c=508.7\pm0.3$ K and $\beta=0.47\pm0.02$.

3. CONCLUSIONS

The value of the static critical exponent β depends on the choice of the hyperfine field distribution parameter (H_{peak} or H_{mean}) and the method for the determination of P(H). It is not clear which one strictly corresponds to the critical exponent β of the magnetization.

ACKNOWLEDGEMENT

We sould like to thank Dr M. Kopcewicz for kindly supplying the $Fe_{40}Ni_{40}Si_{13}B_7$ samples.

REFERENCES

1) S.N. Kaul, J. Magn. Magn. Mater. 53 (1985) 5.
2) B. Window, J. Phys. E4 (1971) 401.
3) G. Le Caer and J.M. Dubois, J. Phys. E12 (1979) 1083.
4) N. Saegusa and H. Morrish, Phys. Rev. B26 (1982) 10.
5) A.K. Bhatnagar, Hyperfine Interactions, 24-26 (1985) 637.
6) C.L. Chien, Phys. Rev. B18 (1978) 1003.
7) J. Hesse and A. Rubartsch, J. Phys. E7 (1974) 526.
8) K. Brzóka, M. Gawroński, K. Jezuita and J. Szlanta, Proceedings of the 3rd International Conference on Physics of Magnetic Materials, Szczyrk 1986, in: Acta Phys. Polonica, in print.

AMORPHIZATION BY SOLID STATE REACTION OF Co-Sn MULTILAYERS. MÖSSBAUER SPECTROSCOPY AND MAGNETIC MEASUREMENTS

P. GUYOT, P. GUILMIN, P. DELCROIX

Laboratoire de Physique du Solide (U. A. au C. N. R. S. n° 155), Université de NANCY - I, B. P. 239 54506 VANDOEUVRE LES NANCY CEDEX (France)

1. INTRODUCTION

Amorphization by solid state reaction at interfaces of crystalline multilayers has been observed in many systems[1]. In Co-Sn multilayers, this reaction occurs at room temperature and was observed by electron microscopy and electrical resistivity measurements[2]. The mechanism of the amorphization process is not yet fully understood, particularly, is the amorphous phase homogeneous or not ?

2. MULTILAYERS PREPARATION AND EXPERIMENTS

Crystalline cobalt-tin multilayers made of 150 bilayers with individual thickness about 65 Å (15 Å Co, 50 Å Sn) were evaporated in ultra-high vacuum, and deposited onto a kapton substrate held at 77 K. Details of preparation are developed elsewhere[2]. The multilayers are kept in liquid nitrogen until their measurements by Foner magnetometer or by Mössbauer spectroscopy. The thickness of Co with regard to Sn is calculated so that the composition after total interdiffusion varies between 40 % and 48 % of cobalt.

2.1 Magnetic measurements

We used a Foner magnetometer with an applied field up to 20 kOe. We measured, at 4.2 K, the variation of the magnetic moment, $\bar{\mu}$, versus annealing time (fig. 1a).Between the two measurements, the multilayer was heated from 4.2 K to room temperature, annealed during fixed time and then quenched at 4.2 K.

We observed strong decrease of $\bar{\mu}$ in the first period, followed by a slower variation of $\bar{\mu}$. It is quite surprising that $\bar{\mu}$ is still about 0.5 μ_B after large annealing time,whereas $\bar{\mu}$ vanishes in co-evaporated $Co_x Sn_{1-x}$ alloys of same composition.

Magnetization has been also measured in the temperature range of 4.2-250 K at a constant field of 100 Oe. For short annealing times (fig. 1b) a typical reversible ferromagnetic behaviour is observed. By contrast, for large annealing times, an irreversible behaviour is observed at low temperature.

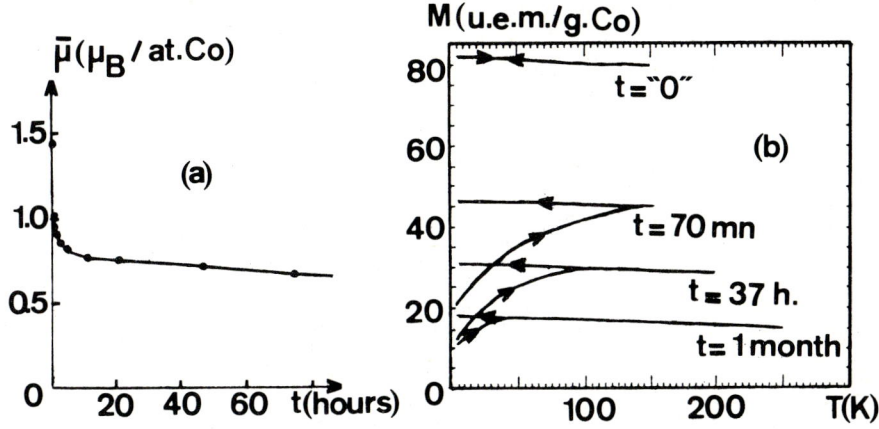

FIGURE 1
(a) Average magnetic moment at 4.2 K extrapolated to H = 0
(b) Subsequent thermomagnetic behaviour
For different annealing times, at 300 K, t and for the same sample of overall composition 43 at.% Co

2.2 Mössbauer spectroscopy

^{119}Sn Mössbauer spectroscopy is done at 77 K. We assume that the spectra are the superposition of the metallic tin line and of a spectrum of amorphous $Co_x Sn_{1-x}$ alloy. We observe (fig. 2) the fast vanishing of the first contribution. Simultaneously the second contribution, roughly analysed as a doublet by comparison with the spectra of $Co_x Sn_{1-x}$ co-evaporated amorphous alloy[4], increases. Moreover, in this second contribution, relative intensity of each line slowly varies.

Some spectra were done at 4.2 K ; they exhibit no broadening of the doublet lines, so that we can conclude that the amorphous phase is never magnetic.

3. DISCUSSION

Mössbauer data show unambiguously that amorphization process begins with the disappearance of metallic tin. It is not surprising, because Co is an anomalous fast diffuser in Sn. Moreover this fact inhibits the nucleation of defined compounds and allows the growth of an amorphous phase.

From the evolution of the doublet line of Mössbauer spectra, we assume a slow cobalt concentration increase in this amorphous phase ; however it seems that small clusters of cobalt atoms remain, even for large annealing times. This conclusion is consistent with magnetic measurements and could explain the persistence of a cobalt magnetic moment. It is also in concordance with EXAFS measurements[5].

It seems that the amorphization can be achieved at 350 K ; indeed magnetic moment disappears after this additional annealing.

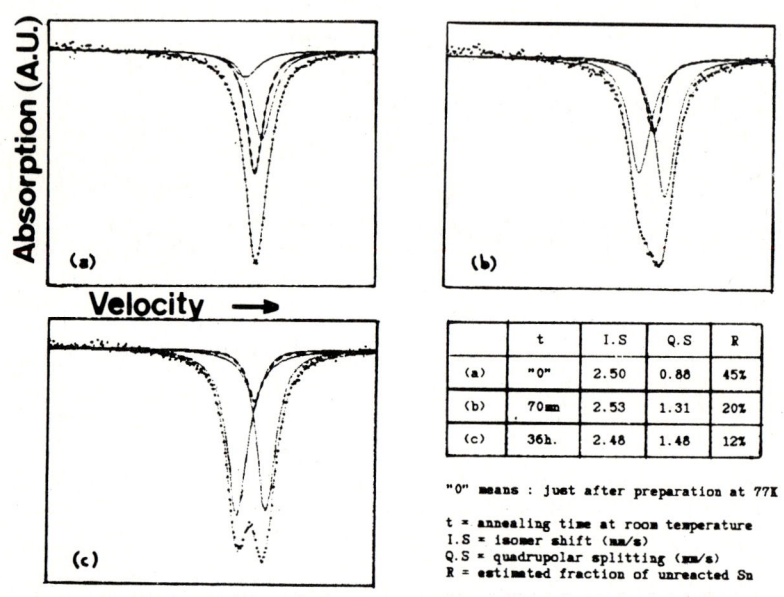

	t	I.S	Q.S	R
(a)	"0"	2.50	0.88	45%
(b)	70mn	2.53	1.31	20%
(c)	36h	2.48	1.48	12%

"0" means : just after preparation at 77K
t = annealing time at room temperature
I.S = isomer shift (mm/s)
Q.S = quadrupolar splitting (mm/s)
R = estimated fraction of unreacted Sn

FIGURE 2
Mössbauer spectra taken at 77 K versus annealing time t for a solid-state reacted multilayer, with overall composition of 40 at % Co, at 300 K
----- metallic tin contribution (R)

REFERENCES

1) W. L. Johnson, Prog. Mat. Science 30 (1986) 81

2) P. Guilmin, P. Guyot and G. Marchal, Phys. Lett. A 109 (1985) 174

3) D. Teirlinck, M. Piecuch, J.-F. Geny, G. Marchal and Ph. Mangin, J. App. Phys. 53 (1982) 7734

4) H. Nabli, M. Piecuch, J. Durand and G. Marchal, J. Phys. C 8 46 (1985) 229

5) C. Brouder, G. Krill, E. Dartyge, A. Fontaine, G. Marchal and P. Guilmin, J. Phys. C 8 47 (1986) 1065

FMR STUDY OF AMORPHOUS $(Sm_xFe_{1-x})_{80}B_{20}$ THIN FILMS

W.A. KACZMAREK[a], J. PIETRZAK[a], and H. RATAJCZAK[b]

[a]Institute of Physics, A. Mickiewicz University, PL 60-780 Poznań, Poland
[b]Institute of Molecular Physics, Polish Academy of Sciences, PL 60-179 Poznań, Poland

The results of FMR and resistivity measurements carrried out in a wide temperature range confirm that the presence of Sm^{3+} ions weakens the exchange interactions and is responsible for changes in the magnetic parameters.

1. INTRODUCTION

Magnetic properties of amorphous Re-FeB alloys have recently attracted interest due to their unusual variations with composition and temperature of heat treatment[1-3]. These variations follow both the structural relaxation, at lower, and crystallisation, at higher temperatures. In the first stage of crystallisation nucleation appears and then microcrystallites can grow even at the temperatures characteristic for relaxation. The crystallization process occurs frequently as a transitions through different phases.

In this paper we report some results of our preliminary study of FMR and resistivity changes under heat-treatment in a wide temperature range (77 to 800 K) in amorphous $Sm_xFe_{80-x}B_{20}$ (x = 3, 6 and 9) thin films.

2. EXPERIMENTAL

The amorphous SmFeB thin films were obtained by the flash evaporation technique in the vacuum of 10^{-7} Pa. They were deposited onto glass substrates cooled with liquid nitrogen. The thickness of the films ranged from 30 to 31 nm. The films were covered with SiO layer immediately after deposition. The resistivity on heating was measured in a vacuum furnace with the four terminal method. The FMR experiments were made with a K-band conventional reflection cavity spectrometer (22.95 GHz) and a variable temperature gas flow cryostat. If general conditions are satisfied then the area under the absorption curve[5], F is proportional to $1/2\ I\ (\Delta B_{pp})^2$, $(I = Y'_{m1} + Y'_{m2}$: amplitude of the recorded signal, ΔB_{pp} : its peak to peak linewidth)[4]. For ordinary ferromagnetics the F(T) dependence is similar in course to the $M_s(T)$ one.

3. RESULTS and DISCUSSION

From the temperature dependences of resistivity we were able to determine accurately the crystallization temperatures T_{cr}, as 680 K for the sample with Sm concentration x = 6 and 700 K for x = 9. The estimated structural relaxation regions are from 350 to 480 K for both alloys. The temperature dependences of $B_{\parallel}$, ΔB_{pp} and I for $Sm_6Fe_{74}B_{20}$ film are shown in Fig. 1 a). The same results for $Sm_9Fe_{71}B_{20}$ films are shown in Fig. 1 b). Fig. 1 presents the results of FMR measurements of the same films (the magnetic field is applied in the film plane ($B_{\parallel}$)). The spectra are composed of two resonance lines : a strong one denoted by A and a very weak one B. Similar temperature dependences $B_{\parallel}(T)$ were observed for both lines. As no essential differences in T_{cr} for x = 6 and x = 9 were found, the changes in temperature dependences of the measured parameters presented in Fig. 1 a and b are the consequence of different T_c values, for x = 6 $T_c \cong$ 750 K and for x = 9 $T_c \cong$ 585 K. For x = 9 the FMR spectra were taken below T_{cr} whereas for x = 6 also above T_{cr}. Analysis of $B_{pp}(T)$ and I(T) confirms significant exchange coupling among iron ions, which weakens with increasing concentration of Sm. Magnetization of the film was calculated from the resonance formula for a thin film, namely from $(\omega/\gamma)^2 = B_{\parallel}(B_{\parallel} + B_{eff})$ where $B_{eff} = \mu_0 M_s - B_k$, having assumed that g = 2 and the magnetocrystalline anisotropy field B_k = 0. For room temperature the values $\mu_0 M_s$ calculated in this way for x = 3, 6 and 9 are 1.85 T, 1.61 T and 1.45 T. In the low-temperature range the FMR line broadening is related with the appearance of non-collinear structure (glass-like behavior). For x = 6 anomalous behavior of ΔB_{pp} near $T_{cr} \cong$ 610 K is related with the process of crystallization, such an anomaly is not observed for x = 9. An increase in concentration of Sm^{3+} ions results in a linear increase of ΔB_{pp} from $5 \cdot 10^3$ Am^{-1} (x = 3, T = 300 K) to $15 \cdot 10^3$ Am^{-1} (x = 9, T = 300 K). On the other hand the presence of small amount of Sm^{3+} ions is responsible for magnetization and T_c greater than those of $Fe_{80}B_{20}$ [5].

On the grounds of the results presented we may conclude that within the whole temperature range studied, we observed a profound influence of Fe^{3+} ions on electric and magnetic properties of the films studied. Thus, the most probable localization of the ions as well as the course of the crystallization process are related with the presence and development of α-Fe type clusters which has been observed in the whole range of temperatures considered.

ACKNOWLEDGEMENT

This work has been partly supported by the fund of Polish Academy of Sciences CPBP 01 04 II 1.13.

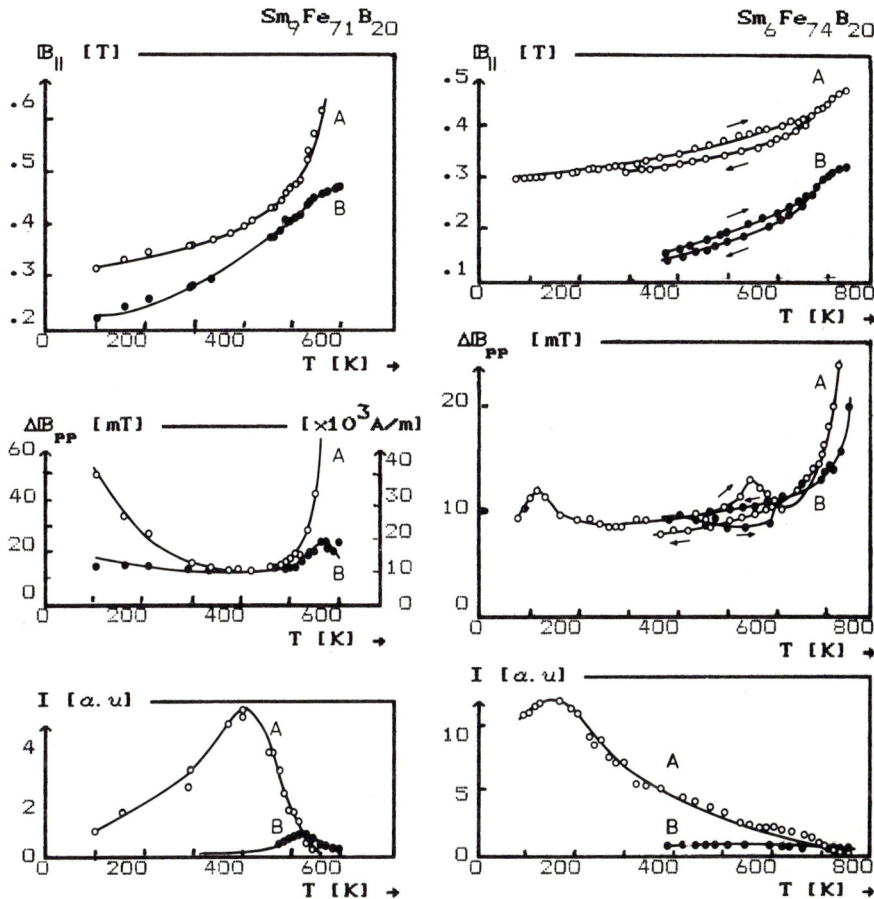

Fig. 1 a), b). Temperature dependence of resonance field $B_\parallel$, linewidth ΔB_{pp} and amplitude of the recorded FMR signal I.

REFERENCES

1) M. Fähnle, J. Magn. Magn. Mat. 45 (1984) 279.
2) K. Kakuno, T. Kubota and T. Yamada, Jap. J. Appl. Phys. 7 (1985) L481.
3) G.L. Whittle, A.M. Stewart and A.B. Kaiser, phys. stat. sol. (a) 97 (1986) 199.
4) C.P. Poole, Electron Spin Resonance (Mir, Moscow, 1972).
5) Chr. Janot, Revue Phys. Appl. 21 (1986) 635.

SURFACE MAGNETIC PROPERTIES OF AMORPHOUS TERBIUM-IRON THIN FILMS

M. AESCHLIMANN, G.L. BONA, F. MEIER, M. STAMPANONI, A. VATERLAUS and H.C. SIEGMANN
Laboratorium für Festkörperphysik ETH-Hönggerberg, CH-8093 Zürich, Switzerland

The magnetic surface properties of perpendicular recording media such as a - TbFe have been investigated with spin-polarized photoemission. The observations are interpreted by a 3-layer model: The virtually non-magnetic surface is formed by Tb-oxide, due to the naturally occuring surface segregation of the rare earth Tb. It is followed by an iron-rich subsurface superimposed on the regular bulk. This subsurface must be thicker than 10 Å and is recognized by a Curie temperature much higher than the one of bulk TbFe. It also exhibits large saturation magnetization and low structural anisotropy. At H = 0 the magnetization of the subsurface lies in-plane. Below the Curie point bulk and subsurface are magnetically exchange coupled producing non-trivial surface magnetic structures.

Amorphous rare-earth-transition metal alloys (RE-TM) with a strong magnetic anisotropy perpendicular to the surface are now one of the most promising media for erasable optical data storage applications [1,2]. However, amorphous RE-TM films show only Kerr rotation angles smaller than 1° leading to low signal-to-noise ratios. It is well known that surface segregation and oxidation of the rare earth [3] produce a variation in the magnetic behaviour at and near the surface, giving rise to a variety of characteristic features affecting the read-out with the Kerr effect which has a probing depth of 200 Å only. The purpose of this work has been to better understand the special magnetic properties of the surface. We compared the Kerr hysteresis loops of amorphous $Tb_{20}Fe_{80}$ samples to spin-polarized threshold photoemission (SPP) loops taken on the same film. The probing depth of the SPP-technique is about 10 Å, suitable for studying the magnetic surface and subsurface properties.

The experiment was performed under ultrahigh vacuum conditions with an equipment already described [4]. The amorphous $Tb_{20}Fe_{80}$ alloy film was about 1000 Å thick and had a protective Al_2O_3-coverage which was sputtered off with 800 eV Ar^+-bombardment. After this cleaning process the sample was placed in a magnetic field perpendicular to the film surface. The spin polarization of the photoemitted electrons is defined as P = (N↑ - N↓)/N↑ + N↓) where N↑(N↓) is the number of the photoelectrons with spin magnetic moment parallel (antiparallel) to the magnetization. For photon energies near the photo-threshold $h\nu$ = 4.0 eV, P is mainly determined by the magnetic iron 3d electrons, because the 4f electrons in terbium lie well below E_F [5]. Also the Kerr effect is mainly due to the interaction of the polarized light with the Fe-subnetwork. Fig. 2 shows the typical rectangular Kerr loop taken at a temperature much higher than the compensation point, which is - if it exists at all - far below room temperature for the bulk composition $Tb_{20}Fe_{80}$. The magnetization curves of the surface measured with the SPP technique are shown in Fig. 1. They are rather different from the Kerr loops. Comparing the SPP- and Kerr-results at the same temperature T = 295 K (see Fig. 1b and 2) the SPP loop shows a discontinuity for an external field of H = 3.4 kOe which is exactly the coercivity H_c found in the Kerr-measurement.

The surface magnetization observed by SPP can be interpreted by a 3-layer model shown in Fig. 3. It is well established and confirmed by XPS and Auger electron spectra that on the surface of the film an enrichment in Tb is built up in a short time caused by spontaneous segregation [3]. The first layer is formed by the reactive RE on the surface which produces an inert nonmagnetic oxide or hydroxide. The migration of Tb atoms leaves behind a Fe-rich ferromagnetic subsurface layer of thickness d which lacks the perpendicular uniaxial anisotropy of the bulk. Accordingly, a state is favored where the spins lie in-plane (see Fig. 3). At $T > T_{curie}$ the bulk becomes paramagnetic. Then the iron-rich subsurface layer is not longer influenced by the bulk and exhibits its intrinsic magnetization curve similar to that of an iron film with the field perpendicular to the surface. Only the saturation field H_{sat} = 11 kOe is smaller than that of clean iron (H_{sat} = 22 kOe). Below T_{curie} of the bulk, the ferromagnetic subsurface layer is exchange coupled to the bulk and this produces anomalous surface magnetic structures. We will first discuss the case where the coercivity of the bulk is much higher than our maximum external field, i.e. the bulk is

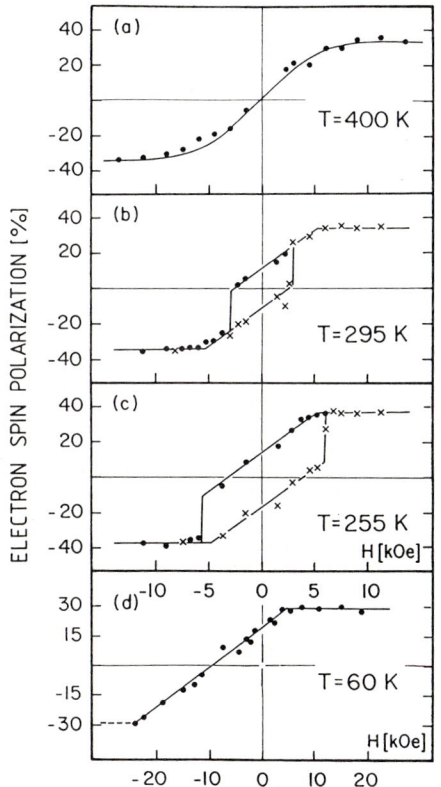

FIGURE 1
Surface hysteresis loops measured with spin-polarized photoemission at different sample temperatures

completely oriented in one direction (Fig. 1d). Neglecting domain wall terms (see Fig. 3) we can write the total energy σ per unit area:

$$\sigma = [H \cdot M_{sat} \cdot d + \frac{A_{21}}{q}] (1 - \cos\Theta) + 2\pi M_{sat}^2 \cdot d \cdot \cos^2\Theta$$

The first term is the magnetostatic energy of the subsurface, in the second term A_{21} is the bulk-subsurface exchange and q = interlayer distance. The magnetic moments of the subsurface may enclose an angle Θ with the z-axis. The last term is the shape anisotropy of the subsurface. Without the second term A_{21} (i.e. without bulk-subsurface exchange) a simple Stoner-Wohlfarth magnetization curve results, without hysteresis. With the exchange coupling there is a shift in the field axis proportional to A_{21} to the left or to the right depending of the sign of A_{21} i.e. the 3d spin direction of the bulk. Fig. 1d shows the case where the direction of the bulk 3d-spin is parallel to the positive external field. When $H > H_{coer.}$ of the bulk the sign of A_{21} changes and we go from the magnetization curve shifted to the right immediately to the magnetization curve shifted to the left and vice versa, dependent on the (M-H) loop of the bulk. This is shown in Fig. 1b and 1c. Comparing Fig. 2 to Fig. 1b the SPP loop in fact jumps from the curve shifted to the right to the one shifted to the left at an external field equal to the bulk coercivity in agreement with our model. From a more detailed model calculation (6) the exchange stiftness A_{21} acting between neighboring bulk and subsurface Fe 3d-moments is found to be 10 times smaller than in the bulk.

The minimum thickness d of the subsurface layer must be larger than the escape depth of the photoelectrons. Independent of the magnetic state of the bulk (T > T_c in Fig. 1a, T < T_c in Fig. 1b) the saturation polarization measured with SPP is in both cases the same - a fact which clearly shows that the bulk

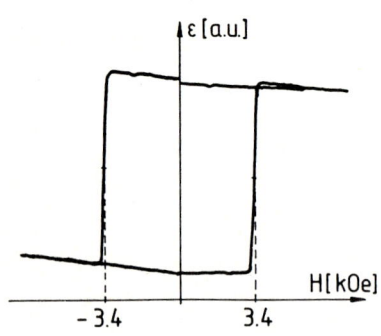

FIGURE 2
Rectangular Kerr loop taken at T=295 K

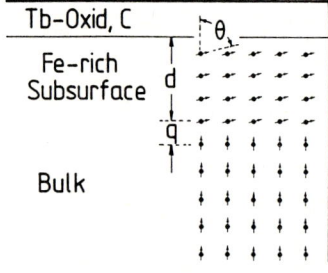

FIGURE 3
3-layer model. The bulk is remanently magnetized perpendicular to the surface, whereas the subsurface layer has the easy direction in-plane

contribution to the photocurrent is negligible.

We thank E. Marinero from IBM San José California for the samples. Support by IBM San José and the Swiss National Science Foundation is gratefully acknowledged.

1) P. Chaudhari, J.J. Cuomo, and R.J. Gambino, Appl. Phys. Lett. 32 (1973) 337.

2) A.E. Bell, Appl. Optics 25 (1986) 3985.

3) R. Allen and G.A.N. Connell, J. Appl. Phys. 53 (1982) 2353.

4) M. Campagna, D.T. Pierce, F. Meier, K. Sattler, and H.C. Siegmann, Adv. Electron. Electron Phys. 41 (1976) 113.

5) G.A.N. Connell, J. Magn. Magn. Mat. 54-57 (1986) 1561.

6) H.C. Siegmann, P.S. Bagus, and E. Kay, Proc. Nato Advanced Research Workshop, University of Sussex, Brighton, UK, Sept. 15-18, 1986, to be published.

MECHANISM OF CHEMICAL SHORT RANGE ORDERING DURING STRUCTURAL RELAXATION IN Co-Fe BASED AMORPHOUS ALLOYS

Takayuki KOMATSU and Kazumasa MATUSITA

Department of Materials Science and Technology, Technological University of Nagaoka, Nagaoka, 940-21, Japan

The resistivity changes during structural relaxation in $(Co_{1-x}Fe_x)_{75}Si_{10}B_{15}$, $(0 \leq x \leq 1)$ and $(Co_{0.75}Fe_{0.25})_{100-x}(Si_{0.4}B_{0.6})_x$, $(20 \leq x \leq 30)$ amorphous alloys were examined to clarify the mechanism of chemical short-range ordering (CSRO) between Co and Fe atoms. The compositional dependence of reversible resistivity changes in pre-annealed Co-rich amorphous alloys coincided well with that of the probability of formation of Co_3Fe unit. It was found that both the values of activation energy and pre-exponential factor for the resistivity changes were distributed widely and the degree of their distributions decreased with increasing Fe and metalloid contents. We propose that the CSRO during structural relaxation in Co-Fe-Si-B amorphous alloys corresponds to the increase in the number of Co-Fe pairs in a random arrangement of trigonal prisms $(Co,Fe)_3(Si,B)$.

1. INTRODUCTION

Since structural relaxation affects various physical, chemical and mechanical properties, it is quite important to clarify the relaxation phenomenon in amorphous alloys not only for technical applications but also for understanding the amorphous structure. It is generally considered from many studies that two types of local atomic rearrangements occur during structural relaxation, namely topological short-range ordering (TSRO; irreversible) and chemical short-range ordering (CSRO; reversible). The purpose of the present study is to examine the compositional dependence and kinetics of resistivity changes during structural relaxation in various Co-Fe based amorphous alloys, in order to obtain more detailed information about CSRO.

2. EXPERIMENTAL

The $(Co_{1-x}Fe_x)_{75}Si_{10}B_{15}$, $(0 \leq x \leq 1)$ and $(Co_{0.75}Fe_{0.25})_{100-x}(Si_{0.4}B_{0.6})_x$, $(20 \leq x \leq 30)$ amorphous alloys were prepared in the form of ribbon, about μm thick and 1.5 mm wide, by rapid quenching using a single-roller casting apparatus. The measurements of electrical resistivity were made using a four-point probe method. As-quenched samples were spot-welded carefully by small copper wires. All measurements were made at the reference temperature 77K. The relative resistivity changes, $\Delta\rho/\rho_0$, were calculated as functions of annealing temperature T_a and time t_a, where ρ_0 is the initial resistivity ($t_a=0$) and $\Delta\rho$ is equal to the resistivity changes caused by annealing, $\Delta\rho=\rho(T_a,t_a)-\rho_0$.

3. RESULTS AND DISCUSSION

The resistivity changes caused by isochronal annealing cycles in pre-annealed $Co_{60}Fe_{20}Si_8B_{12}$ amorphous alloy are shown in Fig.1 as a typical example. Figure 1 shows clearly the reversibility of resistivity changes. Similar behaviors of resistivity changes were observed in other Co-rich amorphous alloys, and it was found that the amount of reversible resistivity changes depended strongly on the Co/Fe ratio and on the amount of metalloid (Si,B) content. In particular, the compositional dependence of reversible resistivity changes in pre-annealed $(Co_{1-x}Fe_x)_{75}Si_{10}B_{15}$ amorphous alloys coincided well with that of the probability of formation of Co_3Fe unit. It has been generally interpreted that the reversible resistivity changes during structural relaxation in Co-Fe and Fe-Ni based amorphous alloys are due to the CSRO between transition metal atoms[1-3]. The present results support strongly that the CSRO occurs during structural relaxation in Co-Fe-Si-B amorphous alloys and causes the

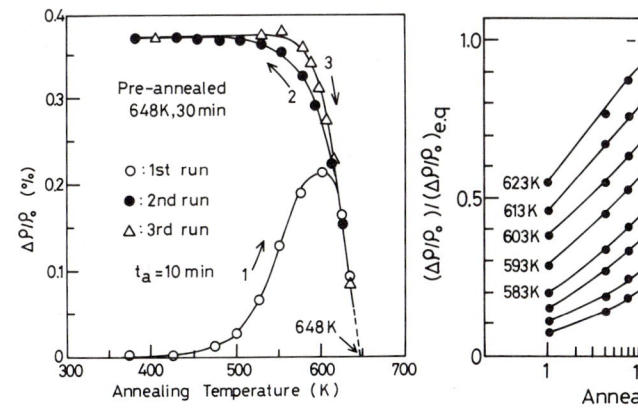
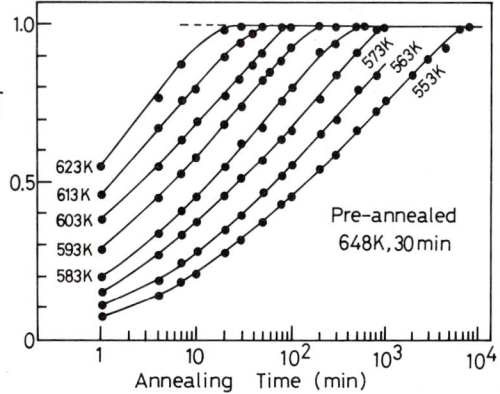

Figure 1
Resistivity changes caused by isochronal annealing in $Co_{60}Fe_{20}Si_8B_{12}$ amorphous alloy

Figure 2
Normalized resistivity changes in isothermal annealing in $Co_{60}Fe_{20}Si_8B_{12}$ amorphous alloy

reversible resistivity changes.

The normalized resistivity changes in isothermal annealing in pre-annealed $Co_{60}Fe_{20}Si_8B_{12}$ amorphous alloy are shown in Fig.2. We analyzed these data using a log-normal distribution model in the relaxation times which was developed by Nowick and Berry[4]. The temperature dependences of mean relaxation time τ_m and distribution parameter β for the relaxation times were obtained and the values of mean activation energy $E_{m,a}$, mean pre-exponential factor $\tau_{m,0}$ and distribution parameters of activation energy $β_E$ and pre-exponential factor $β_0$ were estimated. The obtained kinetic parameters for Co-Fe-Si-B amorphous alloys are summarized as follows. The values of $E_{m,a}$ and $\tau_{m,0}$ are around 2.0 eV and 10^{-16} s, respectively and these values are almost independent on the Co/Fe ratio, but the values of $E_{m,a}$ increase with increasing metalloid content. The values of β are ranging from 3.5 to 4.7 and decrease with increasing annealing temperature. It was found that both E_a and τ_0 are distributed widely and the degree of their distributions decrease with increasing Fe and metalloid contents. These results imply that Co atoms move more easily due to thermal annealing than Fe atoms and local environments of Co atoms fluctuate largely from site to site than those of Fe atoms, and it is considered that these features of Co atoms in Co-Fe based amorphous alloys are the reason why CSRO occurs significantly in Co-rich amorphous alloys. As a conclusion, we propose that CSRO during structural relaxation in Co-Fe-Si-B amorphous alloys corresponds to the increase in the number of Co-Fe pairs in a random arrangement of trigonal prisms $(Co,Fe)_3(Si,B)$ and is affected by a directional chemical bonding between transition metal and metalloid atoms.

ACKNOWLEDGEMENT

This work was supported by a Grant of the Nippon Sheet Glass Fundation for Materials Science.

REFERENCES

1) R.Yokota, M.Takeuchi et al. J. Appl. Phys. 58 (1984) 3037.

2) E.Balanzat, J.T.Stanley et al. Acta Metall. 33 (1985) 785.

3) T.Komatsu and K.Matusita, J. Mat. Sci. 21 (1986) 1693.

4) A.S.Nowick and B.S.Berry, IBM Journal Oct. (1961) 297.

MEDIUM RANGE ORDER AND MAGNETIC PROPERTIES OF AMORPHOUS $Tb_{.50}Cu_{.50}$

D. GUIMARAES*, M. SANQUER**, R. TOURBOT**, M.C. BELLISSENT-FUNEL***
and B. BOUCHER**

* Universidade de Aveiro - Dep. de Fisica - 3800 AVEIRO - Portugal
** DPhG-SRM - CEA Saclay 91191 Gif-sur-Yvette - France
*** Lab. Léon Brillouin - CEA Saclay - 91191 Gif-sur-Yvette - France
  (Lab. Commun CEA-CNRS)

The Tb-poor alloy $Tb_{.22}Cu_{.78}$ has been extensively studied[1,2]. In this paper, we present the results concerning a Tb-rich amorphous alloy obtained by sputtering.

1. MAGNETIC PROPERTIES

The susceptibility (fig.1) obeys the Curie Weiss law for T > 120K with an effective moment of $9.62\mu_B$/Tb (th.cor.value $9.72\mu_B$) and assymptotic Curie temperature θ = +56K, suggesting strong positive interactions. Below 18K the initial susceptibility decreases sharply (fig.1 and 2).

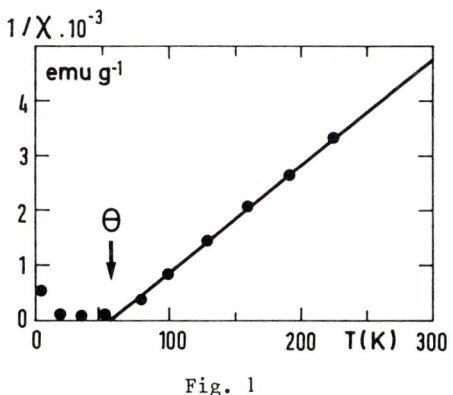

Fig. 1

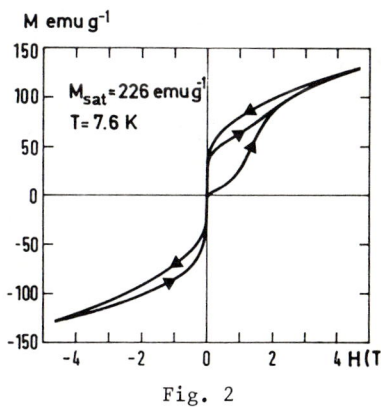

Fig. 2

At low temperature, initial magnetization curve exhibits a large opposition to magnetization. It begins like a spin glass and reaches only 56% of the saturation magnetization for H ∼ 4,5T. The hysteresis loop (fig.2) is composed of two wings, whose areas are not large, with a remanent magnetization of about $2\mu_B$/Tb. A magnetic field (0,15T) which is weak compared to that needed to obtain saturation is sufficient to reverse the magnetization. At higher temperatures (fig.3), the hysteresis loop disappears, the initial susceptibility becomes larger but for high field the field dependence of M seems to remain the same.

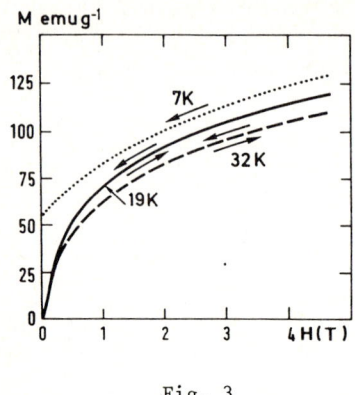

Fig. 3

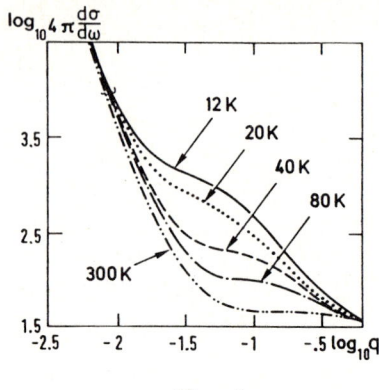

Fig. 4

2. SMALL ANGLE NEUTRON SCATTERING

The nuclear and magnetic medium range order of Tb_xCu_{1-x} amorphous alloys have been discussed in [3]. In this paper we give in details the results for the alloy x = 0,50 (fig.4).

At 300K, a temperature very much higher than the magnetic ordering temperature, we observe a scattering which can be described by the relation:

$$4\pi \frac{d\sigma}{d\omega} = 7\ 10^{-5}\ q^{-4} + 36.4\ e^{-R_1^2/5\ q^2} + 126\ e^{-R_2^2/5\ q^2} \quad (q=4\pi\sin\theta/\lambda) \quad barns/TbCu \quad (1)$$

The origin of this scattering is purely nuclear. The first term on the right hand side corresponds to domains of a few thousand Angströms whose concentration x differs by few percent ($\Delta x \sim 5\%$ for a domain of R $\sim$ 3000A). The two other terms are due to bubbles (assumed spherical with radius R_i). We are able to determine the composition of these bubbles from the absolute values with two criteria : the concentration of the matrix has to be as close as possible to the nominal composition and the volume fraction of the bubbles as small as possible. So we conclude that 16% of the volume is occupied by small bubbles (S.B.) of copper with R_1 = 5A and 4% of the volume of the sample is occupied by large bubbles (L.B.) of a Tb-poor alloy with $R_2 \sim 60A$. We can express the segregation

$$Tb_{.50}Cu_{.50} \rightarrow .20Cu + .75Tb_{.64}Cu_{.36} + 0.05Tb_{.4}Cu_{.6}$$

There are few L.B., but many S.B. of copper. If we assume that the latter are regularly spaced, we obtain an average distance of about 18-20A.

At low temperature a magnetic scattering is superimposed; by subtracting high temperature pattern, we obtain only the magnetic scattering. We account for this scattering with three terms showing the same analytical form as in (1) :

$$4\pi \frac{d\sigma}{d\omega} = 3.7 \; 10^{-5} \; q^{-4} + 350 \; e^{-R_3^2/5 \; q^2} + 2249 \; e^{-R_4^2/5 \; q^2} \quad \text{barns/TbCu} \quad (2)$$

The coefficient of q^{-4} is weak and very difficult to determine in detail because of lack of measurements at low q (fig.4) and the fact that we are obliged to subtract a strong nuclear scattering. The value of the q^{-4} coefficient leads to a resulting magnetization inside the large (few thousand Angströms) domains of about $0.1\mu_B$/Tb. The second term is due to the magnetization of volume ($R_3 \sim 9A$) included between copper bubbles and corresponds to a magnetic difference by $3 \; \mu_b$/Tb from the surrounding vol. This gives the order of magnitude of magnetization of this volume. The third term is related to a few L.B. whose compositions are a little poorer in Terbium. The magnetization of these bubbles presents at the center ($R_4 \sim 25A$) a difference of $1.6\mu_B$/Tb from the surrounding volume. So the sample can be represented by a set of small volumes (S.V.) (equal to 80% of the size of the sample) of diameter 18A with a magnetization of about $3\mu_B$/Tb, whose directions vary from one volume to another, by a set of L.B. (60A) having a weaker magnetization (down to $1.5\mu_B$/Tb) and by a multitude of S.B. of copper which, of course, do not show any magnetization.

3. CONCLUSION

At low temperature, it is difficult to turn the magnetization of the S.V. because the local anisotropy and the exchange between ions are large; M(H) increases slowly with H. Under high field (4,5T), the magnetization of these different S.V. are oriented towards the direction of the field. The remanent magnetization corresponds roughly to two thirds of the magnetization of a S.V.; the resulting magnetizations of S.V. remain "frozen" as a block when the field is reversed. Then the applied field slightly modifies the relative directions of the moments, the shape of the hysteresis loop corresponds to this type of rearrangement. For $18 < T < \theta$, the H dependence of M is reversible, the local anisotropy is weaker. Above θ only the magnetizations of S.V. which at these temperatures are decoupled from the surrounding volume, seem to subsist and we observe a discrepancy relative to the Curie Weiss law up to about 120K.

4. REFERENCES

1) B. BOUCHER, P. CHIEUX, P. CONVERT, M. TOURNARIE
   J. Phys. F : Mat. Phys. 13 (1983), 1339

2) B. BOUCHER, P. CHIEUX, P. CONVERT, R. TOURBOT, M. TOURNARIE
   J. Phys. F : Mat. Phys 16 (1986) 1821

3) B. BOUCHER, M. SANQUER, R. TOURBOT, J. BIGOT, P. CHIEUX P. CONVERT, M.C. BELLISSENT-FUNEL, Proceeding of LAM6- Garmisch-Partenkirchen RFA, 24-29 August 1986

INFLUENCE OF QUENCH RATE ON THE STRUCTURAL RELAXATION IN Fe-B GLASSES

Z. ALTOUNIAN
Physics Department, McGill University, 3600 University Street,
Montreal, Quebec, Canada H3A 2T8

Quench rate dependence of reversible and irreversible structural relaxation in Fe-B glasses, at the eutectic composition, was studied through differential scanning calorimetry measurements. The irreversible part of structural relaxation is strongly quench rate dependent, whereas the reversible part is independent of quench rate and hence is an intrinsic property of the glassy state.

1. INTRODUCTION

Metallic glasses relax upon thermal annealing to a structure closer to a local minimum of free energy. This structural relaxation is accompanied by changes in many physical properties. Relaxation has been classified as of two types : irreversible relaxation and reversible relaxation. A great variety of properties have been used for the study of relaxation in glasses[1]. The interpretation of many of these properties, the simplest being the Curie temperature for magnetic glasses, are not straightforward and in some cases the value of the property depends on the state of the glass and the nature of the measurement[2]. For this reason we have chosen the more general and direct technique of measuring the relaxation enthalpy to investigate quantitatively the quench rate dependence of both the irreversible and reversible part of structural relaxation in Fe-B metallic glasses.

2. EXPERIMENTAL

Fe-B ingots of the eutectic composition, $Fe_{82.5}B_{17.5}$ were melt spun[3] under identical conditions except the quenching rate which was controlled by varying the wheel speed. The crystallization temperatures, T_x, enthalpy changes and Curie temperatures, T_c, were determined by means of a calibrated differential scanning calorimeter (DSC). The change in apparent specific heat for irreversible relaxation, $\Delta C_{p,irr}$ was obtained by heating the as quenched ribbons at a heating rate of 40K/min to a temperature 100K below T_x. A similar scan was taken immediately. The enthalpy difference between the two scans is the irreversible part of structural relaxation (up to T_x-100), as the enthalpy difference between the second scan and any subsequent scan is zero. The pre-annealed ribbons were annealed for 1 hour in the DSC for different annealing temperatures T_a. A DSC scan was obtained at 40K/min from room temperature to at least T_a+100. The difference between this scan and a second identical scan gives the reversible contribution to the relaxation enthalpy.

3. RESULTS AND DISCUSSION

Glassy Fe-B ribbons were obtained for wheel speeds between 40m/s (thickness 26μm) and 70m/s (thickness 16μm). The constancy of T_x (744.9±0.5K, at 40k/min) for all samples suggests that there is no variation in the concentration of quenched-in nuclei in the ribbons[2]. The random scatter of the data for T_c (610±1.5K) outside measurement errors may be due to slight variations of the boron concentration, x, in each ribbon. Near the eutectic composition $\frac{dt_c}{dx} \sim 22K/at.\%B^3$, which gives a maximum variation of about 0.1 at.%B among the samples.

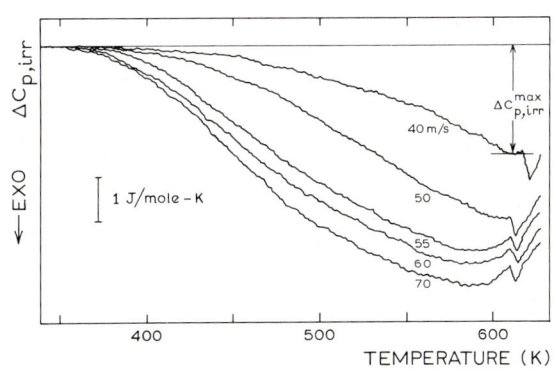

FIGURE 1

DSC scans of $\Delta C_{p,irr}$ for different quench rates.

Figure 1 shows the DSC scans of the irreversible part, $\Delta C_{p,irr}$ of the structural relaxation. The relaxation enthalpy increases with quench rate indicating that samples spun at lower quench rate are already somewhat relaxed. In addition, Figure 1 also confirms the observation that the relaxation rate increases with quench rate[4]. The anomaly at ~ 610K is due to the slight variation in T_c between the two scans. The maximum of $\Delta C_{p,irr}$, $\Delta C_{p,irr}^{max}$ and/or the enthalpy of the irreversible relaxation, ΔH_{irr}, can be used as a measure of the irreversible relaxation process. Figure 3 shows the variation of ΔH_{irr} with quench rate. ΔH_{irr} for the sample

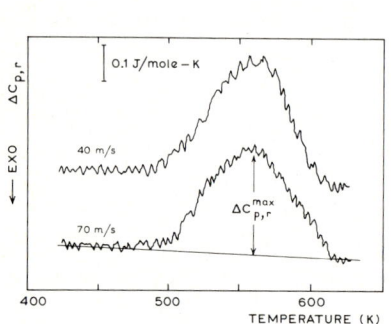

FIGURE 2

DSC scans of $C_{p,r}$ (T_a = 500K) for samples with lowest and highest quench rates.

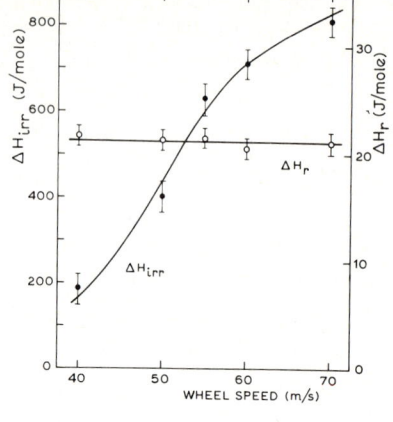

FIGURE 3

Relaxation enthalpies as a function of quench rate.

spun at 70m/s is 4.3 times larger than that spun at 40m/s.

The reversible relaxation enthalpies, ΔH_r, were measured corresponding to anneal temperatures ranging from 400K to 530K. Higher T_a requires scanning to at least T_a+100 which may promote surface crystallization[3] and thus decrease ΔH_r [5]. Figure 3 shows typical DSC scans for samples at the two extremes of quench rate for T_a = 500K. The maximum value, $\Delta C_{p,r}^{max}$ as well as ΔH_r can be used as a quantitative measure for the reversible relaxation process. For ternary metalloid containing glasses ΔH_r is an order of magnitude larger[5] than that for binary Fe-B glasses. As shown in Figure 3 ΔH_r and $\Delta C_{p,r}^{max}$, for a given T_a are identical, within experimental errors, for <u>all</u> samples irrespective of quench condition. This suggests that ΔH_r and/or $\Delta C_{p,r}^{max}$ are an intrinsic property of the glassy state.

REFERENCES

1. See for example, A.L. Greer, J. Non-Cryst. Solids 61 & 62 (1984) 737.
2. A.L. Greer, Acta Metall. 30 (1982) 171.
3. Z. Altounian, J.O. Strom-Olsen and M. Olivier, J. Mater. Res. 2 (1987) 54.
4. F.E. Luborsky and J.L. Walter, Mater. Sci. Eng. 35 (1978) 225.
5. R. Brüning, M.Sc. thesis (McGill, 1986).

MAGNETIC DOMAIN STRUCTURE BY MEASURING MAGNETIC FORCES

J.J. Sáenz and N. García

Dpto. de Física de la Materia Condensada, C-III, Universidad Autónoma de Madrid, Cantolanco, 28049-Madrid Spain.

P. Grüutter, E. Meyer, H. Heinzelmann, R. Wiesendanger, L. Rosenthaler, H.R. Hibder and H.J.-Güntherodt*

Institut für Physik, CH-4056 Basel, Switzerland.

We present in this work a new method to obtain information about local surface magnetic properties, based on the idea of measuring magnetic forces with the recently developed Atomic Force Microscope (AFM). The influence of the experimental conditions and the involved magnetic forces on the performance of the AFM as a magnetic profiling device is discussed. Preliminary experimental results are reported.

We present in this work a way of observing magnetic domain structure based on the idea of measuring magnetic forces with an Atomic Force Microscope[1] (AFM). In the AFM technique, a sharp tip attached to a tiny cantilever is used to map the contours of a sample surface. Instead of measuring the current as in the Scanning Tunneling Microscopy[2] (STM), the force between tip and sample is used as the control parameter. Forces are detected by measuring the deflection of the lever.

In the magnetic version of the AFM, images are obtained by measuring the interaction force between a single domain magnetic microtip and a magnetic sample. The main features of the force acting on the tip as it goes across a domain wall can be illustrated simply by considering two infinite antiparallel domains separated by a sharp domain wall (fig. 1a). The interaction force can be estimated by assuming a direct interaction between the permanent tip and sample magnetic moments per unit volume. Because of the dipolar behavior of the magnetic forces, they manifest themselves only when the tip approaches a domain wall separating two regions of different magnetizations.

To simulate the influence of these forces in an AFM experiment, consider that the base of the lever is kept fixed with respect to the sample. When

---

*Work supported by the Swiss National Science Foundation, the Kommission zur Förderung der wissenschaftlichen Forschung and the Eidgenössische Volkswirtschaftsstiftung.

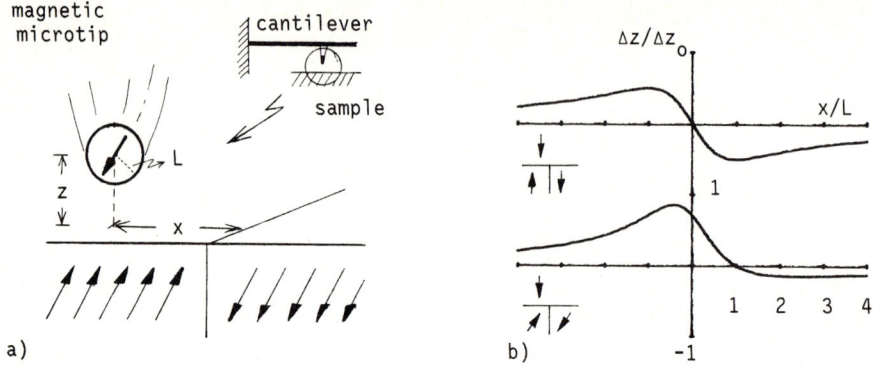

FIGURE 1
a) Scheme of the magnetic version of the AFM. b) Calculated tip displacements for two different spin configurations.

the tip approaches a domain wall, it will suffer a normal displacement given by $\Delta z = F/K$, being K the force constant of the lever. A very important point associated with the long range force is that, as long as the radius L of the single domain particle at the end of the tip is large compared to the gap distance between tip and sample z, the tip force does not depend on z. Then, the magnetic AFM will not be sensitive to the surface roughness, as long as the surface roughness is small compared to L. A measure of the lateral resolution (the minimum domain size which can be distinguished) is the width of the region in which the force changes as the tip goes across a domain wall. This characteristic length is of the order of the tip radius L (see Figure 1b). For typical force constants $K \cong 0.01$ to $1 Nm^{-1}$ we estimated a lower limit to the AFM resolution of $L_{min} > \cong 50 - 100$ Å, or forces $F_{min} > \cong 10^{-11} - 10^{-10}$ N. Provided the domain wall width would be of the order or higher than L, it would be possible to get information, not only of the domain distribution on the sample, but also of the domain wall itself.

We have tried to measure magnetic forces with our AFM, which is operated in air. In order to simulate the situation presented in Fig. 1, we have chosen a ferromagnetic Ni foil as a sample and an electrochemically etched and rolled Ni lever with an integrated ferromagnetic tip. As an example of various measurements, Fig. 2 shows the obtained line scans which are very similar to the asymmetric line shown in Fig. 1b. We have checked very carefully the topography of this Ni foil by STM measurements also performed in air and no correlation between the height of the structures in the topographic and

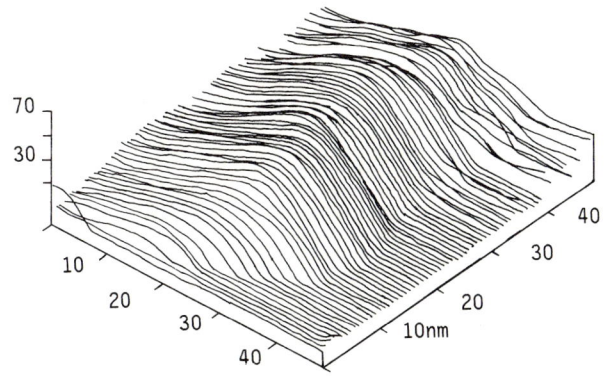

FIGURE 2
Experimental deflection of the lever versus the tip position on the Ni sample. The displayed scans are equiforce lines. In this case the interaction force was attractive.

magnetic images is observed.

We have shown theoretically that with our magnetic version of the AFM, it is possible to observe domain walls and the domain distribution for a magnetic sample. Lateral domain resolutions of the order of 0.01μ can be achieved with this technique. We have presented preliminary experimental results which show the sensitivity of our AFM to magnetic interaction.

REFERENCES

1) G. Binnig, C.F. Quate and Ch. Gerber, Phys. Rev. Lett. 56 (1986) 930

2) G. Binnig, H. Rohrer, Ch. Gerber and E. Weibel, Phys. Rev. Lett. 50(1983) 120; G.Binnig and H. Rohrer, IBM J. Res. Dev. 30 (1986) 355.

AMORPHOUS TO CRYSTALLINE TRANSITION IN FINE IRON-COBALT ALLOY PARTICLES

S. WELLS and S.W. CHARLES
Department of Physics, University College of North Wales, Bangor, Gwynedd, LL57 2UW, UK

Iron-cobalt alloys produced in a high-purity argon atmosphere by the borohydride reduction of iron and cobalt salts in aqueous solution in which oxygen has been removed undergo a spontaneous transition from the amorphous to crystalline state with the evolution of sufficient heat to cause the alloys to glow red-hot.

1. INTRODUCTION

The preparation of amorphous particles by direct chemical methods and their subsequent annealing by heating strongly in vacuo or in an inert atmosphere has previously been reported[1,2,3]. However, in this work we report that annealing occurs in certain iron-cobalt alloys at room temperature in a high purity argon atmosphere.

2. PREPARATION

An equal quantity of deoxygenated 1M sodium borohydride solution was added to a 1M solution of iron and cobalt salts[4] under an atmosphere of high purity argon. After washing with de-oxygenated water and acetone, the sample was dried in a stream of argon using a magnet to hold the fine particles. During drying all alloys except those with the highest fraction of iron ($\geq$ 90 at %) spontaneously glowed red-hot.

3. RESULTS AND DISCUSSION

X-ray diffractometry confirmed that for those alloys with iron concentration $\leq$ 80 at %, a transition occurred from an amorphous to crystalline state (Fig 1.). Mossbauer spectra taken of the samples after the transition had occurred (Fig. 2), revealed the presence of lines consistent with crystalline cobalt/iron alloys and other weaker lines consistent with crystalline metal borides[5]. The boron concentration lay between 1 and 8 at % depending on the Fe/Co ratio. There was no conclusive evidence for the presence of any significant concentration of metal oxides from either the Mossbauer or X-ray diffraction studies. For iron-rich alloys ($\geq$ 90 at %), for which a transition did not take place, Mossbauer and X-ray studies showed that these alloys possessed significant crystalline character, However the

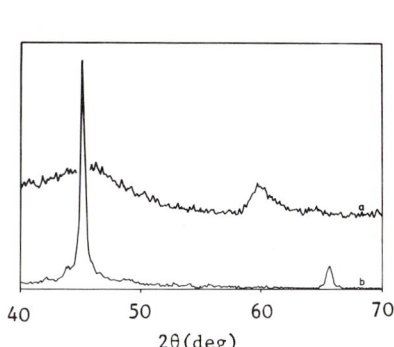

FIGURE 1

X-ray diffraction of a Fe:Co alloy (a) before and (b) after transition.

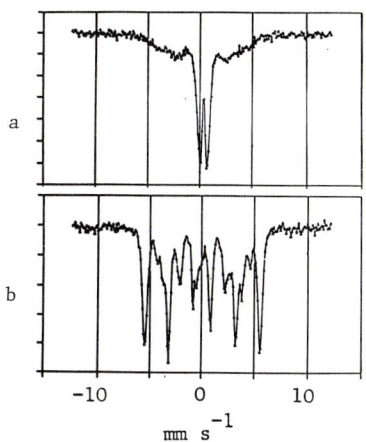

FIGURE 2

Mossbauer spectra of a 40:60 Fe:Co alloy (a) before and (b) after transition.

amorphous state for these alloys could be achieved by increasing the molar ratio of borohydride used in the reaction. During the transition a large increase in particle size from ≈0.05μm to ≈0.5μm occurred (Fig. 3).

A large increase in the values of the saturation magnetization occurred after the transition to near that but somewhat lower than the bulk values. (Fig. 4). The values of the saturation magnetization of annealed samples obtained by Watanabe et al[6] were lower than those reported here. The coercivity of the particles dropped markedly after the transition for the

FIGURE 3

S.E.M. of a 40:60 Fe:Co alloy (a) before and (b) after transition

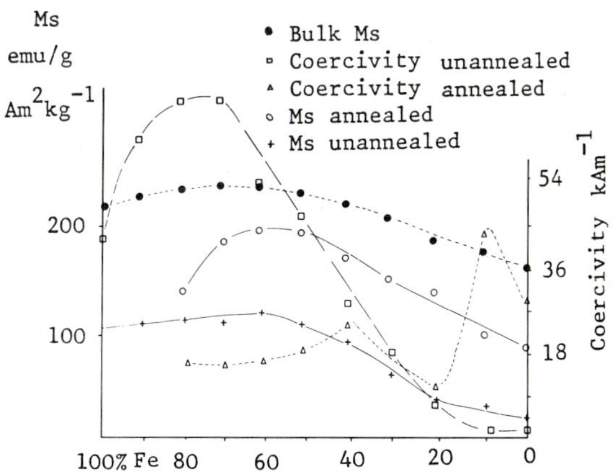

FIGURE 4

Magnetic properties of Fe:Co alloys.

iron-rich alloys whilst there was an equally marked rise for the cobalt-rich alloys. (Fig. 4). The inflection points in the coercivity curve of the crystalline sample at 80 and 90 at % Co occur at phase transitions in the iron cobalt system[7]. The coercivity of similar alloys prepared by the borohydride reduction in a magnetic field of 0.1T [6,8] had very much larger coercivities 160 kAm^{-1} rising to 185 kAm^{-1} on annealing at around 400°C.

The transition from the amorphous to crystalline state would be expected to release an energy not too dissimilar to that of the latent heat of fusion of the Co/Fe alloys ($\simeq 2.6 \times 10^5$ Jkg^{-1}). In view of the specific heat capacity of these alloys at room temperature ($\simeq 400$ Jkg^{-1}), the sudden release of this energy would raise the temperature of the alloys many hundreds of degrees accounting for the red-hot glow. It has previously been reported[8] that a broad exothermic peak (1.2×10^5 Jkg^{-1} in maximum calorific value) in the range 200 to 300 °C is observed for similar alloy systems. An energy release comparable to the heat release already described would also occur as a result of the reduction of surface energy ($\simeq 10^2$ Jm^{-2}) during the sintering process in which an approximate ten-fold increase in particle size resulted.

ACKNOWLEDGEMENT

We are indebted to Dr. Steen Mørup for his measurements of the Mossbauer spectra, and to the EEC and SERC for financial support.

REFERENCES

1) J. van Wonterghem, S. Mørup, S.W. Charles, S. Wells and J. Villadsen, Phys. Rev. Lett. 55 (1985) 410.
2) J. van Wonterghem, S. Mørup, C.J.W. Koch, S.W. Charles and S. Wells, Nature 322 (1986) 622.
3) I.D. Dragieva, M. SL. Slavcheva and D.T. Buchkov, J. of Less-Common Metals 117 (1986) 311.
4) A.L. Oppegard, F.J. Darnell and H.C. Miller, J. Appl. Phys. 32 (1961) 1845.
5) D.T. Buchkov, S. Nikolov, I.D. Dragieva and M. SL. Slavcheva, J. Magn. Magn. Mat. 62 (1986) 87.
6) A. Watanabe, T. Uehori, S. Sitoh and Y. Imaoka, IEEE Trans. Magn. MAG - 17 (1981) 1455.
7) W.C. Ellis and E.S. Greiner, Trans A.S.M. (1940) 415.
8) Y. Taketomi, A. Watanabe, Y. Tokuoka, H. Sugihara and Y. Imaoka, J. Magn. Magn. Mat. 31-34 (1983) 905.

FLASH ANNEALING OF COLD-DRAWN AMORPHOUS MAGNETOSTRICTIVE WIRES

R. Malmhäll[+], K. Mohri[§], F.B. Humphrey[*], T. Manabe[§], H. Kawamura[§] and J. Yamasaki[§]

Cold-drawn amorphous magnetostrictive $Fe_{77.5}Si_{7.5}B_{15}$ wires have been found to exhibit bistable hysteresis loops (large Barkhausen jump) after flash current annealing under applied axial tension. At optimum bistable switching conditions, current pulse amplitude I_p and duration t_p follows the equation $I_p^2 t_p = A d^4$, where $A = 1.6 \times 10^{15}$ A^2 s/m^4. Maximum H^* was 1.1 Oe for 50 µm diameter wire, similar to field-tension oven anneal case.

1. INTRODUCTION

In-rotating-water rapid quenched amorphous magnetostrictive alloy wires have been shown to exhibit complete large Barkhausen jump, leading to sharp and stable induced voltage pulses, when excited by an external ac magnetic field[1,2]. Potential applications for such properties of the wires are as rotary encoders, security sensors, non-contact switching, magnetic field and electric current sensing.

The bistable magnetization (M) reversals in as-quenched (AQ) wires has only been observed in wires of 6 cm length or longer[3], but it was recently demonstrated that three times shorter pulse generation elements can be made by field-tension oven annealing of cold-drawn (CD) amorphous Fe-B-Si wires[4]. However, from a manufacturing point of view the current flash annealing process[5] can be more easily implemented.

2. EXPERIMENTAL

Amorphous $Fe_{77.5}Si_{7.5}B_{15}$ wires were cold-drawn in several steps from 125 µm down to 50 µm diameter. Flash annealing was performed on 10 cm long wires in air by passing a current pulse of amplitude I_p and duration t_p through the wires.

+ Royal Institute of Technology, S-100 44 Stockholm, Sweden
§ Kyushu Institute of Technology, Kitakyushu 804, Japan
* Carnegie-Mellon University, Pittsburgh, PA 15213, USA

Part of this work have been supported by the Swedish National Board for Technical Development, grant STU 86-4867.

3. RESULTS AND DISCUSSION

The hysteresis loops of CD wires in as-drawn and flash annealed states are presented in Fig.1. The loops for AQ and field-tension oven annealed states are shown for comparison. The AQ wire exhibits bistable M reversal with a switching field H^* of 0.08 Oe, whereas the CD wire is not bistable. However, the bistable switching in CD wires was re-established after flash and oven annealing as shown in Fig.1c and d. There is very little difference between H^* and squareness ratio M_r/M_s obtained by the kinds of annealing. It can be noted that the values for H^* and M_r/M_s are much larger than for AQ wire.

The flash annealing process was investigated further by varying I_p for a given t_p at different σ_{an}. The combinations of t_p and σ_{an} which resulted in a bistable M behavior are shown in Fig.2 for 50 µm and 70 µm CD wires. In the regions marked 'bistable', it was possible to find values of I_p resulting in bistable switching. Bistability could not be obtained for $\sigma_{an} = 0$.

In general, it was observed that H^* is decreasing with increasing I_p at fixed t_p and σ_{an}. The current amplitude, which gave the maximum value of H^*, at various σ_{an} is shown in Fig.3, where the square of I_p is plotted as a function of the inverse of t_p. From this figure, it is seen that $I_p^2 t_p$ = constant, the constant being about four times larger for 70 µm wire than 50 µm. If it is assumed that the electric energy in the current pulse is converted into energy of heat, it is expected that $I_p^2 t_p = A d^4$, where d is the wire diameter. The constant A was calculated from experimental data to be $1.6 * 10^{15}$ A^2 s/m^4.

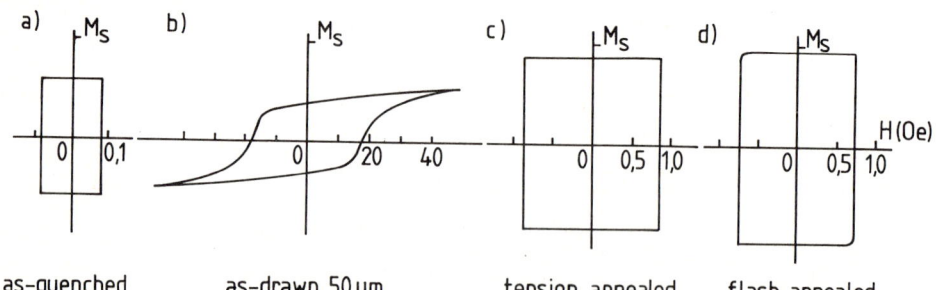

FIGURE 1
Hysteresis loops of amorphous magnetostrictive Fe-B-Si wires, a) before and b) after cold-drawing, and for cold-drawn wires c) after oven tension annealing (350 C, 30 min, 150 kg/mm²) and d) flash current tension annealing (0.45 A, 50 ms, 100 kg/mm²).

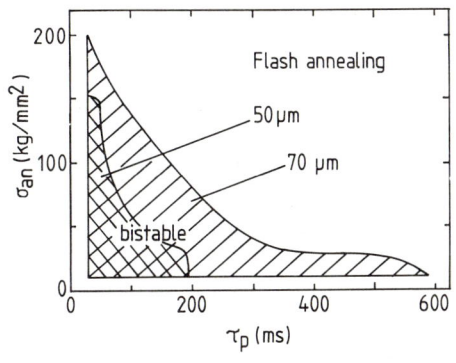

FIGURE 2
Flash current pulse width t_p and applied axial tension σ_{an} for obtaining bistable magnetization reversals in cold-drawn wires.

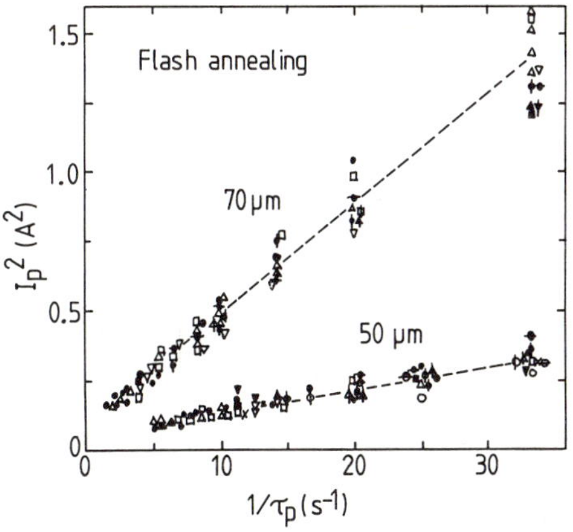

FIGURE 3
Flash current amplitude I_p and pulse width t_p for various tensions as given in Fig.2.

4. CONCLUSIONS

The improved bistable switching properties obtained for oven annealed cold-drawn amorphous magnetostrictive $Fe_{77.5}Si_{7.5}B_{15}$ wires have been reproduced by flash current annealing. An applied axial tension was found to be a necessary condition. The maximum H^* was 1.1 Oe for 50 μm diameter wire, which is similar to field-tension oven annealed wire. The current pulse amplitude I_p and duration t_p at optimum bistable switching conditions was observed to follow $I_p^2 t_p = A d^4$, where $A = 1.6 \times 10^{15}$ A^2 s/m^4, indicating that the temperature rise due to the current pulse is the dominant effect of flash annealing.

REFERENCES

1) K. Mohri, F.B. Humphrey, J. Yamasaki and K. Okamura, IEEE Trans. Magn., MAG-20 (1984) 1409.

2) K. Mohri, F.B. Humphrey, J. Yamasaki and F. Kinoshita, IEEE Trans. Magn., MAG-21 (1985) 2017.

3) F. Kinoshita, R. Malmhäll, K. Mohri, F.B. Humphrey and J. Yamasaki, IEEE Trans. Magn., MAG-22 (1986) 445.

4) R. Malmhäll, K. Mohri, F.B. Humphrey, T. Manabe, H. Kawamura, J. Yamasaki and I. Ogasawara, IEEE Trans. Magn., MAG-23 (1987) Sept.

5) T. Jagielinski, IEEE Trans. Magn., MAG-19 (1983) 1925.

AMORPHOUS Tb(Fe,Co) THIN FILMS BY FACING TARGETS SPUTTERING

Nobuyoshi KITAMURA, Toyoaki HIRATA and Masahiko NAOE

Tokyo Institute of Technology, Faculty of Engineering,
Department of Physical Electronics, 2-12-1, O-okayama,
Meguro-ku, Tokyo 152, JAPAN

Amorphous Tb(Fe,Co) thin films were prepared by a Facing Targets Sputtering (FTS) system in changing magnetic fields for plasma confinement. When the magnetic field was 160 Gauss, the growing films were plasma-exposed and were bombarded by high energy particles from targets. A C/N ratio was 46 dB, but it was higher than that obtained by the conventional magnetron sputtering. Then, the magnetic field was increased to 220 Gauss. In this case, the growing film was plasma-free and was not bombarded by them. The film surfaces were very smooth and any columnars were not remarkable in the Transmission Electron Micrograph (TEM) of the cross section in the cleaved films. As a result, the C/N ratio of the MO disk with these smooth Tb(Fe,Co) thin film deposited by the FTS system was 57 dB for the bit length of about 4 μm.

1. INTRODUCTION

Amorphous Tb(Fe,Co) thin films are well known as one of the most suitable media for magneto-optical recording. They have been studied by many workers using conventional sputtering methods[1,2]. But, they are easily damaged during deposition by the bombardment of high energy particles from the plasma, even when they are deposited at low rate under the low input. For mass production of them for recording media, the development of high-rate deposition technique is of practical importance. On the other hand, the FTS system can confine the plasma in the space between two facing targets and can achieve a high-rate deposition with a minimum influence of the effects by the plasma bombardment[3]. Then, in this study, the plasma-free FTS system was used to prepare Tb(Fe,Co) films in order to investigate the influence of the plasma bombardment to the film properties. Consequently, the magneto-optical disk with a Tb(Fe,Co) thin film by FTS showed a C/N ratio of higher than 50 dB for the bit length of more than 1.4 μm primarily due to the microstructual uniformity in the film by FTS.

2. EXPERIMENTAL

Specimen Tb(Fe,Co) films were prepared by two types of FTS system as shown in Fig.1(a) and (b). For FTS-I shown in Fig.1(a), the magnetic field of 160 Gauss is applied perpendicularly to the target plane. In this case, the surface of

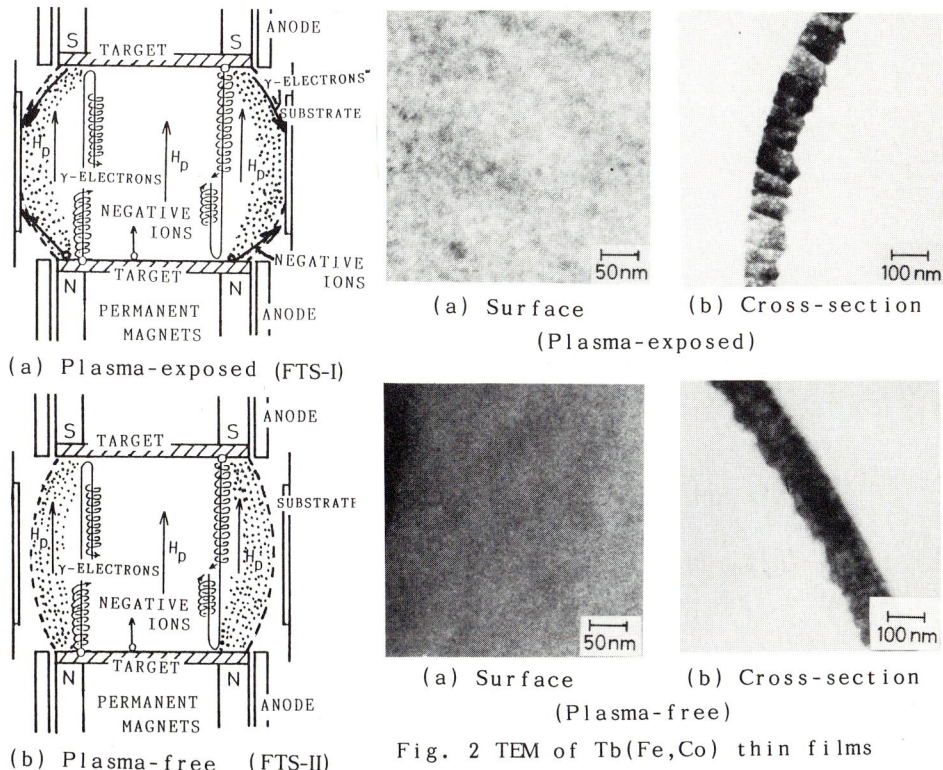

(a) Plasma-exposed (FTS-I)

(b) Plasma-free (FTS-II)

Fig. 1 FTS system

(a) Surface (b) Cross-section
(Plasma-exposed)

(a) Surface (b) Cross-section
(Plasma-free)

Fig. 2 TEM of Tb(Fe,Co) thin films

the substrate is exposed to the plasma bombardment because of relatively weak magnetic field. On the other hand, in the FTS-II case, films can be deposited under the plasma-free condition at the same substrate position. This is because the stronger magnetic field of about 220 Gauss is applied in order to confine the plasma completely between the facing targets. The read/write characteristics were evaluated in the films prepared on 5.25- and 12-inch pregrooved disks made of glass or PMMA.

3. RESULTS AND DISCUSSION

Films 100 nm thick were prepared at the deposition rate of 300 nm/min in the FTS-I and II cases shown in Fig.1 (a) and (b). The microstructures of the films were carefully observed by TEM. They are shown in Fig.2. In the FTS-I case, which is plasma-exposed condition, the film seems to be composed of uneven grains probably due to bombardment by high energy particles. In a plasma-free FTS-II case, the film surfaces were very smooth and any columnars were not remarkable in the cross section in the cleaved film. Figure 3 shows the

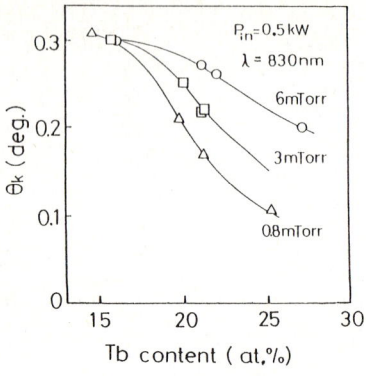
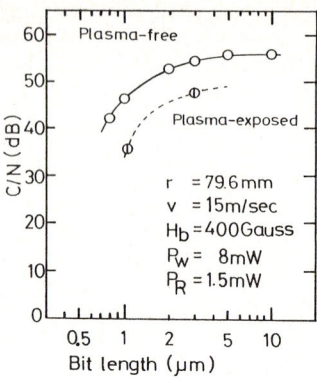

Fig.3 Dependence of θ_K on Tb content Fig.4 Dependence of C/N ratio on bit length

dependence of θ_K on Tb content in the films prepared at various P_{Ar}. It can be found that θ_K becomes larger as P_{Ar} becomes higher. Though the causes of this phenomenon are not clear, one of them is considered that the substrate is apt to be exposed to high energy particles in the lower P_{Ar}. For each P_{Ar}, θ_K took the same maximum value near 15 at.% of Tb content. Figure 4 shows the correlation between a C/N ratio and bit length on the disks. In the FTS-I condition, the C/N ratio became 46 dB for the bit length of 3 μm. This C/N ratio was higher than that obtained by the conventional magnetron sputtering. However, in the FTS-II case, the C/N ratio became higher than 50 dB for the bit length of 1.4 μm and 57 dB for the bit length of 4 μm. Therefore, the C/N ratio in the FTS-II case showed higher values compared with that in the FTS-I case. These results indicate that the plasma-free condition may be very important to obtain the amorphous Tb(Fe,Co) thin films.

4. CONCLUSION

The higher-quality thin films were obtained even at such a high deposition rate as 300 nm/min by the FTS system. These films were able to be deposited in FTS-II case, where the substrate was plasma-free. Evaluation of the media showed C/N ratio as high as 57 dB for the bit length of about 4 μm. Consequently, the FTS system is considered to be quite promissible for mass production of the magneto-optical recording media.

REFERENCES

1) Y.Togami, K.Kajimura, K.Sato and T.Teranishi, J.Appl.Phys., 53, (1982) 2334.

2) S.Miura, N.Imamura and T.Kobayashi, Jpn.J.Appl.Phys., 15, (1976) 933.

3) M.Naoe, S.Yamanaka and Y.Hoshi, IEEE Trans. on Mag., MAG-16, (1980) 646.

TEMPERATURE DEPENDENCE OF BISTABLE MAGNETIC PROPERTIES
IN AS-QUENCHED AND COLD-DRAWN AMORPHOUS Fe-B-Si WIRES

R. Malmhäll[+], K. Mohri[§], F.B. Humphrey[*], T. Manabe[§],
H. Kawamura[§], J. Yamasaki[§] and D-X Chen[+]

Amorphous magnetostrictive $Fe_{77.5}Si_{7.5}B_{15}$ alloy wires were studied in the temperature range 77K to 480 K. The temperature coefficient of the critical switching field was 1.2×10^{-3} K^{-1} and 2.4×10^{-3} K^{-1} for as-quenched (125 μm) and tension annealed cold-drawn (50 μm) wires, respectively, and of minimum domain propagation field 2.0×10^{-3} K^{-1} for 50 μm wire. Induced pickup voltage in the two cases was roughly constant.

1. INTRODUCTION

Amorphous magnetostrictive wires have been shown to exhibit rapid bistable magnetization reversals, when excited by an external ac magnetic field. This was first observed for as-quenched wires[1,2]. Recently, it has also been demonstrated that cold-drawn Fe-B-Si wires, after suitable annealing treatment, can be made bistable[3,4]. Potential applications of bistable wires are as pulse generation elements in rotary encoders, etc. and as sensors for magnetic field and electric current. Depending on the application, it is important to know how their magnetic properties vary with temperature.

2. EXPERIMENTAL

As-quenched (125 μm diameter) and cold-drawn (50 μm) amorphous $Fe_{77.5}Si_{7.5}B_{15}$ alloy wires were obtained from Unitika Co., Japan. The cold-drawn wires were subjected to annealing at 350 C for 30 min with an applied axial tension of 150 kg/mm^2. The temperature dependence of magnetization, switching field and domain propagation field, as well as induced voltage pulses in a 230 turn, 5mm long pickup coil, were measured in the range 77-480 K. The voltage pulses were determined in an applied sinusoidal axial field with amplitude 2 Oe and frequency 60 Hz.

+ Royal Institute of Technology, S-100 44 Stockholm, Sweden
§ Kyushu Institute of Technology, Kitakyushu 804, Japan
* Carnegie-Mellon University, Pittsburgh, PA 15213, USA

Part of this work have been supported by the Swedish National Board for Technical Development, grant STU 86-4867.

3. RESULTS AND DISCUSSION

The temperature (T) dependence of the hysteresis loop for tension annealed cold-drawn (CD) wire is shown in Fig.1. The switching field (H^*) decreases from 1.3 Oe to less than 0.5 Oe with increasing T, whereas the squareness ratio (M_r/M_s) remains close to unity in the whole T-range. In the case of as-quenched (AQ) wire, M_r/M_s (= 0.5) is also found to be constant with T.

The domain wall propagation characteristics, obtained by the Sixtus-Tonks experiment[5], were determined for various T. The result is presented in Fig.2 for CD wire, where the domain wall velocity (v) is plotted as a function of applied axial magnetic field (H). The wall mobility (μ) and the wall propagation field (H_o) can be obtained from the slope and by the intersection with the H-axis, respectively, in Fig.2. It is found that H_o increases almost linearly with decreasing T, and that μ is essentially constant.

The temperature dependence of H^*, H_o and saturation magnetization M_s for AQ and CD wires are summarized in Fig.3, where M_s is plotted in normalized scale and H^* for AQ wire is also shown amplified 10 times. Clearly, H^* for CD wire has a stronger T-dependence than for AQ wire. As expected, $M_s(T)$ is the same in both cases.

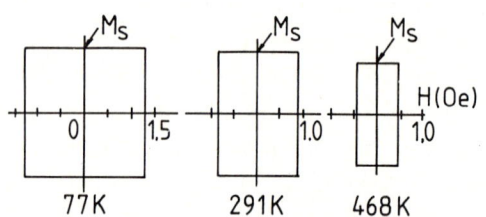

FIGURE 1
Hysteresis loops for tension annealed cold-drawn Fe-B-Si wire at various temperatures.

The variation of induced voltage pulse amplitude (e_p) and width (t_p) with temperature is given in Fig.4. It is observed that e_p for both wires decreases at higher T, but more pronounced for the CD wire. The pulse width t_p for AQ wire has a similar T dependence as e_p, whereas t_p for CD wire is found to have a minimum around 280 K.

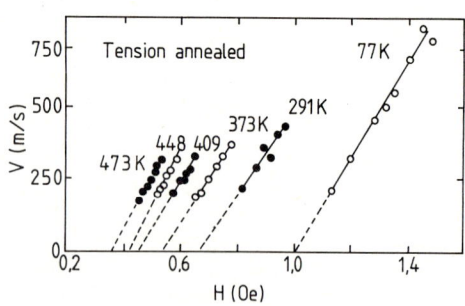

FIGURE 2
Domain wall velocity as a function of applied axial magnetic for different temperatures.

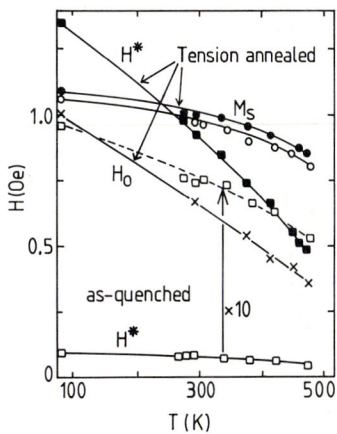

FIGURE 3
Switching field H^*, domain propagation field H_0 and saturation M_S versus temperature for as-quenched and tension annealed cold-drawn wires.

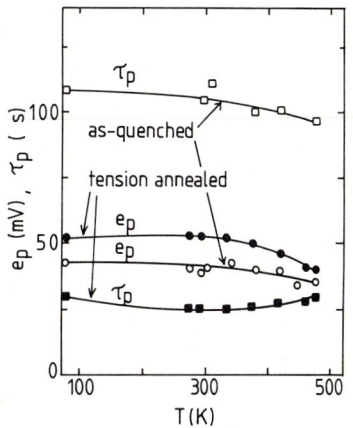

FIGURE 4
Induced voltage pulse amplitude e_p and width t_p as a function of temperature for as-quenched and tension annealed cold-drawn amorphous Fe-B-Si wires.

4. CONCLUSIONS

The magnetic properties of as-quenched and tension annealed cold-drawn amorphous magnetostrictive $Fe_{77.5}Si_{7.5}B_{15}$ wires were studied in the temperature range from liquid N_2 to about 200 K below the Curie temperature. It is found that the as-quenched wire has a weaker temperature dependence than the cold-drawn wire for some of the properties studied, such as the domain nucleation field and induced voltage pulse width.

REFERENCES

1) K. Mohri, F.B. Humphrey, J. Yamasaki and K. Okamura,
   IEEE Trans. Magn., MAG-20 (1984) 1409.

2) K. Mohri, F.B. Humphrey, J. Yamasaki and F. Kinoshita,
   IEEE Trans. Magn., MAG-21 (1985) 2017.

3) R. Malmhäll, K. Mohri, F.B. Humphrey, T. Manabe, H. Kawamura,
   J. Yamasaki and I. Ogasawara,
   IEEE Trans. Magn., MAG-23 (1987) Sept.

4) R. Malmhäll, K. Mohri, F.B. Humphrey, T. Manabe, H. Kawamura and J. Yamasaki, Flash annealing of cold-drawn amorphous magnetostrictive wires, this volume.

5) K.J. Sixtus and L. Tonks, Phys. Rev., 37 (1931) 930.

RANDOM MAGNETISM AND PHASE TRANSITIONS IN HOMOGENEOUS AND MULTILAYERED
AMORPHOUS SYSTEMS*

D.J. SELLMYER

Behlen Laboratory of Physics, University of Nebraska, Lincoln, Nebraska,
68588-0111, U.S.A.

A review is given of recent progress made in understanding magnetic transitions in random-anisotropy amorphous systems. Scaling analyses performed on several rare-earth-based glasses indicate that an Ising-type of spin-glass phase transition occurs. Some of the difficulties in this picture are described, along with recent theoretical advances. Finally, the production and control of perpendicular magnetic anisotropy in amorphous multilayered systems are discussed.

1. INTRODUCTION

The purpose of this article is to review the recent advances made in understanding the magnetic phase transitions and properties exhibited by amorphous solids containing significant random magnetic anisotropy (RMA). Systems of this nature are of interest for fundamental reasons concerning the development of a proper description of the phase transition or glass "transformation" picture of magnets with random interactions. In addition, there may be significant data-storage applications for rare earth-transition metal (RE-TM) glasses, materials with which we will be concerned in this paper.

Research on RMA systems up to about 1983 has been reviewed in the book by Moorjani and Coey.[1] More recent reviews include those by Sellmyer and Nafis[2] and by Rhyne.[3] Here we shall focus on the significant progress, both experimental and theoretical, made since about 1985.

2. PHASE-TRANSITION ASPECTS

  2.1. Scaling-Analysis

It is well known that RE-TM glasses exhibit transitions at low enough temperature to a phase in which the RE moments are frozen in scattered directions (speromagnet). The spin structure is spin-glass-like and the origin of this scatter is the RMA term in the model Hamiltonian

---

*Research supported by the National Science Foundation under Grants DMR 8605367 and INT 8419546.

$$H = - \sum_{ij}{}' (J_o + \Delta J_{ij}) \vec{J}_i \cdot \vec{J}_j - \sum_i D (\hat{n}_i \cdot \vec{J}_i)^2 , \qquad (1)$$

where J_o is assumed to be an average ferromagnetic exchange, ΔJ_{ij} are the exchange fluctuations, D an average local anisotropy arising from electric field gradients of neighboring atoms, and $\hat{n}_i$ the easy-axis direction for the ith spin. If all three terms of the Hamiltonian are present it becomes very difficult to separate the exchange-fluctuation effects from the RMA effects, so attempts generally are made to study systems where either $\langle\Delta J\rangle$ is negligible compare to D or the reverse. In practice this means that Gd-based glasses are used to simulate the $D \cong 0$ case (there is no orbital angular momentum to first order), and the other rare earths (e.g., Tb, Dy, etc.) are studied in the large-D case.

The similarities in the low temperature properties of spin glasses and speromagnets have stimulated efforts to analyse the magnetic properties near the ordering temperature, Tg, in terms of a spin-glass phase-transition model. The first attempt of this kind was that of Sellmyer and Nafis[2], who determined the critical exponents of a-$Tb_{58}Fe_{18}G_{24}$, G = glass former, from ac susceptibility data. Following the spin-glass theory of Suzuki a <u>singular</u> susceptibility,

$$\chi_s \equiv \chi_o - \chi \qquad (2)$$

was determined, where χ_o and χ are the linear and total susceptibilities, respectively. It was assumed that the critical exponents were defined by

$$\chi_s \propto t^{-\gamma} \quad ; \quad t > 0 \qquad (3)$$

$$\chi_s \propto H^{2/\delta} \quad ; \quad t = 0 \qquad (4)$$

$$\beta = \gamma(\delta-1), \qquad (5)$$

where $t \equiv (T-T_g)/T_g$. The resulting values for γ, δ, and β were 2.3, 3.6, and 0.9, respectively. However, this analysis was subject to uncertainty because there appeared to be a low-field and high-field δ value, and because γ could not be determined very close to Tg, precisely where a critical exponent should be obtained.

Our subsequent work[4] on a-$Tb_{64}Fe_{20}Ga_{16}$ employed a scaling analysis based on

$$\chi_s = t^\beta f(H^2/t^{\gamma+\beta}) \qquad (6)$$

where $f(x)$ is the scaling function. The ac susceptibility was measured at 270 Hz with applied dc fields ranging from 0 to 288 Oe. The resulting scaling plot shows an excellent collapse of the H, t data for $0.002 \leq t \leq 0.13$. The exponents obtained in this case are $\beta = 1.7 \pm 0.1$, $\gamma = 3.7 \pm 0.1$, $T_g \equiv T_c = 140.5 \pm 0.6$ K and, with Eq. (5), $\delta = 3.2$. Another glass, a-$Dy_{60}Fe_{30}B_{10}$[5], gave scaling results which are qualitatively and quantitatively very similar to those obtained for a-TbFeGa.

Dieny and Barbara[6] have examined the critical properties of a-$DyNi_{1.33}$, also from a phase-transition viewpoint. Their analysis is slightly different from that described above, but includes a scaling plot leading to $\beta = 1.2 \pm 0.1$, $\delta = 2.7 + 0.1$ and $\gamma = 2.0 + 0.1$; the field and temperature range for their data included $2.4 < H < 100$ Oe and $0.005 < t < 0.25$. Dieny and Barbara argue that most of the well-studied spin-glasses and a-DyNi fall into two or perhaps three classes, based on the hyperscaling relation $D_f \stackrel{\sim}{=} d\phi/(\phi+\beta)$ for dimension $d = 3$. D_f is the infinite cluster dimensionality and ϕ is the crossover exponent, $\phi = \beta\delta$. These classes are suggested to correspond to different universality classes based on whether $D_f \cong 2.2$ (RMA or Ising like) or $D_f \cong 2.5$ (Heisenberg). These ideas are related to earlier work of Kotliar and Sompolinsky,[7] Yeshurun and Sompolinsky,[8] and de Courtenay et al.[9] These latter authors considered the effects of random anisotropy, mainly of the Dzyaloshinsky-Moriya type, on the crossover from Heisenberg critical behavior ($t \geq 0.1$, $h = H/T_g \geq 0.1$) to Ising critical behavior ($h \leq 0.1$, $t \leq 0.1$). The lack of generality or universality of the critical exponents obtained by various workers in different spin-glass and RMA systems is a serious problem, to which we shall return in the following subsection.

An interesting phenomenon in anisotropic rare-earth glasses has been reported recently by Lee, O'Shea and Sellmyer.[10] These workers introduced RMA in a controlled way by replacing Gd by Tb in a-$Gd_{65-x}Tb_xCo_{35}$. For $x = 0$ the system is a well-defined ferrimagnet with a diverging susceptibility at T_c. Increasing the Tb concentration increases the RMA and for $x = 50$, a clear spin-glass like transition is seen at low temperatures. This transition followed a scaling analysis for the nonlinear *magnetization* defined by

$$M_{n\ell} = \chi_o H - M. \tag{7}$$

It was possible to construct a scaling plot of $m_{n\ell} \equiv M_{n\ell}/t^{(3\beta+\gamma)/2}$ versus $h_\ell \equiv H/t^{(\beta+\gamma)/2}$, with critical exponents: $\beta = 1.30 \pm 0.05$, $\gamma = 3.7 + 0.2$, and $T_c = 104.5 \pm 0.1$ K, for $50 \leq H \leq 700$ Oe. At high fields this system shows a crossover to a standard *ferromagnetic* scaling with $\beta = 0.46 \pm 0.01$, and $\delta = 4.0 \pm 0.1$, for $1 \leq H \leq 80$ kOe. These results indicate that a zero- or low-field spin-glass-like or speromagnetic transition can be converted to a ferromagnetic-like or asperomagnetic transition with the application of a high enough magnetic field.

2.2. Recent Theoretical Advances

In the last two years considerable progress has been made in theoretical understanding of RMA systems. Because of space limitations we can mention briefly only a few results--those which seem to bear most directly on the experiments discussed above.

Feigel'man and Tsodyks[11] have considered RMA systems with long-(but finite-) range ferromagnetic interactions. In the paramagnetic region, not too close to T_c, the anisotropy is not essential and the correlations are ferromagnetic-like. However, very close to T_c the nonlinear susceptibility diverges and the phase transition belongs to the same universality class as that of the Ising spin glass. The structure factor $S(k)$ was calculated and found to be a sum of Lorentzian and squared Lorentzian terms, in agreement with neutron scattering results.[3] In general, the theory gives a good description of the phase-transition results described earlier.

The results of Fähnle et al.[12] may be relevant to the lack of uniform critical exponents observed in RMA systems (and also spin glasses). It is essential to do scaling fits over increasingly small reduced temperature intervals, until no further change in the exponents occurs. In practice, this means that $t^{max} \geq 0.1$ may lead to exponents that are suspect. This suggests that the quoted errors on critical exponents (often $\pm$ 0.1 or 0.2) may be grossly underestimated. In addition, the idea even of determining, say, Heisenberg critical exponents for $t \geq 0.1$ may not be meaningful. Another point made by Singh and Fähnle[12] is that $\gamma_{eff}(t)$ for a $\pm J$ Ising spin glass ranges from about 2.9 to 3.8 as t goes from $t = 0$ to $t \cong 0.5$. Again this points toward a possible overestimate of γ if t^{max} is too large.

A number of theoretical papers have dealt with the questions of magnetism in RMA systems in (a) the weak anisotropy limit, and (b) when a coherent or uniform anisotropy is added to the RMA term. These papers include those of Chudnovsky et al.,[13] and Goldschmidt and Aharony.[14] In particular, these papers point out how a coherent anisotropy, if large enough relative to the RMA, can convert the zero magnetization state into a nearly typical ferromagnetic domain structure. Phase diagrams, to which we shall refer in the following, were obtained, including paramagnetic, speromagnetic, and correlated-spin-glass regions.

2.3. Magnets With Weak RMA

In the course of recent work on weak RMA systems, O'Shea et al.[15] have discovered what appears to be the first observation of "double-transition" behavior induced by RMA. In the a-$Gd_{65-x}Er_xCo_{35}$ system, for $x = 0$, one has a soft ferrimagnetic glass, with properties very similar to the correlated spin-glass (CSG) described by Chudnovsky et al.[13] As Er is added, double transitions appear for x = 2, 4, 6, 8. These appear not to be due to normal exchange fluctuation phenomena, because they do not exist for any other rare earths investigated. The resulting phase diagram is qualitatively the same as that predicted by Chudnovsky et al. in Fig. 3 of their paper. While no explanation exists as to the occurrence of this phenomennon only for Er, it may be significant that Er possesses a small D value in comparison with many of the other rare earths, save Gd.

3. COHERENT MAGNETIC ANISOTROPY IN AMORPHOUS MULTILAYERS

As discussed above, the introduction of a coherent or uniform anisotropy of magnitude D_c can convert a speromagnet into a ferromagnet. In fact, such coherent anisotropy, when perpendicular to a thin film RE-TM glassy alloy, leads to possible perpendicular recording applications such as erasable magneto-optic recording. Though controversial, it has been thought for many years that anisotropic pair correlations have been the source of perpendicular magnetic anisotropy (PMA) in thin films such as a-GdCo.

Recently we have been using a computer-controlled, multiple-gun sputtering system with a rotating table to produce either homogeneous glasses or compositionally-modulated alloys. In the latter case, as the individual layer thicknesses approach one or two atomic diameters, the pair correlations ought to become anisotropic. We have performed experiments on Fe/Nd, Fe/Er and Fe/Ta multilayers, and have found that as the thickness of the Fe layers, t_{Fe}, becomes small enough, the films do indeed exhibit PMA.[16] In the case of Fe/Nd, t_{Fe} must be less than about 9 Å for this to occur. We have been able to determine an interface contribution to the anisotropy, along with a volume contribution. While certain systematics are beginning to emerge for the induction of PMA in amorphous multilayers, as yet we have no microscopic models on which to calculate magnitudes of the various contributions. These results may be understood in the general framework of the above-mentioned theories involving coherent anisotropy.

4. CONCLUSIONS

Significant progress, both experimental and theoretical, has been made in understanding the phase transitions present in homogeneous RMA glasses. There are still outstanding questions regarding the generality and universality of the critical description of various systems. As in the spin glass problem, a detailed understanding of the low-temperature phase is far form realized. However, the basic features of the magnetic properties are understood.

The relationship between microstructure, RMA and PMA is just beginning to be studied. It is likely that the control of PMA along with magnetic and optical properties, with a view towards data-storage applications, will be a fruitful research area for some time to come.

ACKNOWLEDGEMENTS

I am grateful for much assistance and many informative discussions to S. Nafis, M.J. O'Shea, and Z.S. Shan, and for financial support to the National Science Foundation.

REFERENCES

1) K. Moorjani and J.M.D. Coey, Magnetic Glasses (Elsevier, Amsterdam, 1984).

2) D.J. Sellmyer and S. Nafis, J. Appl. Phys. $\underline{57}$ (1985) 3584.

3) J.J. Rhyne, IEEE Trans. MAG $\underline{21}$ (1935) 1990.

4) D.J. Sellmyer and S. Nafis, Phys. Rev. Lett. $\underline{57}$ (1986) 1173

5) S. Nafis and D.J. Sellmyer, to be published.

6) B. Dieny and B. Barbara, Phys. Rev. Lett. $\underline{57}$ (1986) 1169.

7) G. Kotliar and H. Sompolinsky, Phys. Rev. Lett. $\underline{53}$ (1984) 1751.

8) Y. Yeshurun and H. Sompolinsky, Phys. Rev. Lett. $\underline{56}$ (1986) 984.

9) N. de Courtenay, H. Bouchiat, H. Hurdequint, and A. Fert, J. Physique $\underline{47}$ (1986) 1507.

10) K.M. Lee, M.J. O'Shea, and D.J. Sellmyer, J. Appl Phys. $\underline{61}$ (1987) 3616.

11) M.V. Feigel'man and M.V. Tsodyks, J.E.T.P. (in press).

12) M. Fähnle, W.U. Kellner, and H. Kronmuller, Phys. Rev. B (in press); R.P. Singh and M. Fähnle, preprint.

13) E.M. Chudnovsky W.M. Saslow, and R.A. Serota, Phys. Rev. B $\underline{33}$, (1986) 251.

14) Y.Y. Goldschmidt and A. Aharony, Phys. Rev. B, $\underline{32}$ (1985) 264.

15) M.J. O'Shea, K.M. Lee and F. Othman, Phys. Rev. B $\underline{34}$ (1986) 4944.

16) D.J. Sellmyer, Z.R. Zhao, Z.S. Shan, S. Nafis, J. Appl. Phys $\underline{61}$ (1987) 4323; Z.S. Shan, Z.R. Zhao, J.G. Zhao, and D.J. Sellmyer, J. Appl. Phys. $\underline{61}$ (1987) 4320.

MAGNETIC PROPERTIES OF SPUTTERED AMORPHOUS SOLIDS AND MODULATED SOLIDS

C.L. Chien

Department of Physics and Astronomy The Johns Hopkins University
Baltimore, Maryland 21218 USA

Vapor quenching methods, (e.g. high rate sputtering), allow the fabrication of amorphous alloys which have compositions unobtainable by liquid quenching and other techniques. We show that the atomic size difference in the alloy is the single most important factor in determining the composition range of the amorphous alloy. The magnetic properties and short range order of Fe-metal and Fe-metalloid alloys are found to be very different. Superlattices and modulated solids are a new class of artificial solids fabricated by depositing alternating layers of different materials. The magnetic properties of modulated systems, such as the magnetization and its temperature dependence, spin wave excitation, effects due to interfaces, magnetic surface anisotropy, and dimensional crossover effects, will be discussed.

1. Introduction

Amorphous metallic solids are metastable solids which do not appear in the equilibrium phase diagrams[1]. Rapid quenching processes (liquid quenching, vapor quenching, solid state reaction, laser melting etc.)[1-4] are essential to achieving the metastable state. These processes are intrinsically nonequilibrium and have high quenching rates. In each rapid quenching process the competing characteristic times are different. Consequently, each process has a different capability to form amorphous solids. Liquid quenching methods (e.g. melt spinning), with quenching rates of about 10^6 K/sec, can fabricate amorphous alloys only within a narrow composition range near the eutectic points in the phase diagram, (e.g. $Fe_{80}B_{20}$, $Fe_{90}Zr_{10}$). However, the stability of an amorphous alloy has no bearing on whether or not it is quenchable, nor are the amorphous alloys with eutectic compostions the most stable. Many alloy systems, even those with eutectic points, contain no alloy of any combination which can be liquid quenched into an amorphous state.

The effective quenching rates of vapor quenching methods (e.g. sputtering) are more difficult to estimate accurately (as reflected by estimates ranging from 10^{10} to 10^{16} K/sec)[3] although they are much higher than those of liquid quenching. Consequently, vapor quench methods can produce amorphous solids in which the composition and therefore the properties can be continuously changed over a very wide range. $Fe_xB_{100-x}(0 \leq x \leq 90)$ and $Fe_xZr_{100-x}(20 \leq x \leq 91)$ are typical examples[5,6]. Large amorphous composition ranges are important not only for facilitating our understanding of amorphous materials but also for creating alloys with tailored properties. Since vaporquenching is a deposition technique, the sample thickness (tens of Å to thousands of μm) can also be varied, leading the way to modulated solids and superlattices.

Modulated solids, or superlattices involve very thin alternating layers of crystalline or amorphous materials, yielding a bulk solid with a quasi-two dimensional nature[7]. These solids are dominated by interface and thin film characteristics. In this work we will briefly describe our recent results on homogeneous amorphous solids and modulated solids.

---

* Work supported by NSF Grant No. DMR86-07150

2. Fabrication and Characterizations

All of the samples described in this work were made using a vapor-quench method. In particular, we used a multi-gun high-rate magnetron sputtering system. The vacuum chamber was typically evacuated to 5×10^{-8} torr by an oil-free cryopump system. DC or RF sputtering was performed with 4-6 mtorr high purity Ar at rates of about 0.1 μm/min onto various substrates cooled by water or liquid nitrogen. The deposition rates were feed-back controlled by a quartz oscillator system. For modulated solids, a substrate platform was rotated by a servo motor system. The typical sample thickness was 6-25 μm thick.

Sample characterizations were performed by a number of techniques including x-ray diffraction ($\theta/2\theta$, energy dispersive), cross sectional transmission electron microscopy, and Auger depth profiling.

3. Binary Amorphous Alloys

Binary alloys are the simplest possible amorphous metallic alloys. We have studied both binary metal-metal alloys, (e.g., Fe-T_i, where T_i = Zr, Hf, Ta, Mo, Ti, etc., Ni-Nb, Cu-Nb, etc), and metal-metalloid alloys, (e.g., Fe-B, Ni-B, Fe-C, etc.). As examples, Fig. 1 shows the magnetic ordering temperatures T_c of Fe-B and Fe-Zr, and the superconducting temperatures T_s of Fe-Zr. In Fe-B one observes a continuous variation of T_c with Fe concentration. Below $x \sim 40$, T_c vanishes due to the loss of the Fe moment. In amorphous Fe-Zr, one observes the evolution of properties from superconducting at the Zr-rich end to magnetic at the Fe-rich end.

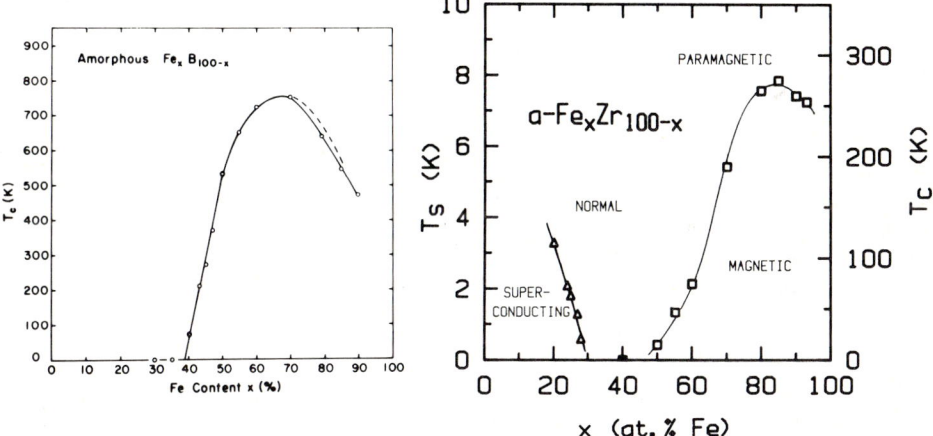

g.1 Magnetic ordering temperatures (T_c) and superconducting temperatures (T_s) of a-Fe-B and a-Fe-Zr.

In Fe-based amorphous alloys, the magnetic properties of the metal-metal alloys are fundamentally different from the metal-metalloid alloys. In the metal-metal alloys, there is an abrupt change in the magnetic moment and hyperfine field at the boundry separating amorphous and bcc crystalline phases, as shown in Fig. 2. The magnetic ordering temperatures of all of the metal-metal alloys, regardless of composition, are below 300 K, whereas all of the crystalline alloys are strongly feromagnetic with T_c in excess of 850 K. Quite the contrary, in the metal-metaloid alloys, there is no discontinuity in the magnetic moment or hyperfine field at the phase boundary. Thus, the properties of the metal-metal alloys and the metal-metalloid alloys extrapolate to different states of amorphous pure Fe. Additional pieces of evidence were found in the quadrupole splitting and isomer shift data, which indicate different short range order in these two types of Fe-based amorphous solids.[8]

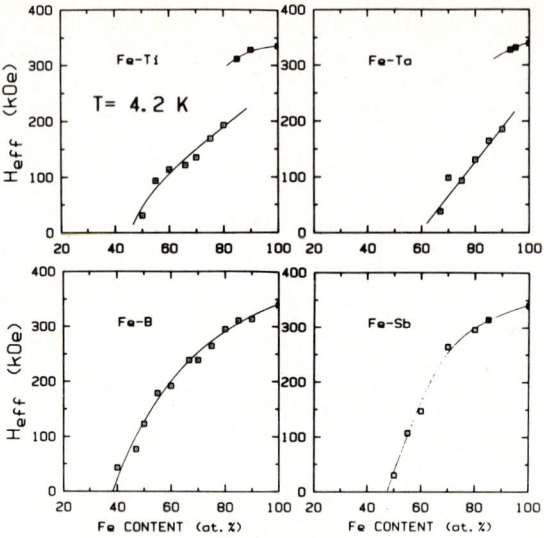

Fig. 2: Magnetic hyperfine fields of a-Fe-B and a-Fe-Ti.

4. Composition Range of Binary Amorphous Alloys

We have determined, for many binary systems, the composition range $x_{min} \leq x \leq x_{max}$, within which the amorphous state can be formed. Some examples are shown in Table 1. In each system there is a single, continuous composition range for the amorphous state regardless of whether there is one eutectic point (e.g. Fe-B) or two (e.g. Fe-Zr) in the phase diagram. Most remarkably, the values of x_{min} and x_{max} can be quantitatively related to the geometrical factors of the atomic species. Specifically, the expressions are[9]

$$x_{max} \sim (1-2\lambda_1) \frac{V_A}{\Delta V} \qquad 1$$

$$x_{min} \sim 2\lambda_1 \frac{V_B}{\Delta V} \qquad 2$$

TABLE I. The constituent atomic radii (r_a, r_b), ratio (z), and composition range $(x_{min} \leq x \leq x_{max})$ of vapor-quenched binary amorphous alloys. The theoretical predictions $[(x_{min})_t \leq x \leq (x_{max})_t]$ are shown in the last column.

$A_x B_{1-x}$	r_A (Å)	r_B (Å)	$z = \frac{r_A}{r_B} = 1+\delta$	Experimental $x_{min} \leq x \leq x_{max}$		Theory $(x_{min})_t \leq x \leq (x_{max})_t$	
Fe-Ti	1.28	1.46	1.14	0.30	0.80	0.28	0.85
Fe-Zr	1.28	1.58	1.234	0.20	0.93	0.19	0.92
Fe-Hf	1.28	1.67	1.305	0.20	0.94	0.16	0.94
Fe-Ta	1.28	1.49	1.164	0.20	0.90	0.24	0.88
Fe-Mo	1.28	1.39	1.086	0.40	0.80	0.41	0.75
Fe-B	1.28	0.78	0.609	0.0	0.90	0.03	0.91
Fe-Cu	1.28	1.27	1.007				
Fe-Nb	1.28	1.46	1.14	0.25	0.85	0.27	0.85
Ni-Nb	1.28	1.46	1.14	0.20	0.80	0.27	0.85
Cu-Nb	1.27	1.46	1.14	0.20	0.80	0.26	0.86

where V_A and V_B are the atomic volumes, $\Delta V = V_A - V_B$ and λ_1 is a free parameter. As shown in Table 1, the calculated values of x_{min} and x_{max} are in close agreement with the experimental results using $2\lambda_1 = 0.07$. This indicates that atomic size is the single most important factor in determining the composition range of the amorphous state[10].

5. Modulated a-$Fe_{70}B_{30}$/Ag

$Fe_{70}B_{30}$ has the highest magnetic ordering temperature of the Fe_xB_{100-x} system, and is highly resistant against crystallization. Thus, it was the ideal choice for a modulated solid composed of amorphous and crystalline layers. Since both the mutual solubility between Ag and Fe and the diffusion at the Fe-Ag interface are very small, the $Fe_{70}B_{30}$/Ag system was the natural choice for an amorphous/crystalline modulated solid[11]. The thickness (d) of the $Fe_{70}B_{30}$ layers ranged from 3 Å to 90 Å, while that of the Ag layers was maintained at four times as large.

The structure of the $Fe_{70}B_{30}$/Ag films was studied by x-ray diffraction and transmission electron microscopy (TEM). Fig. 3 displays a XTEM micrograph of a film with modulation wavelength $\lambda = 51$ Å. The light regions are the amorphous $Fe_{70}B_{30}$ layers (d=10 Å), the dark regions are the Ag layers (d=41 Å). Both regions are continuous and flat over a large lateral distance. The changing contrast within the layers is caused by the crystalline nature of the Ag layers.

Spontaneous magnetizations (M_s) at T = 8 K are presented in Fig. 4 as a function of 1/d. Within our experimental uncertainty, M_s remains practically constant and independent of d in the range from 3 to 3500 Å. There are no magnetically "dead" layers at the interfaces and the interfaces do not cause any changes in the surface magnetization of the $Fe_{70}B_{30}$ layers.

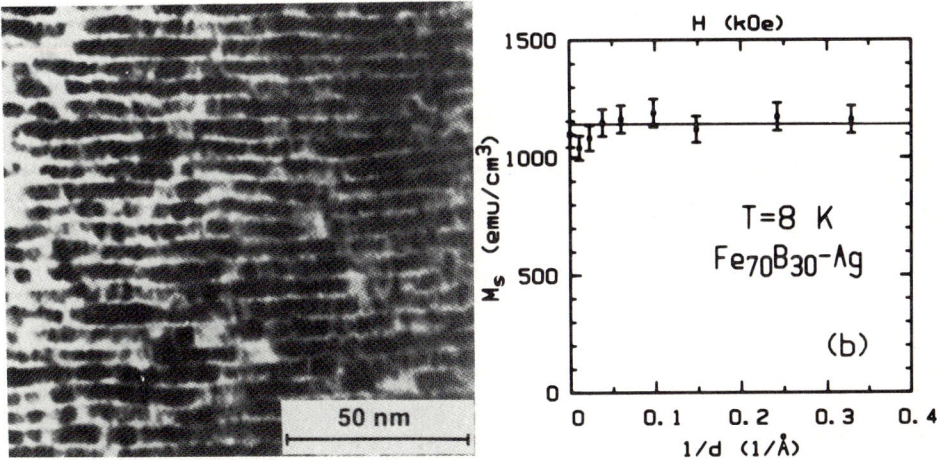

Fig.3　XTEM micrograph of modulated a-$Fe_{70}B_{30}$/Ag.

Fig.4　Spontaneous magnetization of modulated a-$Fe_{70}B_{30}$/Ag as a function of the inverse of the $Fe_{70}B_{30}$ layer thickness.

As the thickness of the magnetic layer is reduced, the temperature dependence of the magnetization may show a dimensional crossover effect. The normalized magnetization of a few samples is plotted in Fig. 5 as a function of $T^{3/2}$. All of the data can be fitted by straight lines, indicating that Bloch's law

$$M_s(T) = M_s(0)(1-BT^{3/2}) \qquad 3$$

is well obeyed. However, as the $Fe_{70}B_{30}$ thickness is reduced below 4.1 Å M_s decreases linearly with temperature, as shown in Fig. 5, indicating a magnetic 3D to 2D dimensional transition.

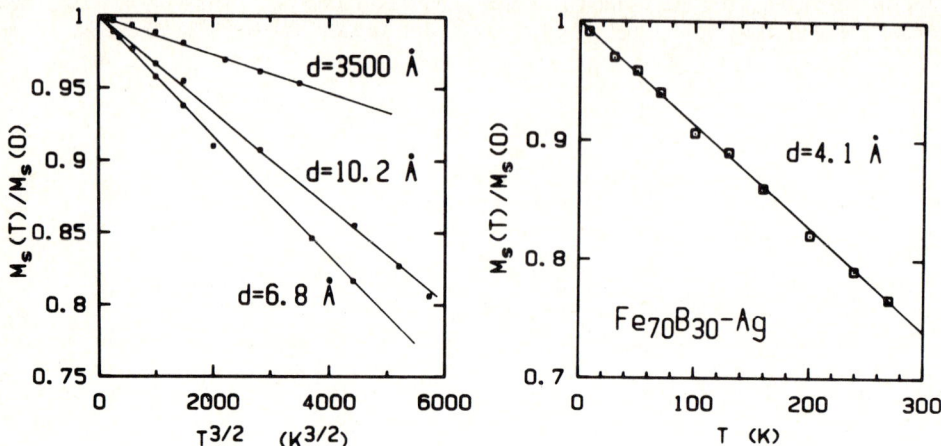

Fig. 5: Temperature dependence of the magnetization of modulated $Fe_{70}B_{30}/Ag$ films. All of the samples show $T^{3/2}$ dependence except that of d = 4.1Å which shows a linear temperature dependence.

A particularily interesting behavior is observed when the spin wave constant B in Bloch's law is plotted agianst the inverse of the $Fe_{70}B_{30}$ thickness as shown in Fig. 6. The value B is linearly dependent on 1/d over the range 6.8 Å to 3500 Å. This can be explained as follows. Each $Fe_{70}B_{30}$ layer consists of interfaces and interior. The spin wave deviations at the interface are larger than that of the interior. As the layer thickness is reduced, the interface contribution becomes progressively more dominant. Therefore, the effective spin wave constant B increases as d decreases. Quantitatively we can consider a simple model in which the spin wave constant

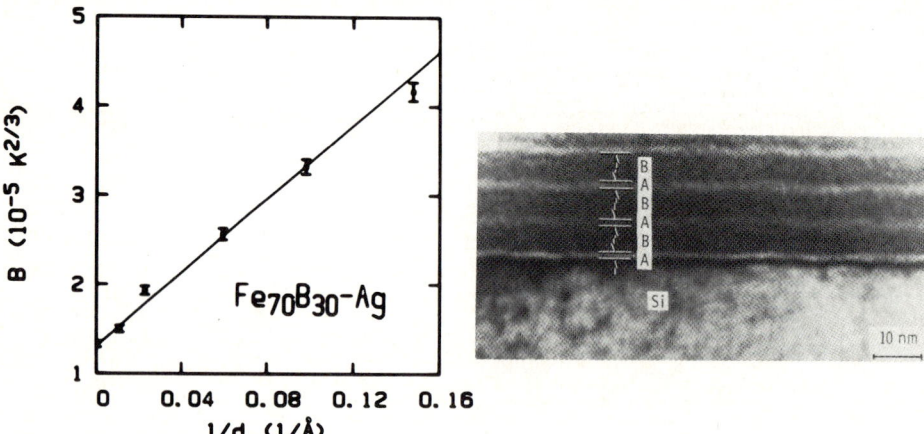

Fig.6 Spin wave constant (B) of modulated $Fe_{70}B_{30}/Ag$ as a function of 1/d.

Fig.7 XTEM micrograph of modulated a- $Fe_{80}Zr_{20}/a$-$Ni_{30}Nb_{70}$.

of the interface layer with thickness d_s (including both interfaces) is B_{surf}, and that of the interior of thickness $d-d_s$ is B_{bulk}. Both the interface and the interior follow Bloch's law. Consequently Eq. (3) is observed, but with an effective spin wave constant:

$$B_{eff} = B_{bulk} + \frac{d_s}{d}(B_{surf} - B_{bulk}).\qquad 4$$

Relation 4 shows that the effective B constant is inversely proportional to d, which is experimentally observed as shown in Fig. 6. The slope of the lines gives $d_s(B_{surf} - B_{bulk})$. Using $d_s \leq 6.8\text{Å}$, one finds that $B_{surf} > 3.2 B_{bulk}$ from Fig. 12. Therefore, the surface magnetization decreases with temperature far more rapidly than that of bulk.

6. Modulated a–$Fe_{80}Zr_{20}$/a–$Ni_{30}Nb_{70}$

In this modulated system, both constituent layers are amorphous. Since we have extensively studied a–Fe_xZr_{100-x} ($20 \leq x \leq 93$) and a–Ni_xNb_{100-x} ($20 \leq x \leq 80$) over a wide composition range, the fabrication, stability, and properties of the two alloy systems are well known. The compositions of both a–$Fe_{80}Zr_{20}$ and a–$Ni_{30}Nb_{70}$ are very stable, with high crystallization temperatures. The compound a–$Ni_{30}Nb_{70}$, although containing Ni, is non-magnetic, whereas a–$Fe_{80}Zr_{20}$ is magnetic with a convenient bulk ordering temperature of T_c=250 K. In the modulated a–$Fe_{80}Zr_{20}$/a–$Ni_{30}Nb_{70}$ solids, the non-magnetic layers serve to seperate the magnetic layers containing Fe, which exhibit thin film magnetic properties. Two XTEM micrographs are shown in Fig. 7, where A = a–$Fe_{80}Zr_{20}$ and B=a–$Ni_{30}Nb_{70}$. (A is about 20 Å and B about 50 Å). Flat growth can be seen close to the Si substrate as well as many layers away.

The magnetic properties of this modulated solid were measured by a SQUID magnetometer, complemented by Mössbauer spectroscopy, and will be described elsewhere[12].

7. Conclusions

Vapor quench amorphous alloys and modulated solids contain extra degrees of feedom upon which the properties of the solids depend. In amorphous alloys, the extended freedom in composition leads to a variety of properties and phenomena. In modulated solids, the choice of different constituent layers and their thicknesses allow one to fabricate materials with new properties which are unattainable in bulk amorphous alloys.

8. References

1. e.g. **Amorphous Metallic Alloys**, edited by F.G. Luborsky, (Butterworth, London, 1983); **Magnetic Glasses**, by K. Moorjani and J.M.D. Coey, (Elsevier, Amsterdam, 1984).
2. P. Duwez, Trans. Amer. Soc. Metals, **60**, 607 (1967).
3. T.W. Barbee, Jr. and D.L. Keith, in **Synthesis and Properties of Metastable Phases**, eds. E.S. Machlin and T.J. Rowland (AIME, 1981).
4. R.B. Schwartz and W.L. Johnson, Phys.Rev. Lett. **51**, 415 (1983).
5. C.L. Chien and K.M. Unruh, Phys. Rev. **B25**, 5790 (1982).
6. K.M. Unruh and C.L. Chien, Phys. Rev. **B30**, 4568 (1984).
7. e.g. **Synthetic Modulated Structured Materials**, eds. L.L. Cheng and B.C. Giessen (Academic Press, New York, 1985).
8. Gang Xiao and C.L. Chien, Phys. Rev. **B35**, (1987).
9. T. Egami and Y. Waseda, J. Non. Cryst. Solids, **64**, 113 (1984).
10. S.H. Liou and C.L. Chien, Phys. Rev. **B35**, 7443 (1987).
11. Gang Xiao, C.L. Chien and M. Natan, J. Appl. Phys. **61**, 4314 (1987).
12. C.L. Chien, S.H. Liou and M. Natan, in **Interfaces, Superlattices and Thin Films**, eds. J.D. Dow, I.K Schuller and J. Hilliard (1987).

RE-ENTRANT MAGNETIC FLUX REVERSAL IN AMORPHOUS WIRES

F.B. Humphrey,
Carnegie Mellon University, Magnetic Technology Center, Pittsburgh, Pa 15213-3674, USA

K.Mohri, J.Yamasaki and H. Kawamura,
Kyushu Inst of Technology, Tobata, Kitakyushu 804, Japan

R. Malmhall,
Royal Institute of Technology, s-100 44 Stockholm, Sweden

I. Ogasawara,
Unitika Research & Development Center, Ujikozakura,Uji 611, Japan

Amorphous magnetostrictive wires, either as-quenched or die-drawn then annealed, can exhibit re-entrant flux reversal with a resultant large Barkhausen discontinuity. This re-entrant reversal occurs without any set field. In the as-quenched wire, the radial stress created by the in-water quenching causes a radial anisotropy of 20 erg/cm^3 in the positive magnetostrictive $Fe_{77}B_8Si_{15}$ wire and circumferential anisotropy of 3 erg/cm^3 in the negative magnetostrictive $Co_{72.5}B_{15}Si_{12.5}$ wire. This anisotropy isolates the center 90 μm and 70 μ, respectively, diameter core where the axial re-entrant reversal takes place. An axial wall mobility of 1400 m/sec.Oe and 6000 m/sec.Oe, respectively, was measured for the two wires using the Sixtus-Tonks method. Wall length varied from 15 mm to 50 mm depending upon the diameter in positive magnetostrictive die-drawn, tension annealed wire. The critical length for re-entrant reversal also varied with wire diameter with the shortest (1.5 cm) observed in 30 μm diameter wire.

INTRODUCTION

Amorphous magnetostrictive wires, either as-quenched or annealed, can exhibit re-entrant flux reversal with a resultant large Barkhausen discontinuity. Even though flux reversal in amorphous systems as a whole seems essentially the same as crystalline systems, amorphous wires are a case where the process that creats the amorphous material also creates a uniquely strained wire with a unique reversal character. This re-entrant reversal can be detected as a sharp voltage pulse induced in a small coil around the wire or as a voltage generated on the wire itself because of the Matteucci effect[1,2]. This characteristic reversal makes the wires useful as pulse generating sensor elements for such applications as rotary encoders, non-contact switchers, magnetic field, electric current, position and security sensors[1,3].

The wire that will be discussed here is made by a melt spinning process that quenches the molten metal in a moving stream of water[4]. As the in-rotating-water rapid quenching technique only allows wires to be produced with diameters down to about 80 μm[5], smaller diameter wire can be obtained by cold drawing the 120-130 μm diameter as-quenched wire. Wires as small as

30 μm in diameter can be obtained in this way. They must, of course, be annealed to recover their unique reversal character[6].

RE-ENTRANT REVERSAL

This unique flux reversal can best be illustrated by referring to the set of hysteresis loops seen in Fig.1. Loops for $Fe_{77}B_8Si_{15}$ wire with positive magnetostriction are shown in (a), $(Fe_6Co_{94})_{72.5}B_{15}Si_{12.5}$ non-magnetostrictive wire in (b) and $Co_{72.5}B_{15}Si_{12.5}$ negative magnetostrictive wire in (c). With the drive field very low as with the first loop of each set, there is no flux change for the magnetostrictive wire (a) and (c) whereas the wire at (b) displays

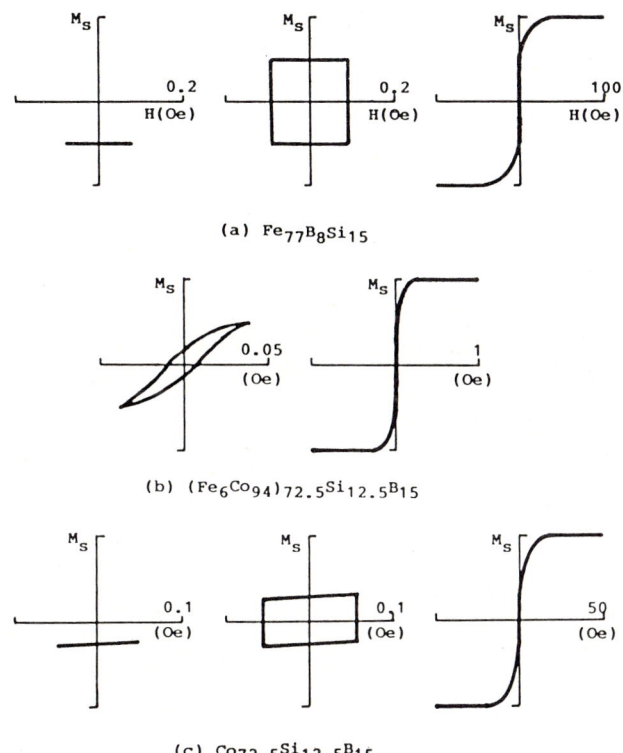

FIGURE 1. Magnetic hysteresis loops for a 20cm length of as-cast amprphous wire with (a) positive magnetostriction, (b) zero magnetostriction and (c) negative magnetostriction. The value of saturation magnetization (M_s) is relative to each set of loops.

the normal minor loop expected of square loop magnetic materials. As the drive field is increased, wire (a) and (c) reach a sharp threshold above which they reverse very abruptly. Reasonable increases in field do not produce more flux change, however, if the field is increased to a very high value as shown by the last loop, it becomes clear that the low field remnance is only a fraction of saturation. The non-magnetostrictive wire (b) saturates at a reasonably low field. Not only does this demonstrate the features of re-entrant reversal ((a) and (c)), but it also shows that magnetostriction is very important.

Magnetostrictive wires exhibit re-entrant reversal and non- magnetostrictive wires do not. The residual stress from the quenching creates an anisotropy to provide the environment for the re-entrant reversal. This has been confirmed by a strain relief anneal. The loops of a magnetostrictive iron wire as seen in Fig.1a before a strain relief anneal look exactly like the non-magnetostrictive case in Fig.1b after the anneal. The residual stress in the wire should be primarily radial tension. When the molten metal stream hits the water, an outside shell solidifies first establishing the diameter of the molten core. As the core then solidifies, it shrinks to create this radial tension. This radial tension should create a strong radial anisotropy for the positive magnetostrictive iron wire and a circumferential anisotropy for the negative magnetostrictive cobalt wire. Evidence for this stress induced anisotropy can be seen in the surface domain structure of the as-quenched wire. Using the Bitter technique to decorate the walls where they meet the surface, the domain patterns can be seen. Unfortunately, only the top 20 μm of surface of the wire can be seen because of the depth of focus of the microscope. The iron wire exhibits a maze like domain pattern with a spacing of about 4 μm for a long wire and wider spacing as the wire length is shortened. A sketch of the surface shell domain pattern in the center of a length of wire is shown in Fig.2. For this case, we can guess from the zig-zag

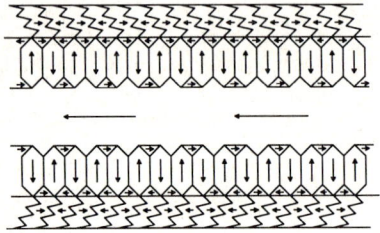
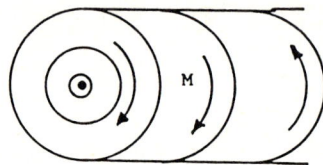

FIGURE 2. Magnetic domain pattern seen on the surface of a positive magnetostrictive amorphous wire including a guess at the domain structure in the strained shell of the wire.

FIGURE 3. Magnetic domain pattern seen on the surface of a negative magnetostrictive amorphous wire and a guess at the domain structure in the strained shell of the wire.

walls on the surface and the radial anisotropy that some kind of closure structure as shown exists under the surface. Since the surface pattern does not change when moderate axial fields are applied, it seems reasonable that this closure structure does not change with field. The cobalt wire undoubtedly has a stable structure as shown in Fig.3 consistent with the circumferential anisotropy of the negative magnetostrictive material. The rings of the bamboo domains are spaced about 20μm along the wire. From the ratio of remnance to saturation, the core size in the 120 μm outside diameter wire can be estimated as 90 μm for the iron wire and 70 μm for the cobalt wire.

WALL DYNAMICS

The classic Sixtus and Tonks experiment[4] was the first to investigate wall motion in wire and is still a good way to measure the dynamics of walls. In this experiment, the wire sample is placed in the uniform field of a long solenoid. A small coil around the wire at one end is connected in series with the solenoid to locally increase the field. Two small pickup coils, connected in series, are wound tightly around the wire to sense the local change in flux. With the wire saturated, a field is generated in the solenoid. At the end, the field will be locally higher so that a reverse domain will be nucleated and a domain wall will propagate along the wire driven by the uniform field of the solenoid. As the wall passes the first pickup loop, a small voltage will be generated because of the local flux change coupling the small loop. The same will happen at the second loop but a short time later because the wall must propagate the distance between the loops. The two voltage pulses are amplified and detected with a digital oscilloscope. The time between the two pulses is then used to calculate the wall velocity.

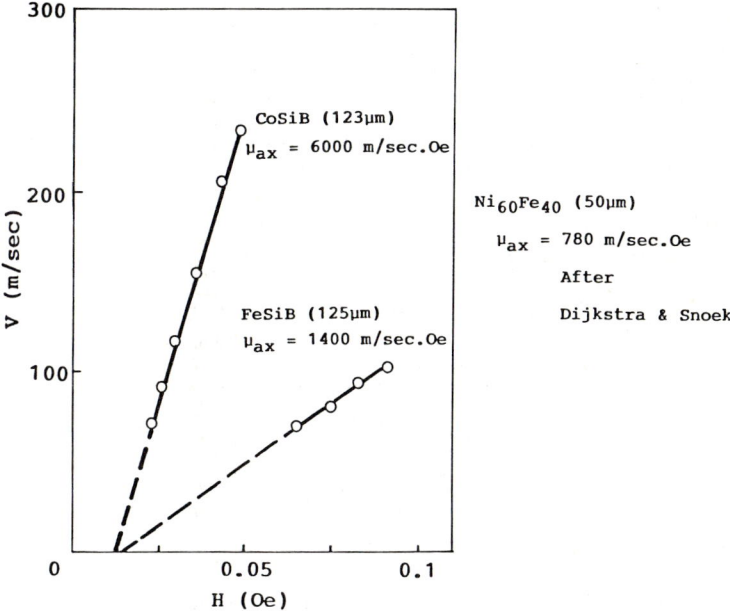

FIGURE 4. Domain wall axial velocity as a function of magnetic field. The wire diameter is included as well as the axial mobility μ_{ax}.

Domain wall velocity as a function of field for the two wires is shown in Fig.4. The mobility (slope) is also included. For comparison, the mobility of a 50 micron diameter wire of permalloy ($Ni_{60}Fe_{40}$) is 780 m/sec.Oe. The differences between the three wires reflect their different conductivity hence their different eddy current damping.

Knowing the wall velocity and the width of the pickup coil signal, the shape of the wall can be inferred. By putting recording heads on each side of the wire and comparing their output, it has been shown that the wall is radially symmetric. It is undoubtedly shaped as a symmetric rounded cone with the base moving in front in the manner of the classic Sixtus-Tonks wall. The wall length (height of the cone) is independent of velocity and wire composition. It is a function of wire diameter as can be seen in Fig.5. This wire has been die drawn to the diameter indicated and then carefully annealed. The die drawn wire as-drawn has a fairly high coercive force (ca.

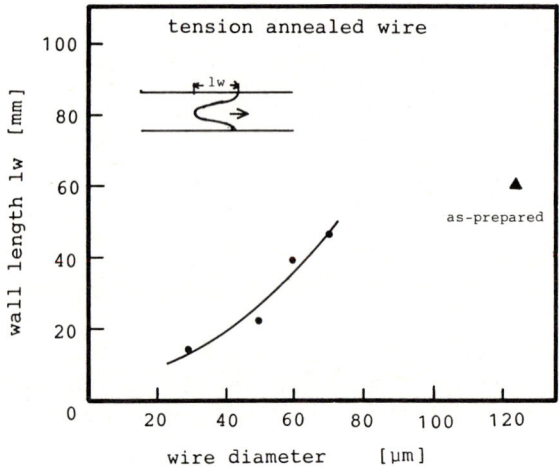

FIGURE 5. Domain wall length as a function of die drawn, tension annealed, FeSiB positive magnetostrictive amorphous wire.

10 Oe) and is not re-entrant. After a fairly low temperature anneal (330 deg C) while under tension (100 Kgm/mm^2) the reversal is re-entrant. It can be seen that the wall length (l_w as indicated on the figure) of the as-cast wire fits the die-drawn data as long as the core diameter is used.

REVERSAL MECHANISM

The Matteucci effect[8] is most convenient to probe the motion of the wall in the finite length wires with spontaneous re-entrant reversal of the kind seen in Fig.1a since the effective pickup loop is closer to the wire providing better resolution along the wire. A slight twist (3 turns/10cm) is included during the tension anneal. The resultant anisotropy contains a small helical component that couples some of the reversing flux to the wire itself resulting in an induced voltage pulse. A 10cm long 50μm diameter twist-strain annealed wire is shown in Fig.6. The voltage as a function of time generated between the two ends of the wire during a reversal is shown as e_{p1}. It can be seen the pulse has an initial and a final high peak (time is increasing to

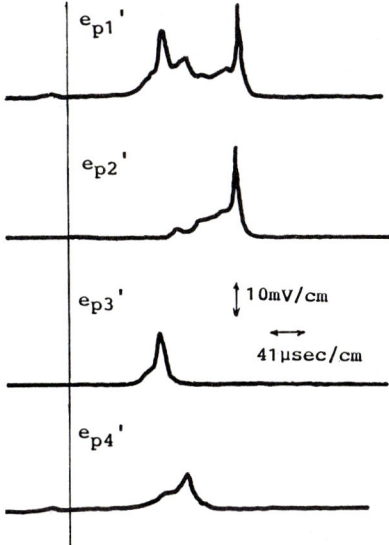

FIGURE 6. Matteucci effect in a 50 μm diameter die drawn, strain and twist annealed FeSiB positive magnetostrictive amorphous wire. A 2 Oe 60 Hz field is used to drive the wire. The voltage e_{p1} is seen across the entire 10 cm length of wire, e_{p2} at the center and e_{p3} and e_{p4} at the ends as indicated. The time and voltage scales are indicated.

the right) with a relatively constant period of flux change in between. The voltage pulse in the center portion of the wire is seen as e_{p2}. It can be seen that it is identical to the final high peak of the total pulse with some of the constant. The voltage pulse at the end can be seen as e_{p3} and e_{p4} for the left end and right end, respectively. It can be seen that the initial pulses of the main pulse are clearly generated by the ends of the wire. It seems clear that this re-entrant reversal proceeds in three steps. There is an initial rapid flux change corresponding to the nucleation of two walls, one at each end of the wire. This is followed by a period of constant flux change as the walls proceed at a constant velocity toward the center. Finally, there is a fast reversal as the walls meet and the enclosed domain collapses.

WIRE ENDS

The static domain structure at the ends of the wire of the wire provide it with the most unique reversal characteristic. Most re-entrant reversals must be set after the re-entrant part of the reversal. Generally, a higher field is used to remove any closure domains and get ready for reversal in the opposite direction. The wire does not require a set field. At threshold, the wall pulls away from the end leaving a new domain structure already set for the next reversal. This structure must be insensitive to all of the damage and variation from cutting the wire. The detailed structure is not clear. As the end is approached, the radial component of flux should greatly increase with the domains opposite the central core decreasing in width and the ones in the direction of the central core increasing. The periodic pattern observed in the center of a length of wire should be disturbed and the wavelength increased. This general character is confirmed by the Bitter patterns similar (like Fig.3 for the wire center), however, the experimental problems of working near the end of the wire are large so that our technique must be improved before the details can be observed.

CONCLUSION

Re-entrant reversal can be obtained in finite length wires with the minimum length dependent upon the wire diameter. Stress is important so that magnetostrictive wires are needed, however, either positive or negative magnetostrictive wires work. Once nucleated, the walls seem to be conical in shape damped by eddy currents.

REFERENCES

1. K. Mohri, F.B. Humphrey, J. Yamasaki and K. Okamura, IEEE Trans. Magn. Mag-20 (1984) 1409.

2. K. Mohri, F.B. Humphrey, J. Yamasaki and F. Kinoshita, IEEE Trans. Magn. Mag-20 (1985) 2017.

3. K. Mohri, IEEE Trans. Magn. Mag-20 (1985) 942.

4. I. Ohnaka, T. Fukusako, T. Ohmichi, T. Masumoto, A. Inoue and M. Hagiwara, Proc. 4th Intl. Conf. on Rapid. Quench. Metals (1982) 31.

5. R. Malmhall, K. Mohri, F.B. Humphrey, T. Manabe, H. Kawamura, J. Yamasaki and I. Ogasawara, IEEE Trans. Magn. 1987 Intermag issue in press.

6. R. Malmhall, K. Mohri, F.B. Humphrey, T. Manabe, H. Kawamura and J. Yamasaki, "Flash Annealing of Cold-Drawn Amorphous Magnetostrictive Wires", this volume.

7. K.J. Sixtus and L. Tonks, Phys. Rev. (1930) 930.

8. C.H. Matteucci, Memorie sur le Magnetisme Developpe par le Courant Electrique, C.R. 24 (1847) 301.

HIGH FATIGUE STRENGTH AND UNIQUE MAGNETIC PROPERTIES OF
AMORPHOUS ALLOY WIRES

A. INOUE, T. MASUMOTO, I. OGASAWARA* and M. HAGIWARA*
Institute for Materials Research, Tohoku University, Sendai 980, Japan
*Unitika Research and Development Center, Unitika Ltd., Uji 611, Japan

1. INTRODUCTION

In order to make the best use of the attractive mechanical properties of amorphous alloys, the fabrication of an amorphous alloy wire with a circular cross section has strongly been desired. In 1980, the authors have succeeded for the first time in producing Pd-[1], Fe[2,3] and Co-[4] based amorphous wires by melt spinning in rotating water in the collaborate research between Tohoku University and Unitika Ltd. Furthermore, the Fe- and Co-based amorphous wires have been found to exhibit high values of tensile fracture strength[2-4], bend ductility[2-4] and fatigue strength[5] as well as unique soft ferromagnetism[6-8] different from that of ribbon-shaped amorphous alloys, in addition to good cold-[2-4] and warm-[9] drawability. Recently, Unitika Ltd. has developed the established equipment and techniques to mass-produce Fe-, Co- and Ni-based amorphous wires with a length as long as several kilometers. Owing to the good mechanical and ferromagnetic properties of the amorphous alloy wires and the energy- and labor-saving process of producing amorphous alloy wires with a fine diameter of 80 to 200 μm directly from melt, amorphous alloy wires have attracted an increasing interest as a new type of engineering material. This paper attempts to review the fundamental characteristics of static and dynamic mechanical strengths and soft ferromagnetism of Fe-, Co- and Ni-based amorphous wires clarified by the present authors for the last eight years.

2. FABRICATION METHOD AND ALLOY SYSTEMS OF AMORPHOUS ALLOY WIRES

The in-rotating-water spinning technique[1,10] was used to produce a long amorphous alloy wire. The details of the equipment and the spinning condition have been presented in some previous references[2-4]. Here it is very important to note that the formation of an amorphous alloy wire by this technique is limited to the following systems because of the limited glass-forming capacity; Fe-Si-B[2], Fe-M-Si-B (M=transition metal)[2], Fe-P-C[3], Fe-M-P-C[3], Co-Si-B[4], Co-M-Si-B[4], Ni-Si-B-Al[11], Ni-M-Si-B-Al[11], Ni-P-B-Al[11], Ni-M-P-B-Al[11], Ni-Pd-P[12], Pd-Si[1], Pd-M-Si[1] and Cu-Zr[13].

3. STATIC MECHANICAL STRENGTHS

Table 1 summarizes the Young's modulus(E), tensile fracture strength(σ_f), Vickers hardness(H_v), tensile fracture strain(ε_f), tensile yield strain($\varepsilon_y=H_v/3E$) of Fe-, Co-, Ni-, Pd-, Pt- and Cu-based amorphous wires[14]. The approximation of $\varepsilon_y=H_v/3E$ is based on the fact that amorphous alloys exhibit little work-hardening. With increasing metalloid content, all the amorphous wires exhibit an appreciable increase in strength, e.g., σ_f and H_v of Fe-Si-B wires increase from 2910 to 3920 MPa and 830 to 1100 DPN which are much higher than those of piano wires with the highest strength in conventional crystalline alloys. The high values of E, σ_f and H_v for Fe-Si-B and Co-Si-B amorphous wires have been clarified to increase further by the replacement of Fe or Co by Nb, Ta, Cr, Mo or W[2,4].

Table 1 Static mechanical properties of Fe-, Co-, Ni-, Pd-, Pt- and Cu-based amorphous wires

Alloy(at%)	E(MPa)	σ_f(MPa)	H_v(DPN)	ε_f(%)	$\varepsilon_y=H_v/3E$(%)
$Fe_{80}Si_{10}B_{10}$	158000	2910	830	2.1	1.7
$Fe_{77.5}Si_{10}B_{12.5}$	164000	3100	935	2.4	1.9
$Fe_{75}Si_{10}B_{15}$	171000	3410	1030	2.8	2.0
$Fe_{70}Si_{10}B_{20}$	187000	3920	1100	2.3	1.9
$Fe_{75}P_{10}C_{15}$	152000	2990	895	2.8	1.9
$Co_{77.5}Si_{12.5}B_{10}$	190000	3580	1140	3.0	2.1
$(Ni_{.75}Si_{.08}B_{.17})_{99}Al_1$	151000	2730	780	2.9	1.7
$(Ni_{.78}P_{.12}B_{.1})_{99}Al_1$	125000	2170	690	2.4	1.8
$Pd_{77.5}Cu_6Si_{16.5}$	87800	1560	390	2.3	1.4
$Pd_{48}Ni_{32}P_{20}$	95700	--	470	--	1.6
$Pt_{60}Ni_{15}P_{25}$	93500	--	385	--	1.3
$Cu_{60}Zr_{40}$	--	1810	440	2.7	--

4. FATIGUE STRENGTH UNDER DYNAMIC BENDING LOAD

The fatigue test was performed on amorphous wires with a gauge dimension of 100 mm at a frequency of 3.2 Hz at 293 K in air where the humidity was controlled. The details of a specially designed bending-type fatigue testing machine have been shown elsewhere[5]. The maximum applied strain(λ) leading to fatigue failure under condition of dynamic bending load was measured as a function of the number of cycles of strain(N_f).

Figure 1 shows the $\lambda-N_f$ curves of amorphous (Fe,Co or Ni)-Si-B and (Fe,Co or Ni)-Cr-Si-B wires[5,11]. The fatigue limit is 0.0032 for $Fe_{75}Si_{10}B_{15}$ and $Co_{72.5}Si_{12.5}B_{15}$ and 0.0050 for $(Ni_{.75}Si_{.08}B_{.17})_{99}Al_1$ and increases significantly by the dissolution of Cr to 0.010 for $Fe_{67}Cr_8Si_{10}B_{15}$ and $Co_{67.5}Cr_5Si_{12.5}B_{15}$ and 0.0070 for $(Ni_{.75}Si_{.08}B_{.17})_{93}Al_1Cr_6$. It is very striking that the enhancement by the dissolution of 5 to 8 %Cr reaches as much as 180 to 300 % and the fatigue limits (0.009-0.0131) of the Cr-containing wires are higher by a factor of 2.3 to 3.3 than those of piano wire and SUS 304 wire. No other effective element to enhance the fatigue strength of Fe-, Co- and Ni-based amorphous wires was found besides Cr.

It is well known that the initiation of fatigue crack usually occurs on the sample surface. Accordingly, it is reasonably believed for the amorphous wires that the crack under the dynamic bending force initiates on the outer surface where the largest strain is applied and propagates to the central region of the wire. The inferrence has been supported from a number of photographs taken from the fracture surface[5]. It may therefore be concluded that the enhancement of fatigue strength by the dissolution of Cr is due mainly to the suppression of the initiation of fatigue crack on the sample surface through the remarkable enhancement of corrosion resistance.

5. MAGNETIC PROPERTIES

Figure 2 shows the changes in coercive force(H_c), saturation magnetization (B_s), residual magnetization(B_r), squareness ratio(B_r/B_s), longitudinal magnetostriction($\lambda_{\shortparallel}$) and initial permeability(μ') of $(Fe-Co)_{72.5}Si_{12.5}B_{15}$ amorphous wires with ≈120 μm diameter as a function of the ratio of Co to Fe+Co. The values of H_c, B_s, B_r and $\lambda_{\shortparallel}$ are the largest for $Fe_{72.5}Si_{12.5}B_{15}$ and decrease continuously with increasing Co content, while the ratio of B_r to B_s remains almost constant(≈0.5) in the entire composition range except $Co_{72.5}Si_{12.5}B_{15}$. It is

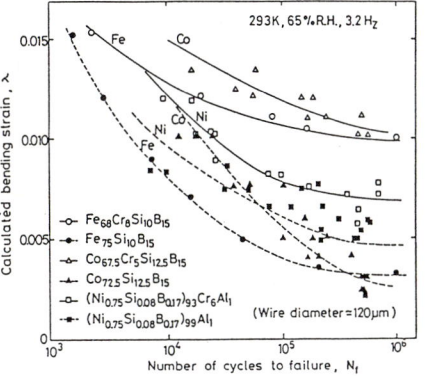

FIGURE 1
Fatigue curves of amorphous (Fe,Co or Ni)-Si-B and (Fe,Co or Ni)-Cr-Si-B wires.

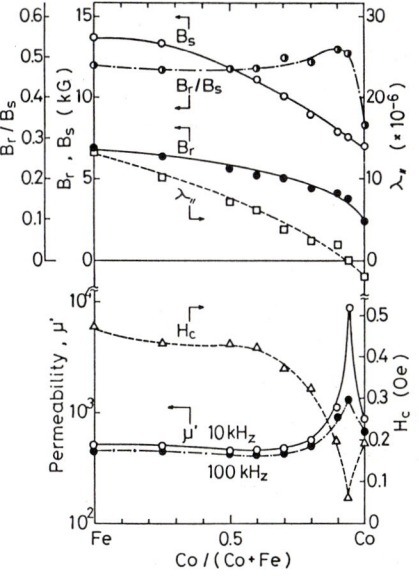

FIGURE 2
Compositional dependence of H_c, B_r, B_s, B_r/B_s, $\lambda_{\shortparallel}$ and μ' for amorphous $(Fe_{1-x}Co_x)_{72.5}Si_{12.5}B_{15}$ wires.

notable for the $(Fe_{.06}Co_{.94})_{72.5}Si_{12.5}B_{15}$ wire that H_c has the smallest value of 4.8 A/m and $\lambda_{\shortparallel}$ is nearly zero. The μ' values at 10 and 100 kHz remain almost constant ($\approx$450) in the compositional range below about 80 %Co and increase steeply to 8800 and 1300, respectively, at 94 %Co having zero magnetostriction.

The large values of H_c, B_s and B_r/B_s for the Fe-Si-B wire imply the appearance of B-H hysteretic loops with large squareness ratios as exemplified in Fig. 3. The hysteretic loop having large values of H_c and squareness ratio has been interpreted[6] as due to a large Barkhausen jump which occurs by an instantaneous inversion of flux, i.e., an instantaneous movement of magnetic domain wall. Additionally, Fig. 3 shows that a critical magnetic field H* leading to a large Barkhausen jump exists and no B-H loop is formed in the field below H*. On the other hand, as is eveident from the data shown in Fig. 3, the large Barkhausen jump does not take place for the Co-Fe-Si-B wire with zero magnetostriction and the change in B-H loop of the Co-Fe-Si-B wire with applied field is similar to that for the Co-Fe-Si-B ribbon. That is, a usual minor loop is formed, resulting in low H_c and high μ' values. It has previously been shown for Fe-based amorphous ribbons[15] that the Barkhausen inversion appears only in the state where the ribbon is twisted by applied stress and the reason for the appearance is due to the increase in the energy density at domain walls resulting from internal stress and magnetostriction. On the contrary, the Fe-Si-B amorphous wires have the outstanding feature that the large Barkhausen inversion appears even in as-cast and undeformed state without applied stress[6]. The appearance of the large Barkhausen inversion is probably because of a unique internal stress state resulted from the unique solidification process for the wire sample.

The H_c and B_r/B_s of Fe-Si-B amorphous wires having different metalloid compositions had nearly constant values of 30 to 40 A/m and 0.5, respectively, and a systematic compositional dependence was not observed.

In addition, Fe-based amorphous wires have been found to exhibit the following unique magnetic properties: (1) Uniaxial anisotropic energy is about ten times larger for the Fe-Si-B amorphous wire than for the amorphous ribbon having the same composition and the value for $Fe_{75}Si_{10}B_{15}$ wire reaches about 2000 J/m^3. (2) By applying tensile stress, AC- and DC-hysteretic loops

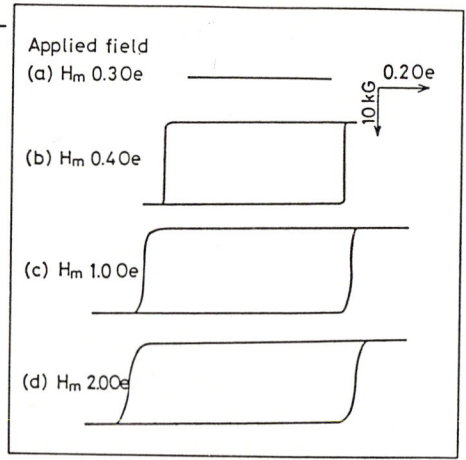

FIGURE 3
Change in hysteretic loop of an amorphous $Fe_{77.5}Si_{7.5}B_{15}$ wire with applied field.

change drastically so as to increase the squareness ratio[6,16]. For instance, B_r and B_s at 50 Hz for $Fe_{81}Si_4B_{14}C_1$ amorphous wire are 0.88 and 1.28 T, respectively, at zero applied stress and increase to 1.51 and 1.63 T, respectively, by applying a tensile stress of 200 MPa. Simultaneously, H_c also decreases by as much as about 40 %. (3) The Fe-Si-B amorphous wires exhibit a large Matteucci effect even in no applied stress state[7]. Whenever an alternating applied magnetic field exceeds a critical value, a sharp voltage pulse generates between both the edges of the amorphous wire.

Figure 4 shows the change in the initial permeability(μ') of $(Co_{.94}Fe_{.06})_{72.5}Si_{12.5}B_{15}$ amorphous wires with wire diameter. The wires with different diameters were prepared by cold drawing of the wire with 122 μm diameter and then annealed for 30 min at 673 K between Curie temperature(641 K) and crystallization temperature(835 K). The μ' increases significantly with decreasing wire diameter and the fine wire with 30 μm diameter has high μ' values even in high frequency ranges. The change in μ' with wire diameter agrees with that in μ' with ribbon thickness and hence the change for the wire appears to be attributed to the difference in eddy current loss with wire diameter. Additionally, the effect of the replacement of Co by transition M metal on μ' of amorphous $(Co_{.94-x}Fe_{.06}M_x)_{72.5}Si_{12.5}B_{15}$ wires with a diameter of about 120 μm was examined with an aim to improving the high-frequency characteristics. The μ' value at 100 kHz remains unchanged for M=Pd and Mn, but decreases for M=Nb, Ta, Cr and Ni. No effective element to enhance the μ' value was found. This is probably because the multiplication of constituent elements results in increase of quenching-induced internal stress.

6. CONCLUSIONS

The features of strength and magnetic properties of engineeringly important Fe- and Co-based amorphous wires were described in section 2 to 5. The amorphous wires possess high strength and hardness, soft ferromagnetism, magnetostriction in the range from zero to a large value and high electrical resistivity, as similar for the amorphous ribbons with the same composition. In addition, the Fe-Si-B amorphous

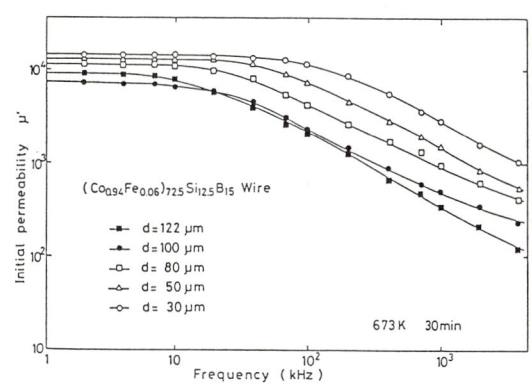

FIGURE 4
Change in μ' of amorphous $(Co_{.94}Fe_{.06})_{72.5}Si_{12.5}B_{15}$ wires having different wire diameters as a function of frequency.

wires have unique characteristics such as geometrical isotropy over the whole cross section, good flexibility leading to easy formation of composite materials, good cold- and warm-drawability, high fatigue strength, huge Barkhausen jump and large Matteucci effect which are not obtained for the amorphous ribbon samples. By taking these advantages inherent to amorphous alloy wires, the Fe- and Co-based amorphous wires have practically been used as sensing elements in magneto-electronic devices which demand these magnetic, electrical and mechanical properties as well as the fine geometry. The practical uses are in new fields where the unique characteristics and morphology of the amorphous alloy wires are simultaneously utilized and hence the importance of the amorphous alloy wires as functional and reinforcing materials is hereafter expected to increase more and more.

REFERENCES

1) T. Masumoto, I. Ohnaka, A. Inoue and M. Hagiwara, Scripta Met. 15 (1981) 293.
2) M. Hagiwara, A. Inoue and T. Masumoto, Met. Trans. 13A (1982) 373.
3) A. Inoue, M. Hagiwara and T. Masumoto, J. Mater. Sci. 17 (1982) 580.
4) M. Hagiwara, A. Inoue and T. Masumoto, Mater. Sci. Eng. 54 (1982) 197.
5) M. Hagiwara, A. Inoue and T. Masumoto, Proc. 5th Int. Conf. on Rapidly Quenched Metals (North-Holland, Amsterdam, 1985) pp. 1779.
6) H.S. Chen, R.C. Sherwood, S. Jin, G.C. Chi, A. Inoue, T. Masumoto and M. Hagiwara, J. Appl. Phys. 56 (1984) 1796.
7) K. Mohri, J. Yamasaki and T. Kondo, Proc. 5th Int. Conf. on Rapidly Quenched Metals (North-Holland, Amsterdam, 1985) pp. 1659.
8) H. Matsuki and K. Murakami, IEEE Trans. Magnetics MAG-21 (1985) 1738.
9) A. Inoue, N. Yano, H.S. Chen, M. Hagiwara and T. Masumoto, Mater. Sci. Eng. 77 (1986) 45.
10) I. Ohnaka, T. Fukusako and T. Daido, J. Japan Inst. Metals 45 (1981) 751.
11) A. Inoue, S. Furukawa, M. Hagiwara and T. Masumoto, Met. Trans. 18A (1987) 621.
12) A. Inoue, Y. Masumoto, N. Yano, A. Kawashima, K. Hashimoto and T. Masumoto, J. Mater. Sci. 20 (1985) 97.
13) A. Inoue, N. Yano and T. Masumoto, J. Mater. Sci. 19 (1984) 3786.
14) A. Inoue, H.S. Chen, J.T. Krause, T. Masumoto and M. Hagiwara, J. Mater. Sci. 18 (1983) 2743.
15) K. Mohri and S. Takeuchi, J. Appl. Phys. 53 (1982) 8386.
16) I. Ogasawara, M. Hagiwara, A. Inoue and T. Masumoto, unpublished research (1985).

MAGNETIC AND CRYSTALLIZATION STUDIES IN AMORPHOUS R-Fe AND R-Co ALLOYS

George HADJIPANAYIS, Anton NAZARETH

Department of Physics, Cardwell Hall, Kansas State University, Manhattan Kansas 66506 USA

The magnetic hysteresis behavior of amorphous rare-earth system is reviewed. Recent results indicate that the RMA models might be more appropriate in explaining the high coercivities. The crystallization kinetcs and products of crystallization are also investigated.

1. INTRODUCTION

Rare-earth amorphous alloys have attracted considerable scientific interest in recent years because of their unique magnetic and electronic properties.[1-3] Most of the early studies have been directed towards the evaporated[4,5] and rapidly-sputtered $R(Fe,Co)$[6-8] and splat cooled $R_{65}(Fe,Co)_{35}$[9-11] and $R_{80}Au_{20}$[12,13] amorphous alloys. These materials show some kind of magnetic ordering at low temperatures. Below the ordering temperature, the non-S-state ion alloys develop a high magnetic anisotropy that restricts magnetic saturation and leads to very high coercive field particularly at cryogenic temperatures. Around the coercive field sharp magnetization jumps are observed together with a strong temperature-dependent magnetic viscosity, suggesting a thermal activation process. The large coercive fields decrease rapidly at higher temperatures and become very small even at temperatures much below the ordering temperature.

Several models have been proposed to explain the magnetic properties of these alloys. The local random anisotropy models[14-16] predict high coercive fields which are due to the intrinsic properties of the microscopic Hamiltonian rather than to macroscopic effects such as domain wall pinning. On the other hand, Callen et al.[17] predicted high coercivities by using the classical approach of interacting single domain particles. Using a different approach, Chi et al.[18] discussed the high coercive fields in $TbFe_2$ in terms of domain wall pinning. This idea was also used to explain the giant intrinsic magnetic hardness of disordered rare-earth intermetallic compounds where it was assumed that fluctuations in exchange and anisotropy can pin the domain walls.

Although there are several models for the magnetic properties of these materials, there seems to be confusion about the origin of the high coercive fields because of fundamental differences in the approach used. Several experiments can be proposed that will aid in selecting among the various competing models. For example, the temperature dependence of the coercivity is different in the different models, so measurements of $H_c(T)$ can determine which model is appropriate for a given situation. A better way to understand the magnetic properties of these materials is to study directly their magnetic domain structure through Lorentz electron microscopy. The magnetic domain structures have not yet been measured in rare-earth amorphous materials with high coercivities.

We have undertaken a study to investigate the origin of anomalous magnetic hysteresis in several rare-earth amorphous systems, by correlating the magnetic properties with the magnetic domain structure and microstructure. We are also studying the crystallization behavior of these systems and its effect on magnetic hysteresis. In the present report we will present some preliminary results of this investigation.

2. CRYSTALLIZATION STUDIES

The DSC studies showed strong exothermic peaks in all samples studied (Fig. 1). The R-Co ribbons are thermally most stable around the equiatomic composition rather than at the eutectic concentration ($Nd_{65}Co_{35}$). In R-Fe the crystallization temperatures are found to be much higher ($T_{cr} \sim 500°C$) and they do not change significantly with Fe concentration (Table I). The reason for this is not yet understood. The Kissinger method[19] was used to determine the activation energy involved in the crystallization of TbFe(Co) and NdFe(Co) alloys (Fig. 2) and the results are shown in Table I. The activation energy is found to be highest for $Nd_{50}Fe_{50}$, the alloy with the highest crystallization temperature. Recently we have also used the Johnson-Mehl-Avrami equation[20] to study the crystallization kinetics in $Tb_{50}Fe_{50}$. The activation energy was found to be E=1.85 eV and the exponent coefficient n=3.5. Values

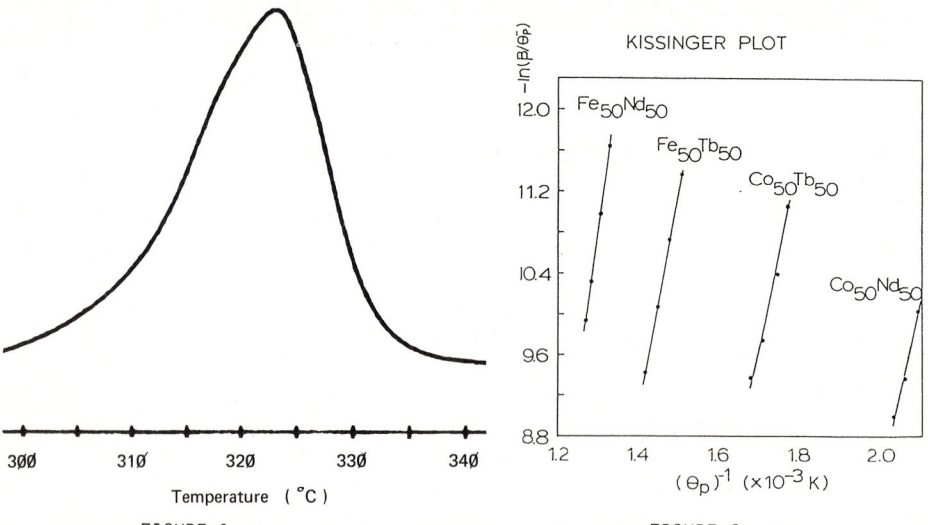

FIGURE 1
DSC studies.

FIGURE 2
Kissinger plot to find E.

of n in the range between 3 and 4 indicate a eutectoid (or interface) controlled growth with a decreasing nucleation rate. These fundamental studies will be extended to other rare-earth systems so that we can understand better the differences in their crystallization.

The phases obtained after crystallization were investigated with ac susceptibility and x-ray diffraction. In Nd-Co four magnetic phases have been identified with ordering temperatures 27, 105, 140 and 190 K. We have identified the phases with ordering temperatures 27 K and 105 K as Nd_2Co and $NdCo_2$ respectively. The former one occurs in Nd rich glasses while the latter in $Nd_{50}Co_{50}$ and in Co-rich glasses. The phases with ordering temperature 140 and 190 K could not be identified with any other existing binary phases. It may be possible that these phases are Nd-Co oxides where oxygen stablizes a ternary phase with relatively high ordering temperature as in $Fe_{14}R_2B$.[21] In Nd-Fe, however the primary phases formed after crystallization are Nd bcc Nd-Fe and Nd_2Fe_{17} (Fig. 3,4). This is consistent with the presence of few

TABLE I

Alloy	T_{cr}(°C)	E(eV)
$Fe_{50}Nd_{50}$	493	2.48 ± 0.01
$Fe_{50}Tb_{50}$	390	1.78 ± 0.01
$Co_{50}Tb_{50}$	283	1.62 ± 0.02
$Co_{50}Nd_{50}$	192	1.61 ± 0.02

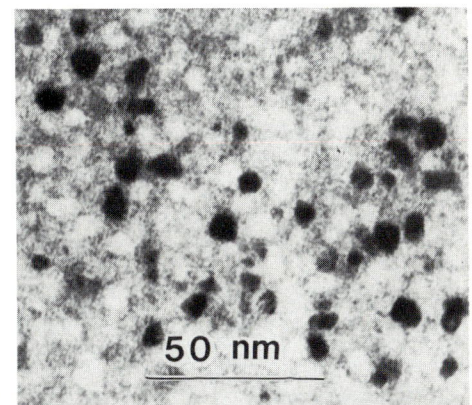

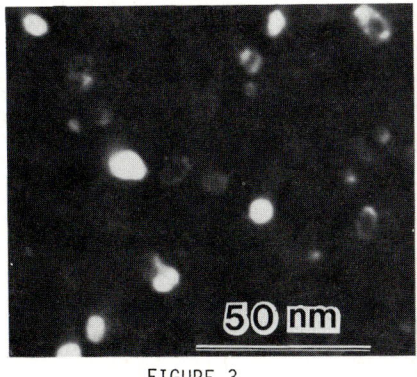
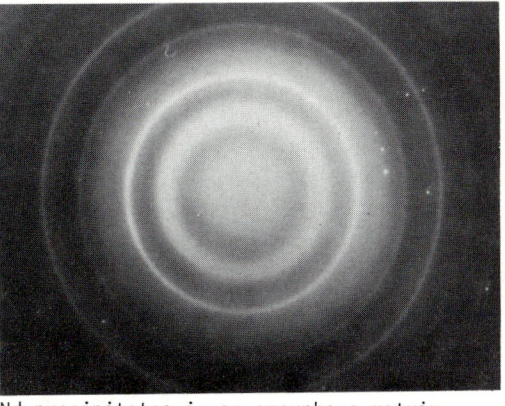

FIGURE 3
(a) BF TEM micrograph showing bcc Fe-Nd precipitates in an amorphous matrix.
(b) DF TEM micrograph showing the bcc Fe-Nd precipitates.
(c) Electron diffraction pattern.

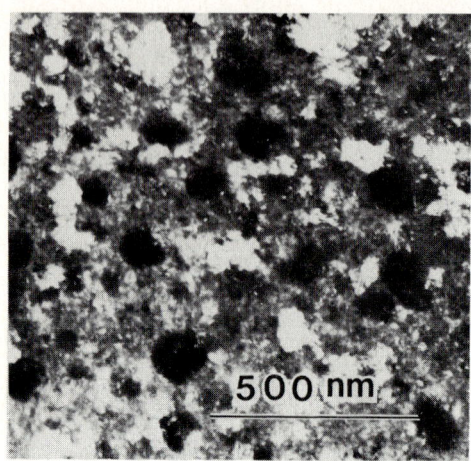

FIGURE 4
DF TEM micrograph showing Nd_2Fe_{17} crystallites together with Nd in a crystallized $Nd_{35}Fe_{65}$ sample.

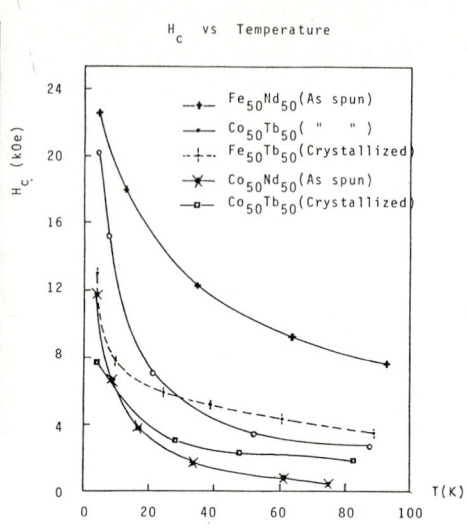

FIGURE 5
Temperature dependence of coercivity.

intermetallic phases in light rare-earth-iron systems. In $Tb_{50}Fe_{50}$ crystallization produces $TbFe_2$ and Tb. Transmission electron microscope data on crystallized samples with high coercivity show fine $TbFe_2$ and Tb crystallities with a size in the range of 50-100 Å.

3. MAGNETIC MEASUREMENTS

The amorphous rare-earth glasses showed the expected behavior; small coercive fields at room temperature which increase to very large values (up to 40 kOe) upon cooling to cryogenic temperatures (Fig. 5). When the samples are crystallized the room temperature coercivity is increased but the temperature coefficient of coercivity is significantly reduced, leading to much smaller values of coercivity at cryogenic temperatures. This behavior shows a distinct difference in the origin of H_c and strongly suggests that the giant hysteresis at cryogenic temperatures is entirely due to the anisotropic nature of the amorphous materials. The coercivities in Nd-Fe are much higher than in Nd-Co because of the much lower ordering temperatures of the latter alloys.[9,22] In Tb-Fe and Tb-Co, the odering temperatures are close and the coercivities at 4.2 K are very similar (30 kOe). The initial magnetization curves show a behavior that is not characteristic of pure uniform domain wall pinning or nucleation of reversed domains. The magnetization curves are rather similar to those of isotropic R-Fe-B magnets.

The high values of coercive field are predicted by all the previously discussed models.[14,18] However, the local random anisotropy models[14] might be

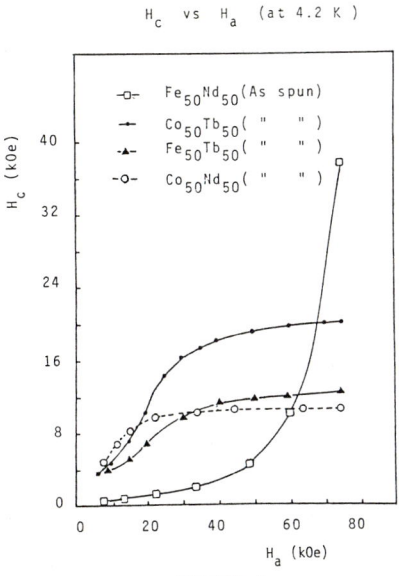

FIGURE 6
Field dependence of coercivity.

more appropriate in explaining the magnetization reversal in these systems. Lorentz microscopy studies of melt-spun $Tb_{50}Fe_{50}$ failed to show the existence of any conventional domain walls down to 100 K. However, the idea of a domain wall in these highly anisotropic materials is not well defined. In $Tb_{50}Fe_{50}$, using the experimentally obtained value of reduced remanence $m_r \approx 0.74$ and applying Callen's model,[17] we obtained an $H_c \sim 40$ kOe which is very close to the experimental value of 30 kOe. (Callen's model predicts $m_r(d)$, $h_c(d)$ where $d = \frac{K}{\lambda M_s^2}$, λ the interaction constant.) The temperature dependence of H_c is presently being investigated according to the predictions of the various models in an attempt to check their validity.

In addition to hysteresis loops two other magnetic measurements were used to investigate the origin of H_c; the field dependence of H_c, $H_c(H_a)$ and the remanence curves. The $H_c(H_a)$ curves are shown in Fig. 6. Most of the samples show a behavior characteristic of nucleation-type magnets;[23] an initial linear increase of H_c followed by saturation at high fields. Only $Nd_{35}Fe_{65}$ showed a different which is very similar to that of uniformly domain wall pinned materials;[23] a small increase at low fields followed by a drastic increase to saturation at a critical field which is close to the coercive field. At the present we do not understand this behavior for $Nd_{35}Fe_{65}$. Lorentz microscopy studies are underway to check whether a domain wall pinning model is more appropriate for this sytem. The remanence curve relationship[24] $M_D(H) = M_R(H_{max}) - 2M_R(H)$ is not obeyed in amorphous $R_{50}Fe_{50}$ alloys. If domain walls were present, they would probably interact uniformly with the random distribution of inhomogeneities (because of the amorphous nature of the sample) and the above relationship would be obeyed. The relationship also is not obeyed for interacting single domain particles, indicating that the local random anisotropy (RMA) models might be more appropriate. This again is consistent with recent small angle neutron diffraction which shows highly correlated "spin-clusters" in these systems.

The room temperature coercive field of Tb-Fe and Tb-Co alloys is found to increase after crystallization. The increase has been attributed to the anisotropic fine $TbFe_2$ crystallites which are present after crystallization. On the other hand, the coercivity of Nd-Fe alloys is decreased after crystallization because of the presence of Nd_2Fe_{17} and Nd-Fe bcc phases which are considered to be soft.

ACKNOWLEDGEMENT

This work was supported by NSF Grant No. DMR-8607023.

REFERENCES
1) R.W. Cochrane, R. Harris, and M.J. Zuckermann, Phys. Rep. 48 (1978) 1.
2) J.J. Rhyne, Handbook on the Physics and Chemistry of Rare Earths, edited by K.A. Gschneidner, Jr. and L. Eyrin (North-Holland, Amsterdam, 1979) p. 259.
3) R. Asomoza, A. Fert, I.A. Campbell and R. Meyer, J. Phys. F: Metal. Phys. 7 (1977) L327.
4) N. Heiman and K. Lee, Phys. Rev. Lett. 33 (1974) 778.
5) R.C. Taylor, J. Appl. Phys. 47 (1976) 1164.
6) A.E. Clark, Appl. Phys. Lett. 23 (1973) 642.
7) J.J. Rhyne, J.H. Schelleng and N.C. Koon, Phys. Rev. B10 (1974) 4672.
8) H.A. Alperin, J.R. Cullen and A.E. Clark, AIP Conf. Proc. Series 29 (1976) 186.
9) J.A. Gerber, D.J. Miller and D.J. Sellmyer, J. Appl. Phys. 49 (1978) 1699.
10) G. Hadjipanayis, S.C. Cornelison, J.M. Gerber and D.J. Sellmyer, J. Mag. Magn. Mat. 21 (1980) 101.
11) G. Hadjipanayis, S.G. Cornelison and D.J. Sellmyer, J. De Physique 41 (1980) C8-642.
12) A. Berrada, J. Durand, N. Hassanain and B. Loegel, J. Appl. Phys. 50 (1979) 7621.
13) D.J. Sellmyer, G. Hadjipanayis and S.G. Cornelison, J. Non-Cryst. Solids 40 (1980) 437.
14) R. Harris, M. Plischke and M.J. Zuckermann, Phys. Rev. Lett. 31 (1973) 160.
15) M.C. Chi and R. Alben, J. Appl. Phys. 48 (1977) 2987.
16) J.D. Patterson, G.R. Gruzalski and D.J. Sellmyer, Phys. Rev. B18 (1978) 1377.
17) E. Callen, Y.J. Liu and J.R. Cullen, Phys. Rev. B16 (1977) 263.
18) M.C. Chi and T. Egami, J. Appl. Phys. 50 (1979) 165.
19) H.E. Kissinger, J. Res. Nat. bur. Stand. 57 (1956) 217.
20) J.W. Christian, The Theory of Transformation in Metals and Alloys, 2nd Ed. Part 1 (Pergamon Press, Oxford, 1975).
21) G.C. Hadjipanayis, R.C. Hazelton and K.R. Lawless, Appl. Phys. Lett 43 (1983) 797.
22) C.P. Wong, S.H. Aly, N. Nazareth, K. Gudimetta, B. Dale and G.C. Hadjipanayis, J. Phys. D19 (1986) 1057.
23) H. Zijlstra, Ferromagnetic Materials, ed. E.P. Wohlfarth (North-Holland Amsterdam) 3 (1982) 37.
24) E.C. Stoner and E.P. Wohlfarth, Phil. Trans. R. Soc. London Ser. A240 (1948) 599.

$Co_{1-x} P_x$ ELECTRODEPOSITED ALLOYS: STRUCTURAL AND MAGNETIC PROPERTIES

Luciano LANOTTE and Flavio PORRECA

Dipartimento di Fisica Nucleare, Struttura della Materia e Fisica Applicata Piazzale V. Tecchio 80, 80125 Napoli, Italy§

In proper conditions $Co_{1-x} P_x$ alloy can be obtained with crystalline, amorphous and intermediate structure as a function of the P content. The behaviours of magnetization saturation, coercive field and magnetostriction, versus the parameter x, validate the changes in the structural properties.

1. INTRODUCTION

The electrochemical deposition of alloy in the composition $Co_{1-x} P_x$ is well known[1,2]. The new results reported here derive from the fact that a substratus of Al, instead of Cu, was used. The $Co_{1-x} P_x$ deposit does not implant in the Al electrode, in fact it is not necessary to use chemical bath in order to dissolve the substratus because this one can be removed mechanically.

2. EXPERIMENTAL

The samples were produced according the Brenner[3] directions on the electrochemical bath. The temperature was fixed at 75 ± 1 °C and the PH controlled in the range 0.5 - 1. The current density, σ, in the cell was changed from $3 A/dm^2$ to $30 A/dm^2$ obtaining different P percentage. The last one was deduced both by means of weight considerations and taking into account literature data[4] about magnetic properties dependence on P content. The sample thickness resulted variable as a function of σ and the time of deposition t (15 min $\leq$ t $\leq$ 45 min), but was included between 100 and 300 μm. All the samples were reduced to the size 2 x 0.3 cm^2. A thermal treatment for 1h at 300 °C in inert gas was performed.

By fluxmetric technique the hysteresis loops (at low frequency, 8 Hz) were obtained and then they were analyzed by means of Data 6000 Signal Analyzer in order to calculate the saturation magnetization, M_s, and the coercive field, H_c. The complete evolution of the hysteresis curves as a function of the magnetizing field H (applied along the longitudinal direction of the ribbons) was registered for obtaining the magnetization curves under the action of an external tensile stress δ. Repeated measurement were performed at several δ values in order to calculate the longitudinal magnetostriction coefficient λ by the relation $\lambda = \lim_{\Delta\delta \to 0} \Delta k_a / \Delta\delta$ where k_a is the magnetic anisotropy energy.

---

§ Unita CISM del MPI e GNSM del CNR

X-ray diffractometry and differential thermal analysis were performed by means of the collaboration with the Department of Materials Engineering and Production (University of Naples).

3. RESULTS AND DISCUSSION

In figures 1, 2 and 3 the results concerning M_s, H_c and λ are reported. It is evident the presence of two critical values of x: $x_A \simeq 4$ and $x_B \simeq 11$. Around these values of P content there is a gradual transition towards more soft magnetism, and for $x \simeq x_B$ also a change in the λ sign occurs.

According to literature data[4] we found that when x<11 the structure is a crystalline one (fig. 4a), while for x>11 the samples have amorphous structure (fig. 4b). In the last case, when $4 < x < 11$ the D.T.A. (fig. 5) shows that a metastable phase is present that evolves in a more stable one when the temperature reaches 400 °C. We think that an intermediate structure ("microcrystalline") between crystalline and amorphous one is obtained in these conditions, as preceeding results[5] in $Ni_{1-x}P_x$ samples support.

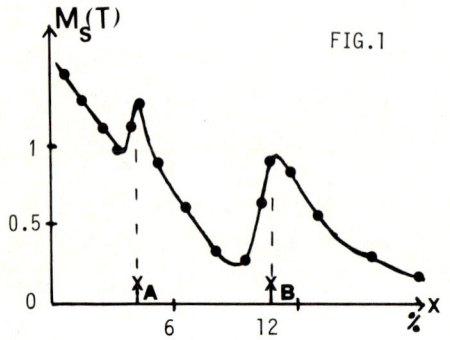

FIG.1

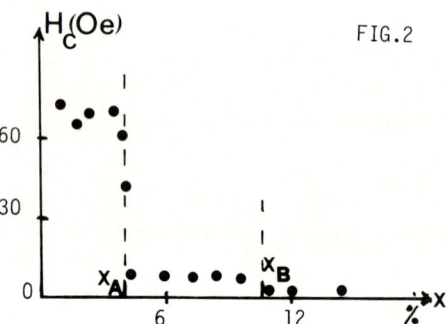

FIG.2

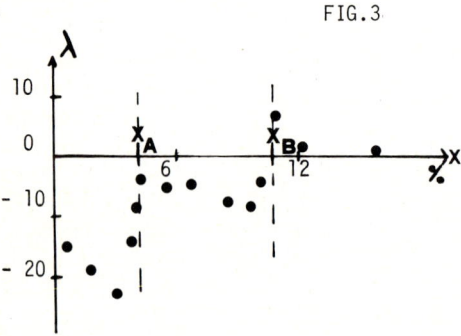

FIG.3

FIG.1   Saturation magnetization M_s versus the P content x (at %)

FIG.2   Coercive field H_c versus the P content

FIG.3   Longitudinal magnetostriction λ versus P content in the sample

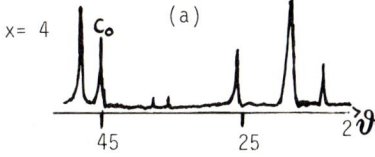

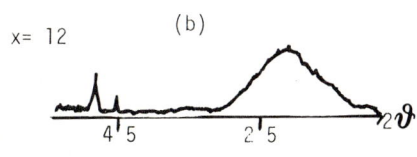

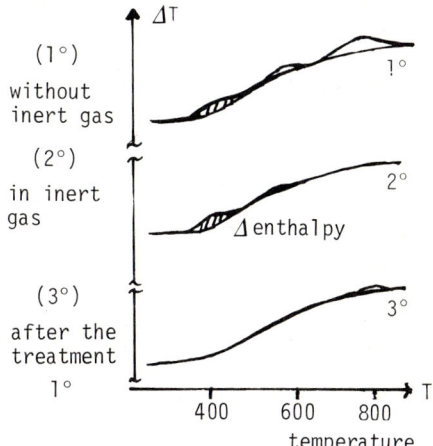

FIG. 4 - X-ray diffrattography results

FIG.5 - Differential Thermal Analysis D.T.A. results

4. CONCLUSIONS

More extensive and accurate structural analysis is going on in order to consolidate the aspect mentioned at the hand of the preceeding section.

However, since the behaviours of M_s, H_c and λ are monotonic from $x \simeq 0$ to $x \simeq 11$ when Cu substrata are used, we conclude that the effects obtained in the range $x_A \leq x \leq x_B$, in the present investigation, are to be related with the fact that the Al substratus does not alloy with the $Co_{1-x} P_x$ deposit, namely it obstacles the formation of ordered structure more than a Cu substratus makes.

ACKNOWLEDGMENT

We would like to thank professors I. de Iorio and V. Tagliaferri for their supports in samples preparation and structure analyse.

REFERENCES

1) J. Logan and M. Yung, J. Non-Crystalline Solids 21 (1976) 151.
2) D. Pan and D. Turnbull, J. Appl. Phys. 45 (1974) 1406.
3) A. Brenner, Electrodeposition of Alloys vol. II (Academic Press, N.Y. - London, 1963) chap. 35.
4) K. Hüller, M. Sydow and G. Dietz, J. Magn. Magn. Mat. 53 (1985) 269.
5) I. de Iorio, L. Lanoote and V. Tagliaferri, J. Metals Nov. issue (1985) 48.

LOW ANGLE DIFFUSE X-RAY SCATTERING IN GLASSY $(FE-NI)_{1-C_B}B_C$ ALLOYS
BEFORE AND AFTER STRUCTURAL RELAXATION AND EMBRITTLEMENT.

D. BIJAOUI, A.R. YAVARI, F. LIVET.

LTPCM (CNRS UA 29), Institut National Polytechnique de Grenoble, BP 75, Domaine Universitaire, 38402 Saint Martin d'Hères Cédex, France.

Using mechanical testing and low-angle X-ray measurements, we show that effects of relaxation-annealing on the structure of $(FeNi)_{1-C_B}B_C$ glassy tapes depend sensitively on the B-content. The hypoeutectic structure remains unchanged after annealing. Eutectic and hypereutectic alloys undergo changes in chemical short range order also becoming brittle after annealing.

1. INTRODUCTION

As part of a study of embrittlement phenomena during relaxation annealing of soft-magnetic Fe-B type glassy alloys[1] we have studied the effect of relaxation on the ductility of $(Fe_{50}Ni_{50})_{1-C}B_C$ glassy alloy tapes prepared by planar-flow-casting[2,3]. We determined the fracture strain ε_f = d at various test temperatures T_t, after annealing for 7200s at T_a = 573K, by bending tapes of thickness d = 25 µm between two parallel plates inside a vacuum furnace until fracture at interplaten distance D_f[4,1]. As can be seen in figure 1, the hyper eutectic glassy alloy with boron content C_B = 23at% fractures at temperatures T_t < 440K and for C_B = 20% fracture occurs at T_t < 330K. On the other hand the annealed hypoeutectic glassy tape with C_B = 17% does not break on bending (ε_f = 1) at test temperatures T_t ≫ 300 K. Glassy alloys with C_B = 0.17 do not undergo thermal embrittlement until the onset of crystallisation. To gain insight into this phenomenon, we performed low angle X-ray scattering measurements on these glassy alloys which we report here.

2. LOW-ANGLE DIFFUSE X-RAY SCATTERING : THEORETICAL AND EXPERIMENTAL RESULTS.

Since Fe and Ni have equal atom diameters and form compounds of similar structure with boron, we use the quasi-binary approximation for ternary FeNiB glasses. Indeed chemical-short-range-order (CSRO) is insensitive to Fe/Ni ratio in this system[5]. With fractions C_B and C_A of boron (B) and transition metal (A) atoms, the scattering intensity I(Q) as a function of wave vector Q is given by[6] :
$$I(Q) = \bar{f}^2 S_{NN}(Q) + 2\bar{f}\Delta f S_{NC}(Q) + \Delta f^2 S_{CC}(Q) \qquad (1)$$
where $\bar{f} = C_A f_A + C_B f_B$, $\Delta f = f_A - f_B$, f_A and f_B are the scattering amplitudes and $S_{NN}(Q)$, $S_{CC}(Q)$ and $S_{NC}(Q)$ are the density and concentration structure factors and their coupling term as defined by Bhatia and Thorton[6]. In the long wave-length limit and at thermal equilibrium, for an atom with volume V,

$$S_{CC}(0) = NkT/(\partial^2 G/\partial c^2)/ \quad S_{NN}(0) = (N/V)kT\ K_T + \delta^2 S_{CC}(0) \quad (2)$$
$$S_{NC}(0) = -\delta S_{CC}(0) \text{ with } \delta = (V_A - V_B)/(C_A V_A + C_B V_B)$$

where G is the free energy, V_i the atomic volumes, K_T the compressibility. These limits are attained as $Q \to 0$ but their values are masked by low angle scattering due to heterogeneities and surfaces[7,8]. Furthermore, glassy alloys depart from thermal equilibrium during cooling at $T < T_g$ and departures from eq.(2) are expected. Lamparter et al[7] and others have shown that $S_{ij}(Q)$ and $I(Q)$ at low Q in metal-metalloid glasses attain nearly constant values I_D in the diffuse scattering range (see fig.2). We will take this constant value in the range $0.5 < Q < 1 \text{Å}^{-1}$, as the low angle limit of $I(Q)$ thus avoiding the effect of scattering from heterogeneities at very low Q. By eq.s(1) and (2) we obtain:

$$I_D = I(0) = \frac{N}{V} kT\ K_T\ \bar{f}^2 + S_{CC}(0)\ (\bar{f}\delta - \Delta f^2) \quad (3)$$

Suppose now an ideal solution where $G(C) = NkT(C_A \ln C_A + C_B \ln C_B)$ and $\partial^2 G/\partial C^2 = NkT/C_A C_B$ with $S_{CC}(0) = C_A C_B$. Now $K_T = 1/2\ E$ with the Young's modulus[9] $E \simeq 1.7 \times 10^{12}$ degrees/cm² in Fe-B glasses[10]. This allows an estimation of the first terms of eq.(3) with $(N/V)kTK_T\ \bar{f}^2 \simeq 0.1\ e^2/\text{Å}^3$ independent of C_B. The second term, of the order of $3e^2/\text{Å}^3$, overwhelms the first term which can be neglected. Figure 3 shows a plot of the second term of I_D versus B-content calculated with $S_{CC}(0) = C_A C_B$, boron to transition metal volume ratio V_B/V_A = 0.4, 0.5 and 0.6 and the values of f_i at $Q = 0.7$ Å^{-1} (in the range of constant $S_{ij}(Q)$). I_D is seen to increase in a monotonic fashion with C_B in ideal solutions with constant V_B/V_A. V_B/V_A is generally reported to be about 0.5 in Fe-B type glasses[5] but may increase as best B sites are filled up with increasing C_B.

We measured X-ray scattering in $(Fe_{50}Ni_{50})_{1-C}B_C$ glassy tapes with C = 0.17, 0.20 and 0.23 at low angles in transmission using molybdenum K_α radiation.

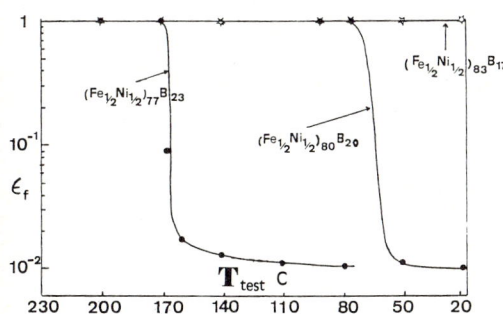
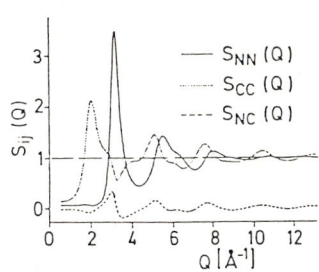

Figure 1
Strain ε_f at fracture versus T_{test} for $(Fe_{50}Ni_{50})_{1-C}B_C$ after 2hrs at $T_a^{test} = 573$ K.

Figure 2 :
Experimental partial structure[7] factors for amorphous $Ni_{80}P_{20}$

After background and other corrections, I(Q) was obtained for $0.05 < Q < 1$ Å^{-1}. For experimental details see[1]. Figure 4 shows I_D versus Q before and after annealing at 573 K for the three glassy alloys which show different behavior. Consistent with previous work[7], all three alloys show intensity values $I(Q) = I_D$ nearly independent of Q in the range near $Q = 0.7$ Å^{-1} as discussed above. I(Q) values at $Q = 0.7$ Å^{-1} (see fig.4) have therefore been replotted in Fig.5 as a function of C_B for comparison with the calculated values of Fig.3 calculated also for $Q = 0.7$ Å^{-1}.

3. DISCUSSION.

Concerning thermal embrittlement, we observe in figures 4 and 5 that the diffuse scattering I_D does not change with annealing in the hypoeutectic glass ($C_B = 0.17$) which also remains ductile. The hypo-eutectic as-quenched structure is thus stable and no significant change occurs in its CSRO with annealing.

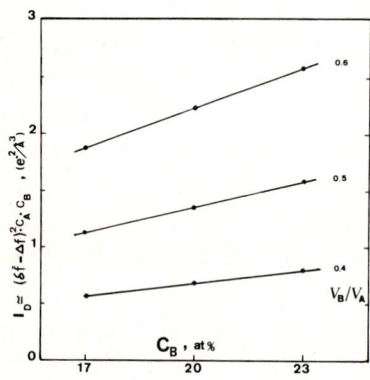

Figure 3 :
Calculated low angle limits of X-ray scattering $I_D C_B$ versus C_B.

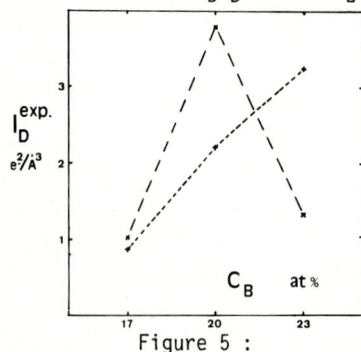

Figure 5 :
Diffuse scattering I_D versus C_B for alloys of Fig.2 before (+) and after (x) annealing

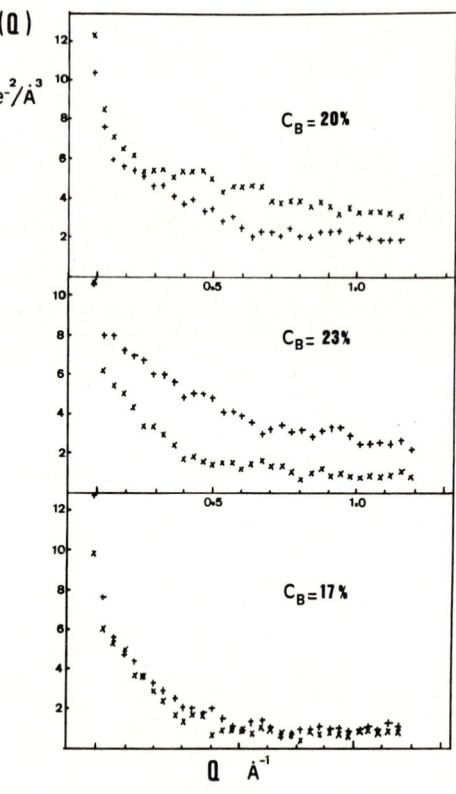

Figure 4
Experimental I(Q) in glassy $(Fe_{50}Ni_{50})_{1-C_B}B_C$ before (+) and after (x) annealing.

On the other hand, I_D, for the near-eutectic (C_B = 0.2) and hypereutectic (C_B = 0.23) glasses are strongly modified, increasing for C_B = 0.2 decreasing for C_B = 0.23 after relaxation that results in embrittlement. In the as quenched state, I_D shows a monotonic increase with C_B (fig.4) similar to the calculated ideal solution curve of Fig.3 but with a stronger slope. This is indicative of some departure from ideality but suggest comparable structures for the three as-quenched alloys. The stronger slope of the experimental curve suggests an increase in V_B occupied by the boron atoms as the Bernal-Polk type interstitial sites in the disordered alloy[11] are progressively filled up. The annealing results in further departure from ideal solution behavior and the 20 at % and 23 at % alloys evolve towards different sub-T_g structures. The lowering of I_D with annealing observed for C_B = 0.23 which is nearly stoichiometric $(FeNi)_3B$, indicates increasing CSRO with $S_{CC}(0)$ values below $S_{CC}(0) = C_A C_B$ (see[6]) and reduced I_D as per eq.(3). For C_B = 0.20, the increase in I_D may be due to occurence of phase separation phenomena during annealing as detected by FIM and atome-probe depth profiles[12] which also contributes to embrittlement[13]. If the hypoeuctectic alloy at low C_B corresponds closely to a stuffed Bernal-Polk type DRP structure[11] and the near stoichometric hypereutectic alloy with $C_B \approx 0.25$ corresponds to a nano-crystalline trigonal-prism packing[14] as in the intermetallic Fe_3B, they are both stabilised (lower G(C)) in relation to the intermediate near-eutectic alloys with C_B = 0.2 and a transformation of the structure of alloys with $C_B \approx 0.2$ into zones poor and rich in boron corresponding locally to each of the more stable structures would become energetically feasible. This would result in G(C) curves ranges in C_B where $\partial G^2/\partial C^2 < 0$ and phase separation is possible. Near zero or negative G"(c) values increase I_D in eq.(3) and as observed experimentally.

REFERENCES.
1) D. Bijaoui, Doctoral thesis, INPG 1987
2) M.C. Narasimhan, US Patent n°4142571, 1979.
3) A.R. Yavari, this conference
4) F.E. Luborsky and J.L. Walter, J. Appl. Phys. __47__ (1976) 3648.
5) Y. Waseda, K.T. Aust and J.L. Walter, J. Mat. Sci., __15__ (1980) 1252.
6) A.B. Bhatia and D.E. Thornton, Phys. Rev. B. __2__ (1970) 3004.
7) P. Lamparter and S. Steeb, Proc. Int. Conf. On Rapidly Quenched Metals RQ6, vol. I, eds : S. Steeb and P. Warlimont, Würsburg 1984, p.459.
8) A.R. Yavari, P. Desré and P. Chieux, Atomic Transport and Defects in Metals by Neutron Scattering, eds. C. Janot et al, Springer Verlag, 1985, p.842-8.
9) H.S. Chen, J.T. Krause and E. Coleman, J. Non Cryst. Solids, __18__ (1975) 157.
10) L.A. Davis, Proc. Int. Conf. on Rapidly Quenched Metals RQ2, Sec.I, eds : N.J. Grant and B.C. Giessen, M.I.T. Press, Cambridge 1975, p. 369.
11) J.D. Bernal, Nature __185__ (160) 68 ; D.E. Polk, Scripta Met. __4__ (1970) 117.
12) J. Piller and P. Haasen, Acta Met., 30 (1982) 1.
13) A.R. Yavari, J. Mat. Res., __1__ (1986) 746.
14) J.D. Dubois and G. Le Caer, Acta Met., __32__ (1984) 2101.

ELASTICITY AND MAGNETOMECHANICAL COUPLING DEPENDENCES OF THE $Fe_{79}Si_{12}B_9$ METALLIC GLASSES ON THE MAGNETIC FIELD*

Zbigniew KACZKOWSKI

Polish Academy of Sciences Institute of Physics
Al. Lotników 32/46, 02-668 Warszawa, Poland

As quenched amorphous $Fe_{79}Si_{12}B_9$ alloy ribbons with the resonant frequencies ranging in the band of 44 kHz were investigated. The magnetomechanical coupling coefficient reaches value of 0.2 and the moduli of elasticity were changing from 140 GPa through minimum of 130 GPa to 155 MPa.

1. INTRODUCTION

Fe base amorphous alloys exhibit relatively high saturation magnetostriction (about 30×10^{-6})[1-3]. This alloys can be used as a core of the ultrasonic transducers, even without thermal treatment, because of their very good piezomagnetic properties.

2. EXPERIMENT

The measurement were carried out on the as quenched $Fe_{79}Si_{12}B_9$ metallic glass ribbons. The strip shape samples were cut out from the ribbons with thickness from 30 to 40 µm. The length was changing from 49.5 to 52 mm and width was in the range from 3.5 to 4.2 mm. The mass of the investigated samples was changing from 34 to 44 mg. The resonant - antiresonant method was used [4] (Fig. 1). The magnetomechanical coupling coefficient (k), the moduli of elasticity at the constant magnetic field (E_H) and the constant magnetic induction (E_B) have been determined from the following relations:

$$k = \left| \frac{\pi^2}{8} \cdot \frac{E_B - E_H}{E_B} \right|^{1/2} = 1.11 \left| \frac{f_a^2 - f_r^2}{f_a^2} \right|^{1/2} = 1.57 \left| \frac{f_a - f_r}{f_a} \right|^{1/2}, \quad (1)$$

$$E_H = c_H^2 \varrho = 4 f_r^2 l^2 \varrho, \quad (2)$$

$$E_B = c_B^2 \varrho = 4 f_a^2 l^2 \varrho, \quad (3)$$

---

*Work supported by Warsaw Technical University Institute of Material Science and Engineering in the frame of the CPBR 2.4.

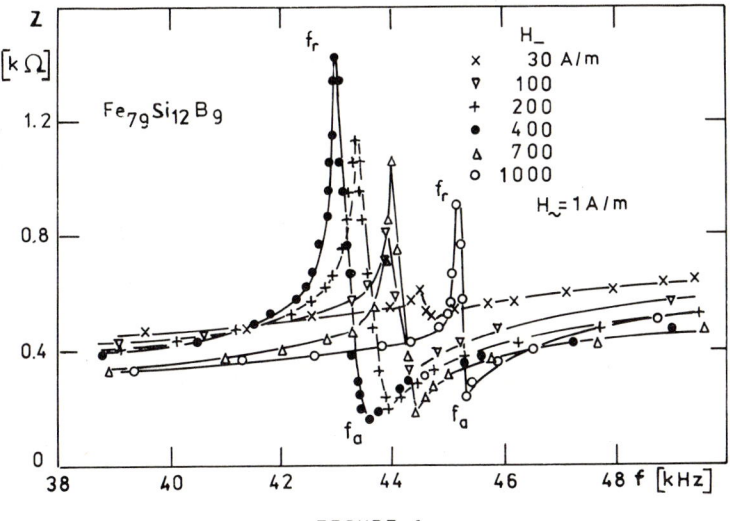

FIGURE 1
Frequency charcteristics of the impedance for the various field

where c_H and c_B are ultrasound velocities at constant magnetic field H and at constant induction B, and f_r and f_a are the resonant and antiresonant frequencies, respectively, ρ denotes the density and l is length of the sample.

3. RESULTS

Measurements carried out at the two times higher frequency than the lower limit of the ultrasound range show that the magnetomechanical coupling coefficient reaches value of 0.2 (Fig. 2). The moduli of the elasticity and the dynamics dependences on the magnetic bias field in the figures 3 and 4 are presented.

4. DISCUSSION AND CONCLUSIONS

Both moduli of the elasticity, i.e. E_H and E_B, exhibit the minimum at the bias field about 400 A/m. In the same region the maxima of magnetomechanical coupling and of the dynamics were observed. The influence of measurement amplitude of the alternating magnetic field is very low. This alloys can be used as material for the core of the ultrasonic transducers working at the higher than 20 kHz frequency ranges. The piezomagnetic properties can be improved after heat treatment in magnetic field [1,5].

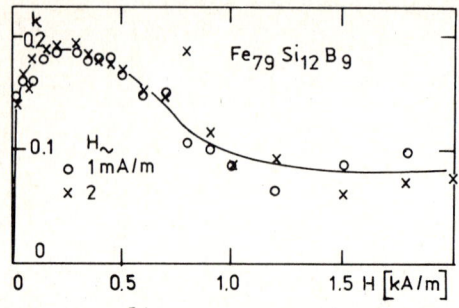

FIGURE 2
Dependence of the magnetomechanical coupling coefficient k on the magnetic bias field H at the amplitude of the AC magnetic field equal to 1 and 2 mA/m

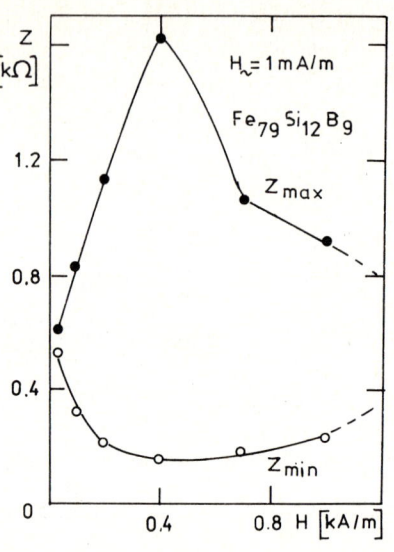

FIGURE 4
Maximum (Z_{max}) and minimum (Z_{min}) values of the impedances for the resonances (f_r) and antiresonances (f_a) (see Fig. 1) frequencies versus magnetic bias field H at the amplitude of the alternating magnetic field equal to 1 mA/m for the half-wave strip resonator made from the as quenched $Fe_{79}Si_{12}B_9$ metallic glass. Differences between Z_{max} and Z_{min} are the diameters of the motional impedance circles connected with the transducer dynamics and efficiency

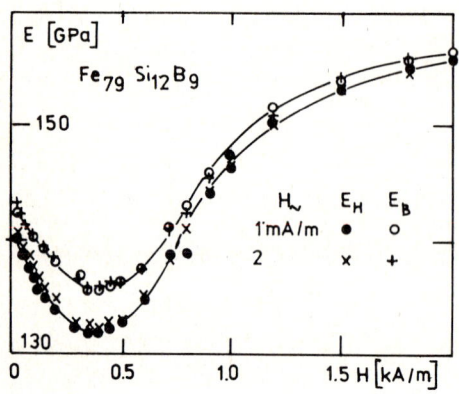

FIGURE 3
Magnetic bias field dependences of the moduli E_H and E_B for the amplitude of the AC field equal to 1 and 2 mA/m

REFERENCES

1) M.A. Mitchel, J.R. Cullen, R. Abbundi, A. Clark, H. Savage, J. Appl. Phys. 50 (1979) 1627.

2) T. Jagieliński, K.I. Arai, N. Tsuya, J. Onuma and T. Masumoto, IEEE Trans. Magn. MAG-13 (1977) 1553.

3) S. Ito, K. Aso, Y. Makino and S. Uedaira, Appl. Phys. Lett. 37 (1980) 665.

4) Z. Kaczkowski, Archiwum Elektrotech. 11 (1960) 635.

5) M. Brouha, J. van de Borst, J. Appl. Phys. 50 (1979) 7594.

MAGNETIC FIELD DEPENDENCE OF THE MAGNETOMECHANICAL COUPLING OF THE $Fe_{78}B_{12}Si_{10}$ METALLIC GLASSES ANNEALED IN PERPENDICULAR MAGNETIC FIELD

Zbigniew KACZKOWSKI and HO Su Nam*
Polish Academy of Sciences Institute of Physics
Al. Lotników 32/46, 02-668 Warszawa, Poland

Piezomagnetic properties of the amorphous $Fe_{78}B_{12}Si_{10}$ ribbons (annealed in the perpendicular to their plane magnetic field) were investigated. The maximum values of the magnetomechanical coupling coefficient k were equal to 0.6 and mechanical quality factor was changing with the bias from 13 to 145.

1. INTRODUCTION

Fe-B-Si metallic glass ribbons, when heat treated near Curie temperature, exhibit high magnetomechanical coupling coefficient k, equal to 0.6 [1]. Thermal treatment in parallel or perpendicular magnetic field improves this coefficient from 0.6 to 0.96.

2. EXPERIMENT AND RESULTS

Ribbons of the $Fe_{78}B_{12}Si_{10}$ amorphous alloy (obtained from Prof. Dr H.Matyja's Laboratory of the Warsaw Technical University Institute of the Material Science) were cut into strips 44 mm long. Their width was equal to 9 mm and thickness was changing from 30 to 40 μm. Mass of the samples was about 52 mg. The samples were annealed in the magnetic field perpendicular to the ribbon plane. Magnetomechanical coupling coefficient was calculated from the resonant (f_r) and antiresonant (f_a) frequencies (Figures 1 and 2), or moduli of the elasticity at the constant magnetic field (E_H) and at the constant magnetic induction (E_B) (Fig. 3). The values of the resonant frequencies were obtained from the motional impedance circles (Fig. 1). The measurement were carried out as the functions of the magnetic bias field for the amplitude of ac magnetic field ranging mainly from 1 to 1 A/m. The maximum values of the magnetomechanical coupling coefficient k were equal to 0.6 (Fig.4). The magnetomechanical and mechanical quality factor and internal friction or mechanical or magnetomechanical damping dependences on the magnetic bias field are given (figures 4 and 5).

---

*On leave of absence of the Academy of Sciences of The DPRK Institute of Physics.

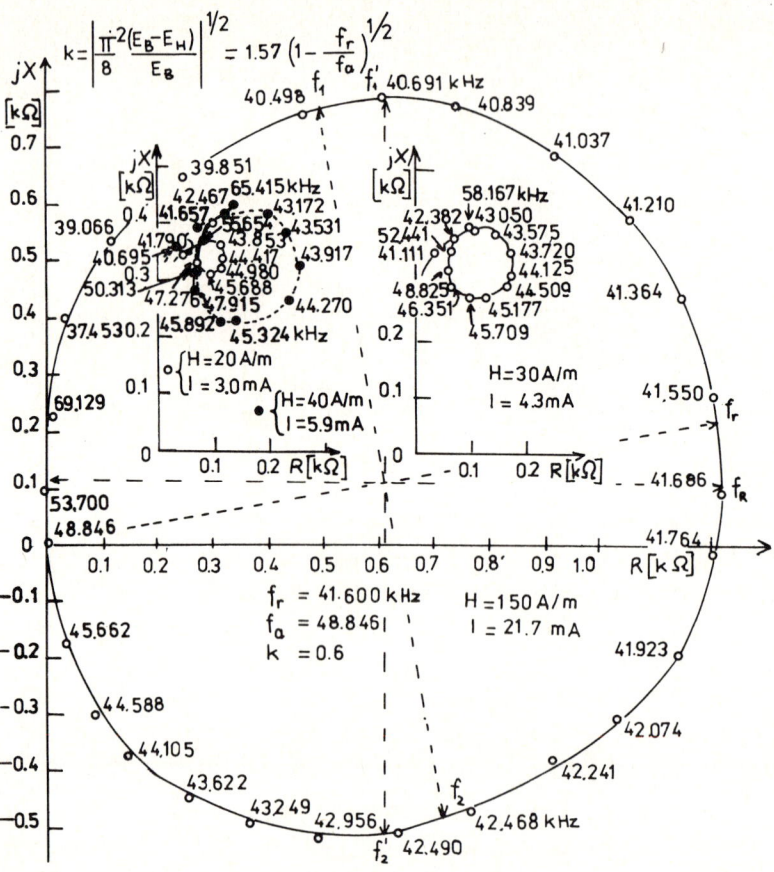

FIGURE 1
Motional impedance circles for the magnetic bias field equal to 20, 30, 40 and 150 A/m

3. DISCUSSION AND CONCLUSIONS

Heat treatment perpendicular to the long axis and laying in the ribbon plane magnetic field induces a weak transverse anisotropy and easily rotatable magnetization. In this case magnetic domains extended entirely across the ribbon width and only 180° Bloch walls exist. When the magnetic field is perpendicular to the ribbon plane effect is similar to the annealing without magnetic field [1].

REFERENCE
1) A. Hernando, A. Garcia-Escorial, E. Ascasibar and M. Vázquez, J. Phys. D: Appl. Phys. 16 (1983) 1999.

$Fe_{78}B_{12}Si_{10}$ metallic glasses annealed in perpendicular magnetic field

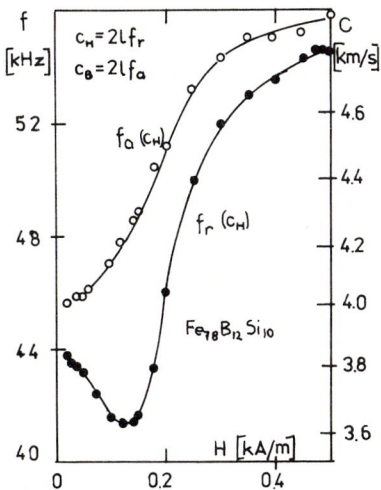

FIGURE 2
Dependences of the resonant (f_r) and antiresonant (f_a) frequencies or ultrasound velocities c_H and c_B on the magnetic field H

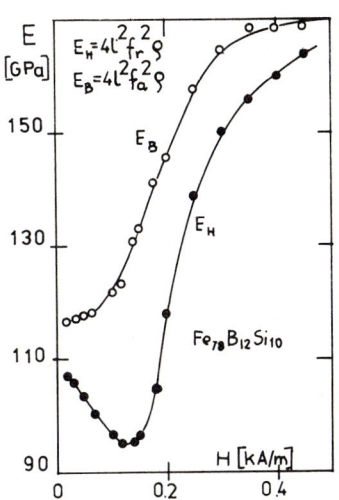

FIGURE 3
Magnetic bias field dependences of the moduli of elasticity E_H and E_B for the $Fe_{78}B_{12}Si_{10}$ metallic glass ribbon

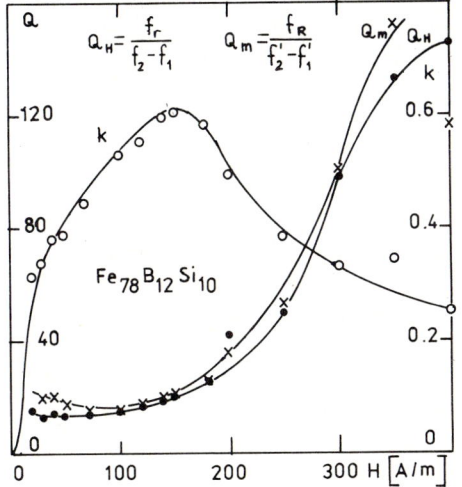

FIGURE 4
Magnetic field dependences of the magnetomechanical coupling coefficient k, mechanical (Q_m) and magnetomechanical (Q_H) quality factor

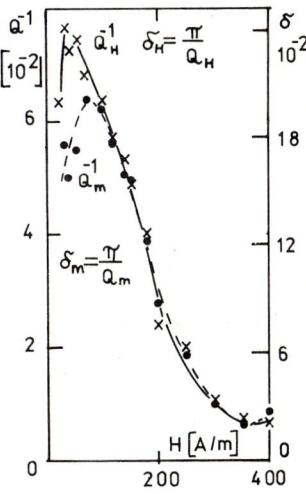

FIGURE 5
Magnetic field dependences of the internal friction (Q_m^{-1}, Q_H^{-1}) and mechanical and magnetomechanical damping (δ_m and δ_H)

SHORT RANGE ORDER AND MAGNETOELASTIC PROPERTIES OF SOME Fe-RICH METALLIC GLASSES

J.M.BARANDIARAN, J.GUTIERREZ, F.PLAZAOLA AND I.ZABALA

Dpto. de Electricidad y Electrónica, Facultad de Ciencias, U.P.V. Lejona, Apdo. 644, 48080 Bilbao (Spain).

Fe Mössbauer spectroscopy and magnetoelastic measurements have been performed in $Fe_{82}M_2B_{16}$ and $Fe_{78}M_2B_{20}$ (M=Cr,Mn,Pd,Zr or Mo) metallic glasses and values of the hyperfine field, isomer shift, spontaneous magnetization, magnetoelastic coupling and damping coefficients have been obtained. Isomer shifts are sensitive to the Boron content but not to the non-ferrous metal. All the others properties are quite similar in all the studied samples and show little variation with respect to the $Fe_{80}B_{20}$ glasses.

1. INTRODUCTION

Fe rich metallic glasses show outstanding magnetic and magnetoelastic properties of great interest in technological applications. Such properties are intimately related to the structural short range order, in particular with the localization of different kinds of atoms in the neighbourhood of the Fe. Addition of small quantities of metallic atoms other than Fe is usually performed in order to improve properties such as thermal stability, resistance to corrosion, etc.

In this work we present magnetic, magnetoelastic and Mössbauer studies in a series of metallic glasses with compositions close to $Fe_{80}B_{20}$, intending to correlate the Short Range Order with the macroscopic magnetic properties.

2. EXPERIMENTAL

Fe rich metallic glasses with compositions given in Table I were used for Mössbauer measurements at room temperature.

For magnetoelastic measurements the real and imaginary part of the ac susceptibility were determined as a function of the frequency (see Fig.1). Typical values of the alternating field were 3mOe RMS and the bias field was in the range 0-150 Oe. Young modulus Y, magnetoelastic coupling coefficient k and damping factor Q^{-1} were determined as in ref. 2.

3. RESULTS AND DISCUSSION

Fig. 2 shows two typical Mössbauer spectra. The lines of the spectra are very broad. As usual for metallic glasses the magnetic splitting of such Mössbauer spectra has to be described by an hyperfine field distribution instead

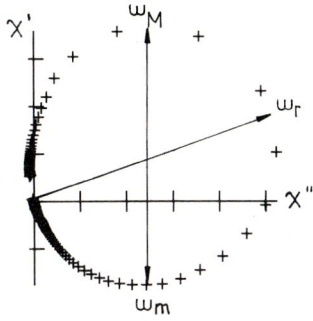

Fig.1. Susceptibility circle of $Fe_{82}Zr_2B_{16}$.

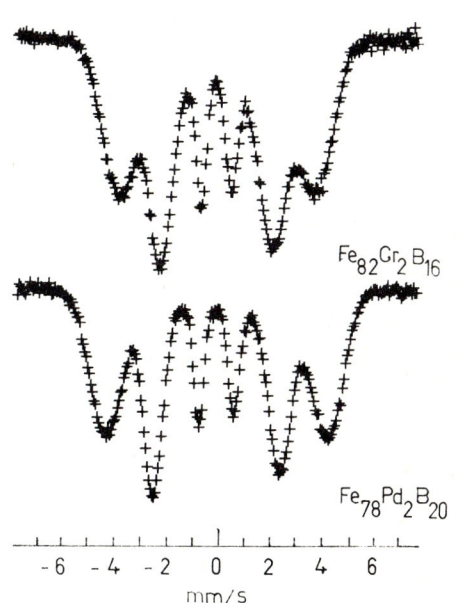

Fig.2. Mössbauer spectra

of a single value. Therefore, the absorption lines were fitted to six Lorentzian peaks, characterized by a mean hyperfine field and a width Γ.

Fig. 3 shows the variation of γ^H, k and Q^{-1} as a function of the bias field for one of the samples. Similar behaviour have been found for the other compositions.

Measurements of spontaneous magnetic polarization (J_s), maximum magnetoelastic coupling factor(k), mean hyperfine field (H_{hf}) and isomer shift at room temperature are summarized in Table 1.

TABLE 1

Sample	J_s(T)	k_{max}	H_{hf}(T)	IS(mm/s)
$Fe_{82}Mn_2B_{16}$	1.3	0.40	24.3	0.061
$Fe_{82}Zr_2B_{16}$	1.5	0.46	24.0	0.067
$Fe_{82}Cr_2B_{16}$	1.4	0.34	23.4	0.069
$Fe_{78}Cr_2B_{20}$	1.4	0.36	23.5	0.047
$Fe_{78}Pd_2B_{20}$	1.5	0.31	26.4	0.041
$Fe_{78}Mo_2B_{20}$	1.3	0.40	22.5	0.052
$Fe_{80}B_{20}$	1.6 (a)	0.49 (b)	25.4 (c)	

(a) Taken from ref.1   (b) Taken from ref.2   (c) Taken from ref.3

In spite of the different metalloid content in the alloy, the measured magnetic and hyperfine parameters (except isomer shift) in the two glasses containing Cromium are very similar. This behaviour indicates that the Boron content of the alloy produces very little influence on the magnetic properties of the studied alloys.

On the other hand, the isomer shift is influenced by the metalloid content and not by the non ferrous metallic component of the alloy, as can be seen in Table 1. The outher shells of these last elements being very different from each other, the observed behaviour may indicate that those metallic elements do not locate in the first neighbour position of Fe atoms preventing the bounding between them. Another explanation could be the hybridation of s-p or s-d orbitals making these electrons indistinguishable from the Fe-ones. Values of the spontaneous magnetization J_s and k_{max} are slightly lowered with respect to $Fe_{80}B_{20}$ glasses by the inclusion of metal atoms other than Fe. The obtained values of k_{max} are high enough for as-quenched metallic glasses, anyway values of k are expected to increase upon annealing.

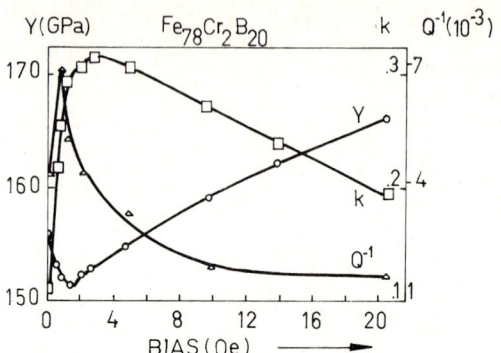

Fig.3. Magnetoelastic parameters Y, k, Q^{-1} versus bias field

Hyperfine field also shows the same trend except for the case of the sample containing Pd in which surprisingly H_{hf} is larger than the $Fe_{80}B_{20}$ one. It is also in this sample where the width of the hyperfine field distribution is narrowest.

ACKNOWLEDGEMENTS

The authors wish to thank Dr.J.A. Leake for supplying the samples used in the experiments. This work has been supported by the University of the Basque Country under contracts 310.05-7/86 and 310.05-65/86.

REFERENCES

(1) F.E.Luborsky in "Ferromagnetic Materials", Vol.1, North-Holland Pub.Co. (1980), 451.
(2) Z.Kaczkowski, E.Lipinski, L.Makinski. IEEE Trans. on Mag. MAG-20 (1984). 1403.
(3) C.L.Chien. Phys. Rev. B18 (1978) 1003.

ENTHALPY AND CURIE TEMPERATURE RELAXATION IN $Fe_{40}Ni_{38}Mo_4B_{18}$ GLASSES

J.M. BARANDIARAN, A.L. GREER* and I. TELLERIA†

Dpto de Electricidad y Electrónica, F. Ciencias, U. P. V., Leioa, SPAIN
*Department of Metallurgy, University of Cambridge, Cambridge, U.K.
†Dpto de Física, Facultad de Química, U. P. V., San Sebastian, SPAIN

We have studied the enthalpy of relaxation and Curie temperature in $Fe_{40} Ni_{38} Mo_4 B_{18}$ metallic glasses using D.S.C. measurements. Results obtained have been interpreted in the framework of activation energy spectrum model modified to be applied to non-isothermal measurements.

1. INTRODUCTION

In recent years Curie temperature[1] (T_c) and enthalpy[2] (H) measurements have been used for monitoring structural relaxation in metallic glasses. Both magnitudes can be simultaneously obtained by differential scanning calorimetry (DSC) and a comparison between them seems to be interesting although, to the best of our knowledge, up to now, none has been reported.

2. RESULTS AND DISCUSSION

Samples of METGLAS® 2826 MB from Allied Co. of nominal composition

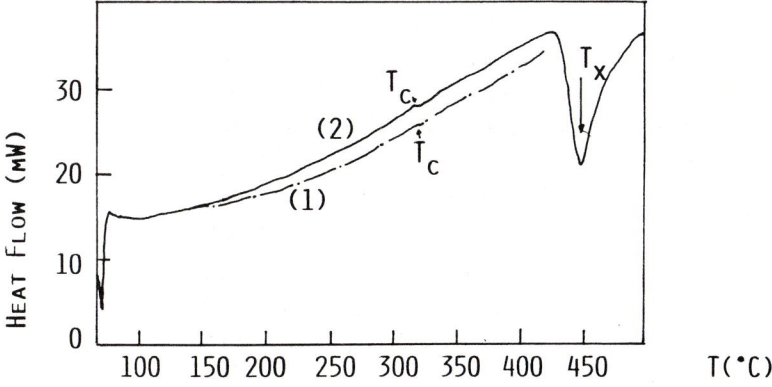

Figure 1.- DSC traces for successives scans, (–·–) 1st scan, (———) 2nd scan. T_x correspond to the crystallization temperature.

$Fe_{40}Ni_{38}Mo_4B_{18}$ have been used in this work, because of the low value of Curie temperature far away the crystallization one. Measurements were made on a Perkin-Elmer DSC-7 using samples of about 30 mg. The temperature axis of the different DSC scans was corrected for the temperature lag using the Curie temperature (Tc) of a Ni standard.

Kinetics of enthalpy relaxation have been studied following the activation energy spectrum (AES) model[3] modified to be applied to non-isothermal measurements[4]. This model assumes that all the processes having activation energies below or equal to

$$E = k_B T \ln(\nu_0 t) \quad [1]$$

contribute to the relaxation during an isothermal annealing of time t at a temperature T. Here ν_0 is a frequency factor and k_B the Bolzmann's constant. The relaxation rate for the quantity X is:

$$\frac{dX}{dT} = c(E)q(E)k_B \ln(\frac{\nu_0}{R}) \quad [2]$$

where R is the heating rate. The difference between the specific heat curves obtained in two DSC successive scans is a direct measurement of the relaxation rate given by [2].

Figure 2 shows the relaxation spectra (c(E)q(E)) using a frequency factor value around 10^{20} s^{-1}. With this ν_0 value the activation energy spectrum ranges from 1.7 to 3 eV. However, these experimental points can be sufficiently well fitted by using bigger frecuency factor values up to 10^{30} s^{-1}. If this high limit ν_0 value is used, the corresponding activation energy

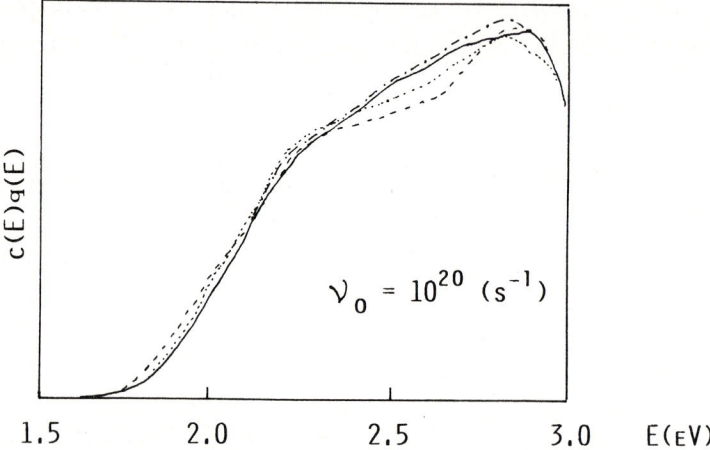

Figure 2.- Relaxation spectra obtained fitting the experimental DSC scans by equation (2). The experimental points correspond to heating rates of : 80(— . —), 40(———), 20(·····) and 10 (-----) K/min.

spectrum ranges between 2.3 and 4.5 eV. This last energy value is close to the apparent activation energy for the crystallization process which start when the relaxation process finishes (~ 4.4 eV).

The scan at a heating rate of 5 °C/min, however, does not fit the general behaviour This fact indicates a change in the sample when enough time is provided. This fact agrees with an anomalous behaviour of the isothermal relaxation of the Curie temperature, which could be attributed to phase a separation ocurring at long times, reflected in changes of the crystallization peaks

Figure 3 shows the relaxation of T_c and H after successive runs at different heating rates. The "as quenched" value of T_c was determined by extrapolation at an infinite heating rate and resulted 293 ± 3 °C. As shown, no direct correlation between both magnitudes exists, indicating a different mechanism involved. Only initial enthalpy relaxation seems to contribute to the increase in T_c, high temperature processes having no noticeable effect.

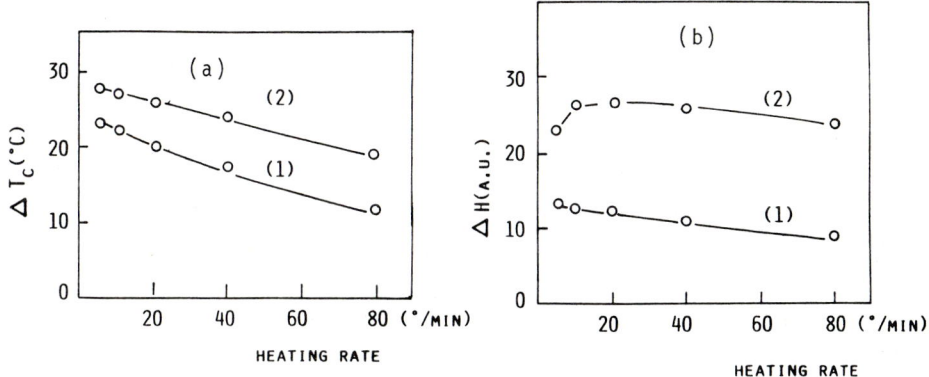

Figure 3.- Curie temperature (a) and enthapy relaxation (b) after sucessive heating up to the Tc(1) and 420 °C(2).

REFERENCES

1) A.L. Greer, Thermochimica Acta, 42 (1980) 193

2) H.S. Chen and E. Coleman , J. Appl. Phys. Letters, 28 (1976) 245

3) I. Tellería, J. Colmenero, A. Alegría and J.M. Barandiarán, Current Topics on Non-Crystalline Solids, pg. 199, World Scientific Publishing Co., Singapore 1986

4) M.R.J. Gibbs, J.E. Evetts and J.A. Leake, J. Mat. Sci., 18(1983)278

MAGNETIC STABILITY OF WATER QUENCHED Co-BASED AMORPHOUS ALLOYS

Han Zhenghe[+], Wang Xinlin and Ke Cheng

Central Iron and Steel Research Institute, Beijing, China

From thermal expansion studies on different Co-based alloys, we find that the free volume fraction in the as-received samples is about 1.5 %, of which only about 40 % can be removed by structural relaxation. Thus, we show that magnetic stability of water-quenched samples can not be improved radically. Neither by prolonging annealing time nor changing the composition. However, magnetic field annealing is found to improve the stability considerably.

1. INTRODUCTION

Water-quenching is a simple and efficient method often used in the heat treatment for Co-based amorphous alloys (1,2). However, magnetic properties of the water-quenched samples are quite unstable (3,4,7). This decay of magnetic properties can be understood as being due to the rebuilding of local-induced anisotropy in these nearly zero magnetostrictive alloys (6). The development of the induced magnetic anisotropy is governed by thermal activation processes which can be explained by the atom pairs model (5). These processes have a broad activation energy distribution because of the random amorphous structure. According to our calculations (7), the activation energy for the processes which cause the Co-based samples to loose their superior soft magnetic properties is under 1.2 eV. Such small activation energy can induce atomic rearrangements at the neighborhood of vacancy-type defects (i.e. free volumes) (8). It is reasonable to assume that the possible existence of large free volumes in a water quenched amorphous sample result in the instabilities of the magnetic properties.

In order to understand the relation between the magnetic stability of water quenched samples and the free volume in them, the amount of free volume must be measured. The free volume fraction in an amorphous sample can be described by the expression

$$V_f = (V_a - V_c)/V_c \qquad (1)$$

[+] Presently at Dept of SSP, Royal Inst. of Tech. Stockholm, Sweden

where V_a is the volume of amorphous sample, V_c is the volume of crystallized sample and the V_f is the free volume fraction in this sample. Let

$$\Delta V_f = \Delta V/V \qquad (2)$$

where ΔV is the change of volume of a sample. Since $\Delta V/V = 3 \Delta L/L$, we can know the change of free volume from measurements of expansion length, ΔL.

The purpose of this work is to study the magnetic stability of water-quenched samples in terms of free volume fraction as well as to search for a practical method to improve the magnetic stability.

2. RESULTS AND DISCUSSION

Fig 1 shows a typical study of the length change of amorphous ribbons during heating. An as-received sample (ribbon) was heated to 510° C, then cooled down to room temperature. The process was repeated three times. In the third sequence, the sample was heated to 700° C when crystallization occurred.

It can be seen that the expansion coefficient of the as-received sample is much more smaller than that of annealed. This fact indicates that during heating, the amount of free volume in as received sample decreases which compensates for some part of thermal expansion. However, there is no significant difference between the second and the third curves. This fact indicates that after the first heating, free volume does not change any more before crystallization.

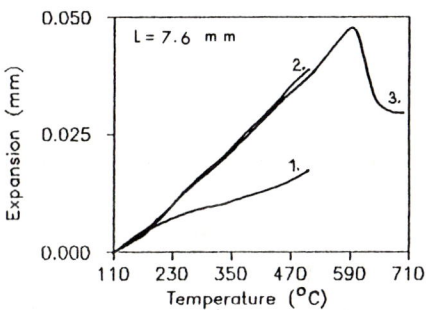

Fig 1. Thermal expansion for ribbon of $Fe_{4.2}Co_{65.8}Ta_7Si_9B_{14}$

The large contraction which occurs at about 590° C indicates that during crystallization, the free volume decreases considerably. From similar measurements on other samples of different composition, we determined the fraction of free volume in the as-received samples is about 1.5, only about 40 % of which can be removed by structural relaxation. Thus, because there is always a large amount of free volume still remaining in amorphous samples, we believe that magnetic stability of water quenched sample can not be improved radically by means of changing composition or prolonged annealing times.

Fig 2 shows one result of these experiments.

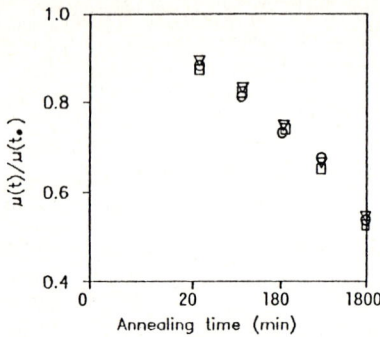
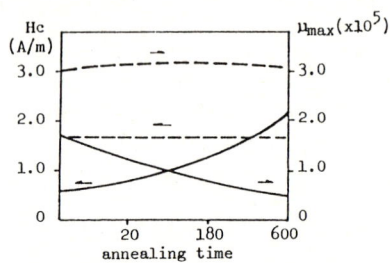

Fig 2. Permeability (20 kHz, 0.5 T) changes during isothermal annealing at 130°C for $(Fe,Co)_{73}Ta_2(Si_xB_{1-x})_{25}$ water-quenched samples.
▽ = 0.36   □ = 0.48   ○ = 0.60

Fig.3. permeability (μ_{max}) and coercivity (Hc) changes during annealing at 130°C for $(Co,Fe)_{73}Nb_2(Si,B)_{25}$ alloys.
— water-quenched
-- magnetic field annealed

However, Fig 3 shows that after water quenching, if the sample is then annealed in a magnetic field, the magnetic properties will be quite stable. The free volume in a field annealed sample does not decrease significantly, but there is a relatively strong induced anisotropy in it which restricts the atom pairs removing and the magnetic stability is improved.

References
1) R.Hasegawa, J.Appl. Phys. 53(11) (1982) 7819
2) Y.Makino et al. J. Appl. Phys. 52(3) (1981) 2477
3) C.E.Fish et al. IEEE MAG-19 (1983) 1937
4) W.Chambron and A.Chamberod, J. Phys. Coll. C5-511 (1981)
5) S.Chikazumi and C.D.Graham, "Magnetism and Metallurgy" Ed. A.E.Berkowitz
6) H.Fujimori et al. "Rapidly Quenched Metals (3)B" Ed. Cantor (1978) 222
7) Han Zenghe et al. Central Iron and Steel Research Inst. Tech. Bull. 6 Supp. 57 (1986)
8) H.Kronmuller and N.Moser, "Amorphous Metallic Alloys", Ed. F.E.Luborsky (1983) 300

ON DIFFUSION IN AMORPHOUS MULTILAYER FILMS $Fe_{1-x}B_x/Fe_{1-y}B_y$

Feliks STOBIECKI

Institute of Molecular Physics, Polish Academy of Sciences,
ul. Smoluchowskiego 17/19, 60-179 Poznan, Poland

Studies of the diffusion in multi- and bilayer Fe-B films performed in a wide concentration range are described. It is demonstrated that the low temperature diffusion process is directly connected to the dependence of free enthalpy on the concentration.

Amorphous multilayer films $Fe_{1-x}B_x/Fe_{1-y}B_y$ ($0.15 \leq x, y \leq 0.62$) to be studied were obtained by r.f. sputtering[1,2]. The samples, 240 nm thick, were composed of 5 alternating $Fe_{1-x}B_x/Fe_{1-y}B_y$ layers (each 40 nm thick) closed between 20 nm thick $Fe_{1-y}B_y$ layers. A series of double-layer films $Fe_{1-x}B_x/Fe_{1-y}B_y$ with each layer of 120 nm in thickness, was also produced. The films were annealed in the temperature range 525-625 K for 0.75 - 24 h in the vacuum of 10^{-6} Torr. Concentration depth profiles of as-deposited and annealed samples were measured by Auger electron spectroscopy combined with Ar ion sputtering (Fig. 1.).

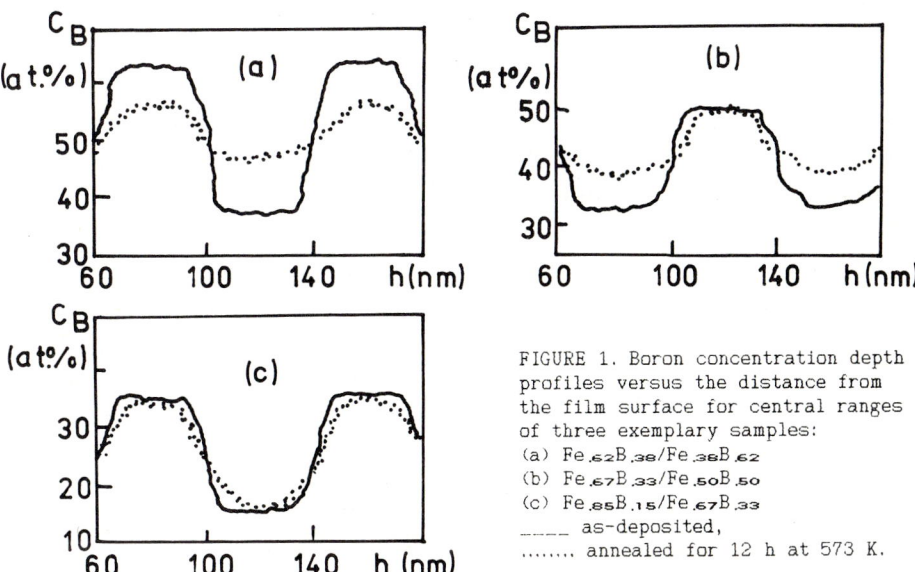

FIGURE 1. Boron concentration depth profiles versus the distance from the film surface for central ranges of three exemplary samples:
(a) $Fe_{.62}B_{.38}/Fe_{.38}B_{.62}$
(b) $Fe_{.67}B_{.33}/Fe_{.50}B_{.50}$
(c) $Fe_{.85}B_{.15}/Fe_{.67}B_{.33}$
——— as-deposited,
········ annealed for 12 h at 573 K.

For the double-layer films $Fe_{1-x}B_x/Fe_{1-y}B_y$ annealed at 573K the concentration dependence of diffusion coefficient was determined from $c_B(h)$ dependence according to well known Boltzmann - Matano method[3] (Fig. 2).

For annealed samples, essential differences in the concentration depth profile for different boron concentrations (x,y) as well a peak in diffusion coefficient at a boron concentration of 50 % were observed, both effects may be explained on the grounds of concentration dependences of free enthalpy (see fig.7, ref. 4), density and free volumes (see ref. 5) for $Fe_{1-x}B_x$ amorphous alloy.

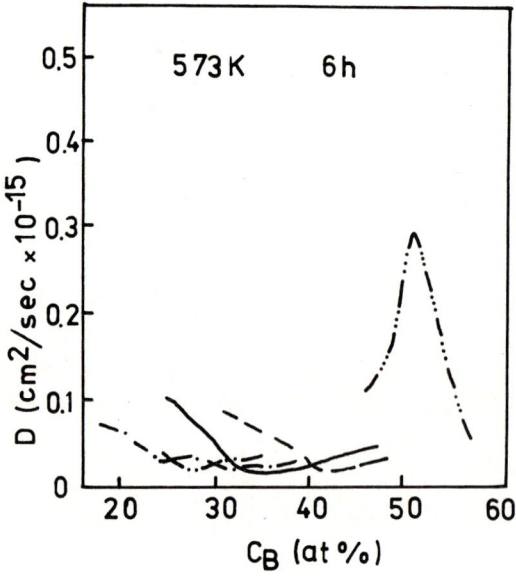

FIGURE 2
Concentration dependence of the diffusion coefficient as measured for several Fe-B layer systems (-...- $Fe_{.62}B_{.38}/Fe_{.38}B_{.62}$; ---- $Fe_{.75}B_{.25}/Fe_{.50}B_{.50}$; ____ $Fe_{.80}B_{.20}/Fe_{.50}B_{.50}$; -.-.- $Fe_{.85}B_{.15}/Fe_{.59}B_{.41}$; -..-.. $Fe_{.75}B_{.25}/Fe_{.67}B_{.33}$).

The analogies should be noticed between the observed diffusion process for different boron concentrations in amorphous Fe-B alloys and the amorphization process through the solid state reaction (see eg. ref. 6).
It has been shown[6,7] that amorphous Fe-B alloys cannot be obtained via a solid state reaction between crystalline Fe and B layers. Results presented in Fig. 3 prove that the aim can be achieved by annealing of multilayer $Fe_{1-x}B_x/Fe_{1-y}B_y$ films, if only X and Y are chosen such that (X+Y)/2 = 0.5 i.e. the mean concentration stays in the region where the minimum of the free enthalpy is located.

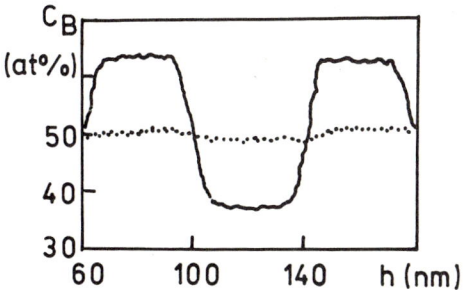

FIGURE 3
Boron concentration depth profiles versus the distance from the film surface for central ranges of $Fe_{.62}B_{.38}/Fe_{.38}B_{.62}$ ____ as-deposited, ... annealed for 6h at 624 K

This work was sponsored by Deutsche Forschung Gemeinschaft and Technical University of Warsaw. The author would like to express his gratitude to Prof. H. Hoffmann for having made it possible to carry out the measurements.

REFERENCES

1) F. Stobiecki, S. Schwarzl, T. Stobiecki, H. Hoffmann, J. de Physique C8 (1980) 171.

2) F. Stobiecki, W. Pamler, R. Reill, K. Roll, H. Hoffmann, J.Non. Cryst. Solids 88 (1986) 209.

3) P.G. Shewmon, "Diffusion in Solids" (Mc Graw-Hill, New York 1963)

4) C. Cuant, M. Notin, J. Hertz, J.M. Dubois, G. Le Caer, J. Non. Cryst. Solids 55 (1983) 45.

5) F. Stobiecki, J. Zweck, W. Pamler, G. Bayreuther, H. Hoffmann, LAM-6 (1986) to be published.

6) B.M. Clemens, Phys. Rev. B, 33 (1986) 7615.

7) W. Reill, H. Hoffmann, K. Roll, J. Mater. Sci. Lett. 4 (1985) 359.

COERCIVE FORCE AND LOSSES IN FeBSi METALLIC GLASS RIBBONS

Bogdan SZYMAŃSKI[a], Marian SOIŃSKI[b] and Ivan ŠKORVANEK[c]

[a]Institute of Molecular Physics, Polish Academy of Sciences, Poznań, Poland
[b]Institute of Electroenergetics, Technical University of Częstochowa, Poland
[c]Institute of Experimental Physics, Slovak Academy of Sciences, Košice, CSSR

In this paper we report some measurements of the effect of annealing on the hysteresis properties of coated and uncoated ribbons of AFB[x] metallic glass.

1. INTRODUCTION

As iron-based metallic glasses offer a great possibility of interesting application they have recently become the subject of thorough investigation. These soft magnetic glasses may replace traditional soft magnetic materials especially in low-frequency devices[1]. So far they have been considered in connection with the problems of finding a suitable insulating coating to decrease eddy-current losses and choosing the appropriate annealing for stabilization of magnetic parameters.

2. EXPERIMENTAL

Measurements were performed on ribbons of AFB metallic glass of 1.9 cm in width and 27 μm in thickness. The glassy state of the samples was checked by X-ray diffraction. The ribbons were coated using MgO based colloidal solution[2]. The reference samples were uncoated ribbons which had been subjected to the same heat treatment as the coated ones. Magnetic properties of the samples were obtained from dynamical and quasi-statical hysteresis loops. In the later case the frequency of magnetization reversal was 0.02 Hz. The value of power loss was determined from the area of the dynamical hysteresis loop. All measurements were done in room temperature.

3. RESULTS AND DISCUSSION

The influence of the insulating coating on the coercive force H_c of AFB ribbons is shown in Table I. For a "suitable" coating (i.e. coating which has good insulating properties and which does not deteriorate hysteresis loop parameters) a slight decrease in H_c was observed. After the annealing recommended

---

[x]AFB - Trade name of metallic glass (with nominal composition $Fe_{78}B_{13}Si_9$) produced by Institute of Materials Science and Engineering, Technical University of Warsaw.

by the producer of the ribbons H_c was observed to decrease in all samples.

TABLE I
Influence of coating on H_c (A/m) of AFB ribbons

Sample	without coating	with coating	
		unsuitable	suitable
as-prepared	7	10	6
after 1 h anneal at 325°C in air	5.5	6	5

The ribbons with the "suitable" coating were taken for further experiments. As the annealing at this relatively high temperature can lead to deterioration of mechanical properties (embrittlement) we decided to carry out the thermal treatment at a lower temperatures. As follows from the paper by Mehrer et al.[3] for precursor alloy $Fe_{80}B_{20}$ the greatest reduction in H_c occurs after annealing at about 250°C. The influence of such thermal treatment on H_c is illustrated in Table II.

TABLE II
The effect of 1 h annealing (in silicon oil bath) on H_c (A/m)

Sample	uncoated	coated
before annealing	6	5
1h at 200°C	5	4.5
1h at 250°C	5	4

So far by H_c we have defined the coercive force along the axis of the ribbon. However, it is well known that in perpendicular to this direction H_c significantly increases. This influences directional losses of the material through the hysteresis loop. For uncoated ribbons the directional losses were investigated in ref. 4. They were found to be of complex nature as the direction of the minimum losses changes with the frequency f of flux reversal. Anisotropy of the losses is much less for coated ribbons than for the uncoated ribbons. Figure presents dynamical hysteresis loops along the axis of the ribbons (ϕ = 0°) and in perpendicular to the axis (ϕ = 90°). The losses along the direction ϕ = 90° are 25 % greather than those in the direction ϕ = 0°, for f = 50 Hz. For f < 10 kHz the increase for losses does not exceed 40 to 50 %.

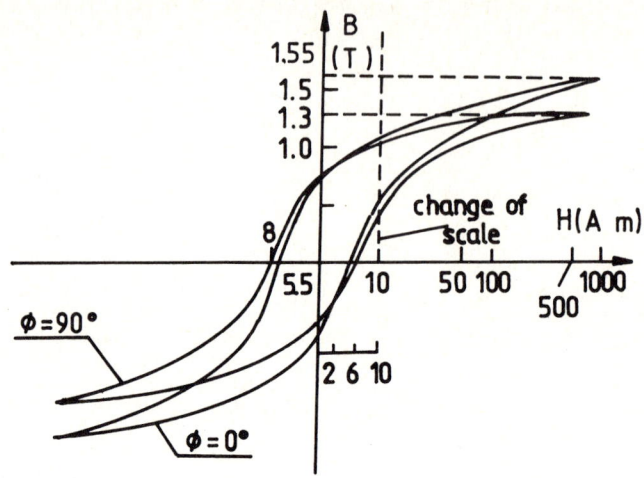

FIGURE
Dynamical hysteresis loops of coated AFB ribbon at f = 50 Hz after magnetic annealing (325°C, 8 kA/m, 1 h)

ACKNOWLEDGEMENT

This work was partly supported by Institute of Materials Science and Engineering, Technical University of Warsaw.

REFERENCES

1) R. Hasegawa, J.Magn.Magn.Mater. 41/1-3/84, 79.

2) J. Szczygłowski and M. Soiński, Proc. 3rd Natl.Symp. "Amorph.Magn.Mater." Kazimierz n.Wisłą, Nov. 1985, p. 45, in polish.

3) H. Mehrer et al., phys.stat.sol./a/ 72/1/82, 215.

4) Y.C. Kuo et al., J.Magn.Magn.Mater. 31-34, 1983, 1563.

NONEXPONENTIAL DECAY AND MAGNETIC RELAXATION IN METALLIC GLASSES

Juan COLMENERO, Isabel TELLERIA and Angel ALEGRIA

Departamento de Física, Facultad de Química, Universidad del País Vasco, Apdo 1072, 20080 San Sebastián (SPAIN)

We have studied the relaxation process of magnetization work in cold-rolled samples of METGLAS 2826® and $Fe_{40} Ni_{40} B_{20}$. Results obtained have been interpreted in the framework of Ngai model for nonexponential relaxations in disordered systems. Activation energy values for the microscopic processes underlying relaxation of magnetization work have also been deduced and interpreted.

1. INTRODUCTION AND THEORY

Nowadays it is well known that annealing of a metallic glass produces atomic rearrangement and results in a change in almost all physical properties. This phenomenon is usually termed relaxation process. One of the main features of this kind of behaviour is the well known log-time kinetic. During the last years, this empirical law has been interpreted in the framework of different theoretical approaches to the relaxation behaviour in metallic glasses, mainly: activation energy spectrum (AES) model[1] and those models involving free-volume concept and both topological and chemical short-range order (TSRO and CSRO)[2,3]. However, log-time kinetics also appear in relaxation processes in very different disordered systems as for example spin glasses[4]. In such systems, log-time kinetics can be understood as an approximation of the *Stretched Exponential* universal behaviour of the relaxation function $\Phi(t)$

$$\Phi(t) \sim \exp[-(t/\tau_p)^{1-n}] \qquad 0 < n < 1 \qquad (1)$$

where τ_p is a macroscopic time scale for the relaxation process and n is an index of slowness. Up to now there are different theoretical appoaches relating this relaxation functional form to microscopic processes. The most direct and elaborated one is the model of Ngai[4] which deduces a second universal relationship conecting τ_p with a microscopic time scale τ_o

$$\tau_p \sim \tau_o^{1/(1-n)} \qquad (2)$$

In the framework of the Ngai model n measures the complexity of the structural rearrangement in the system. The apparent activation energy for the relaxation process, E_p, experimentally deduced from $\tau_p(T)$, is coupled to the microscopic activation energy, E_o, through n as

$$E_p = E_o/(1-n) \qquad (3)$$

This alternative formalism has not been extended, as far as we know, to the relaxation behaviour in metallic glasses. In this work we have applied it to the relaxation of magnetization work in cold-rolled samples of METGLAS 2826® and $Fe_{40}Ni_{40}B_{20}$. As is well known, cold-rolling introduces a great number of defects in both amorphous and crystalline materials. In the case of glassy metal alloys, the exact nature of such defects is not yet clear, although it appears that the level of internal stresses is strongly increased after cold rolling. These internal stresses relax at temperatures well below the glass-transition where the nonequilibrium state of the glass (measured for example by the excess free volume with respect to the metastable supercooled liquid-like state at the same temperature) is hardly modified in the experimental time scale. These relaxation processes can be easily monitored by magnetic measurements and allow you to study the atomic mobility in such a range of temperature.

2. EXPERIMENTAL AND RESULTS

Pieces 5 cm. long were cut from the cold-rolled (~ 14% thickness reduction) ribbons, and annealed for succesive periods of time at a given temperature in a high purity nitrogen atmosphere. The samples were cooled to room temperature and the magnetization work measured after each annealing period by conventional methods.

Normalized relaxation function $\Phi(t)$ at each temperature T was built as

$$\Phi(t) = \frac{Wm(t) - Wm(\infty)}{Wm(0) - Wm(\infty)} \quad (4)$$

where $Wm(t)$ is the value of the magnetization work at time t, $Wm(0)$ is the starting value and $Wm(\infty)$ is the final relaxed limit. In this work, $Wm(\infty)$ was taken as the magnetization work corresponding to the as-quenched samples. Results obtained show that $\Phi(t)$ behaves in all cases studied almost linearly with log t suggesting a *Stretched Exponential* behaviour. In order to fit the experimental data of $\Phi(t)$ to expressions like (1) a master curve for each sample was obtained by reducing all the data to a single reference temperature (T_r). The temperature behaviour of the shift factor, a(T), used to displace the points from other temperatures allows the apparent activation energy for the relaxation process, E_p, to be obtained. Values of E_p corresponding to the two samples studied are shown in the TABLE. Figure 1 shows the master curves $\Phi(t^*)$ for these two samples, where t^* is a reduced time defined as $t^* = t \times a(T)$. The continuous line is in each case the best fit of $\Phi(t^*)$ to expression (1). Values of n and $\tau_p (T_r)$ obtained from these fits are also shown in the TABLE.

Interpreting both n and E_p values, in the framework of the Ngai model, the values of activation energy, E_o, corresponding to the microscopic processes underlying relaxation of magnetization work can be obtained. These values are also shown in the TABLE and are in the range ($E_o < 1eV$) one can expect for migration of quasi-vacancies in the structuraly unrelaxed

TABLE: Values of the parameters used to fit $\Phi(t^*)$ experimental behaviour

SAMPLE	T_r (K)	E_p (eV)	n	$\tau_p(T_r)$ (s)	E_0 (eV)
METGLASS 2826®	481	3.2	0.81	$6.0\ 10^5$	0.6
$Fe_{40}Ni_{40}B_{20}$	450	2.9	0.85	$1.2\ 10^7$	0.4

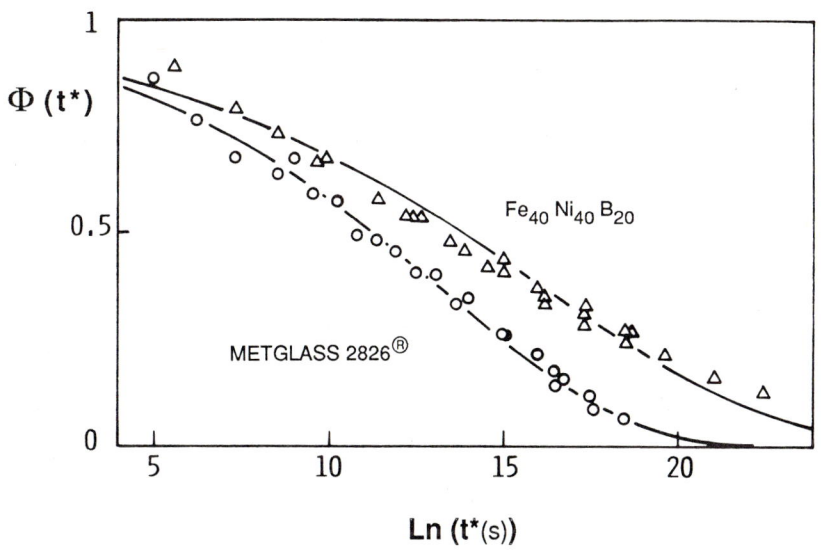

FIGURE 1.- Master curves $\Phi(t^*)$ corresponding to the two samples studied.

state of Fe Ni B glassy alloys[5]. Similar values have also been recently asigned to the reorientation of atom pairs involved in the reversible Magnetic After Effect (MAE) spectrum of glassy alloys[6].

REFERENCES

1) M.R.I. Gibbs, J.E. Evetts and J.A. Leake; J. Mater. Sci. 18 (1983) 278.

2) A. van den Beukel and S. Radelaar; Acta Metall. 31 (1983) 419.

3) A. van den Beukel, S. van der Zwaag and A.L. Mulder; Acta Metall. 32 (1984) 1895.

4) "Relaxations in Complex Systems", K.L. Ngai and C.B. Wright (ed.) Washington DC (1984)

5) J. Horvath, K. Pfahler, W. Ulfert, W. Frank and H. Mehrer; J. Physique 12 (1985) C8-645.

6) F. Rettemeier and H. Kronmuller; Phys. Stat. Sol. (a) 93 (1986) 221.

ANNEALING EFFECT OF Pr-Co FILMS DEPOSITED BY ION BEAM SPUTTERING

Youichi HOSHI* and Masahiko NAOE**

*Tokyo Institute of Polytechnics, Atsugi-shi, Kanagawa-ken 243-02 Japan
**Tokyo Institute of Technology, Meguro-ku, Tokyo 152 Japan

Pr-Co films deposited by an ion beam sputtering metod is annealed in a vacuum ($<2 \times 10^{-6}$ Torr) and in the air, and the changes in magnetic properties and structure of the film have been investigated. The Pr-Co film is oxidized easily even in the vacuum, which results in an increase in perpendicular magnetic anisotropy, though the saturation magnetization of the film changes little by the annealing. Moreover, the oxidation of the film with a crystalline phase leads to the change in crystal structure from the crystalline phase to an amorphous phase along with a remarkable decrease in coercive force Hc.

1. INTRODUCTION

$PrCo_5$ and Pr_2Co_{17} are well known as the permanent magnet materials with the highest value of (BH)max in rare-earth cobalt magnets. Therefore, it can be expected that these films have excellent magnetic properties for high density magnetic recording media. In our previous paper,[1] we reported the magnetic properties and the structure of Pr-Co and Nd-Co films deposited by an ion beam sputtering method and showed that (1) the Pr-Co with crystalline phase has very large coercive force Hc as large as 4 kOe, (2) the film with amorphous phase deposited at substrate temperature below 200 °C has low Hc ($<$100 Oe), (3) the film deposited at a substrate temperature in the range from 200 °C to 300 °C are composed of an amorphous phase with low Hc and crystalline one with high Hc. In this work, annealing the films in a vacuum ($<2 \times 10^{-6}$ Torr) and in the air was performed and the changes in structure and magnetic properties of the Pr-Co film by the annealing were investigated in detail.

2. EXPERIMENTAL PROCEDURE

The ion beam sputtering apparatus with duo-plasmatron type ion source was used for the film preparation. Co and Pr plates of 99.9 % purity were used for the targets. Typical film preparation conditions are listed in Table I. Details of the film preparation method were described in our previous papers[1,2].

The film without over coat or coated with SiO thin film were annealed in a vacuum below 2×10^{-6} Torr at a temperature in the range from 100 °C to 600 °C for 2 hours. Annealing in the air was also performed to confirm the oxidation effect of the film.

3. RESULTS AND DISCUSSIONS

The as-deposited amorphous film has in-plane uniaxial magnetic anisotropy with its easy axis parallel to the incident direction of sputtered particles to the substrate. The anisotropy energy $K_{//}$ takes a maximum value of 2.5×10^5 erg/cm^3 at Co content around 84 at.%. Figure 1 shows the change in the anisotropy energy $K_{//}$ with annealing temperature Ta. $K_{//}$ decreases monotonically with an increase of Ta and takes a value below 2×10^4 erg/cm^3 at Ta of 450 °C. While, the annealing leads to a remarkable increase of perpendicular magnetic anisotropy in the film and the film annealed at Ta of 400 °C has a perpendicular magnetic anisotropy field as large as 1 kOe. Corresponding to the changes in $K_{//}$ and $K_{\perp}$, the hysteresis curve (M-H curve) of the film changes as shown in Fig.2. It is evident from the figure that a large perpendicular magnetic anisotropy is induced in the film annealed at Ta above 300 °C, whereas inplane uniaxial magnetic anisotropy is reduced significantly. However, saturation magnetization Ms of the film changes little by the annealing as shown in Fig.3.

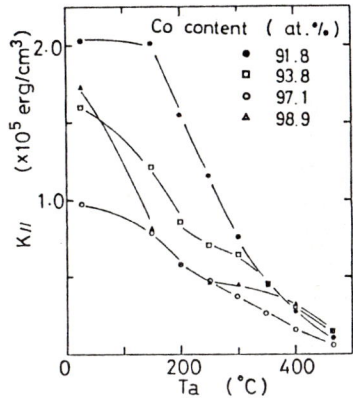

Fig.1 Changes in K with Ta.

Table I Film preparation conditions.

Deposition rate	30	Å/min
Substrate temp.	70~400	°C
Ar gas pressure	2×10^{-4}	Torr
Film thickness	4000	Å
Substrate	glass, and SiO$_2$/Si	
Film composition	0~20	at.% Pr

Figure 4 shows the typical X-ray diffraction diagrams of the films (a) with SiO over coat and (b) without over coat annealed at Ta of 600 °C in vacuum. It is clear from the figure that the films with SiO over coat are crystallized by the annealing, though the film without over coat are not crystallized yet. The films with SiO over coat are not crystallized by the annealing at Ta below 400 °C and their magnetic properties change little. This is quite different from the results observed in the films without over coat shown in Fig.1 and 2. The films crystallized by the annealing have large coercive force as large as 2.5 kOe. Above mentioned results suggest that even if the annealing is performed in a vacuum ($<2 \times 10^{-6}$ Torr), the amorphous film without over coat is easily oxidized

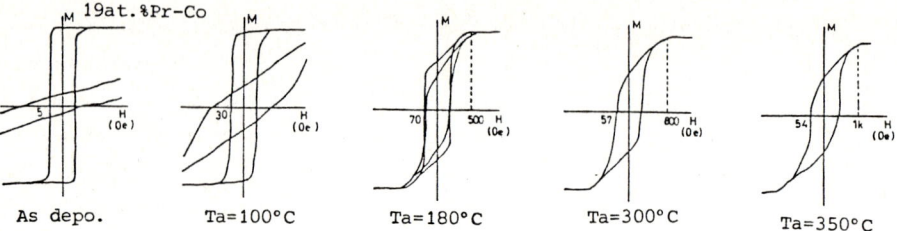

Fig.2 Changes in M-H curves of the film with Ta.

by the annealing, which results in the decrease in $K_{//}$ and the increase in $K_{\perp}$ of the film. Furthermore, the oxidation supresses the crystallization of the film.

The film deposited at substrate temperature Ts of 200 °C, which is composed of high Hc crystalline phase and low Hc amorphous phase, was also annealed in a vacuum (2×10^{-6} Torr). Figure 5 shows the change in M-H curves of the film by the annealing. The high Hc phase decreases by the annealing with an increase of Ta and disappears completely in the film annealed at Ta of 400 °C. Corresponding to the changes in the M-H curves, the X-ray diffraction diagram of the film changes as shown in Fig.6. It is evident from the figure that the crystalline phase with high Hc is changed to an amorphous phase with low Hc by the annealing. Taking the results observed in the amorphous film shown in Fig.4 into consideration, the changes in crystal structure and magnetic properties are thought to be caused by the oxidation of the film. This is agree with the fact that the structure and magnetic properties of the film with SiO over coat changes little by the annealing. Moreover, these changes of the film occur at much lower Ta in

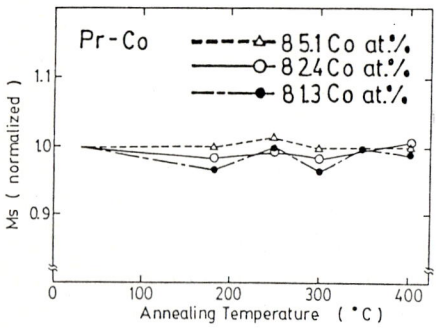

Fig.3 Changes in Ms with Ta.

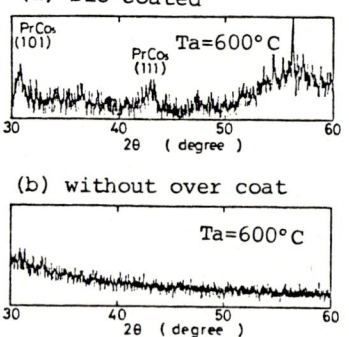

Fig.4 X-ray diffraction diagrams of the film annealed at 600 °C.

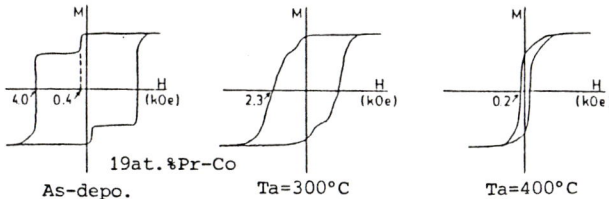

Fig.5 Changes in M-H curves of the film deposited at Ts of 200 °C.

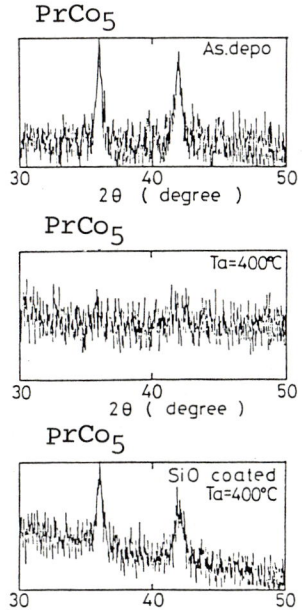

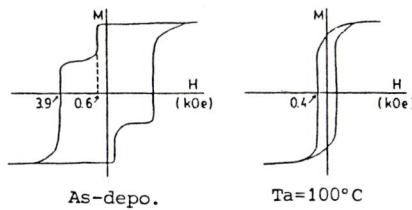

As-depo.   Ta=100°C

Fig.7 Changes in M-H curves of the film by the annealing in the air.

Fig.6 Changes in X-ray diffraction diagrams of the films deposited at substrate temperature Ts of 200 °C by the annealing in vacuum at 400 °C.

the air than in vacuum as shown in Fig.7.

Therefore, it can be concluded that the Pr-Co film is oxidized easily even in a vacuum ($<2\times10^{-6}$ Torr), though the film with crystalline phase has very high Hc, and the oxidation leads to a remarkable increase in $K_\perp$ and decrease in Hc of the film. However, a protective coating such as SiO thin film is useful to prevent the oxidation of the film.

REFERENCES

1) N. Terada, M. Naoe and Y. Hoshi, J. Appl. Phys., 57 (1985) 4170
2) Y. Hoshi, M. Naoe and S. Yamanaka, J. Appl. Phys., 53 (1982) 2344

OBSERVATION OF CROSSOVER BETWEEN REVERSIBLE AND IRREVERSIBLE MAGNETIC AFTER-EFFECTS IN AMORPHOUS $Fe_{40}Ni_{40}P_{14}B_6$

J. RIVAS [1], F. WALZ [2] and H. KRONMÜLLER [2]

(1) Depto. de Física Aplicada, Facultad de Física, Universidad, 15706 Santiago de Compostela, Spain.

(2) Max-Planck-Institut für Metallforschung, Institut für Physik, P.O. Box 800665, 7000 Stuttgart 80, F.R. Germany.

Long time-scale isothermal reluctivity relaxation (T=500 K, $0 < t(s) \lesssim 10^5$) has been measured in as-quenched and pre-annealed specimens of the magnetostrictive amorphous alloy $Fe_{40}Ni_{40}P_{14}B_6$. After times longer than $10^3 - 10^4$ s, a considerable decrease in the relaxation amplitudes has been observed which, is interpreted in terms of the " theory of two-level systems " as being due to a change from reversible to irreversible after-effects.

1. INTRODUCTION

Amorphous ferromagnetic alloys are metastable with respect to both their structure and their magnetic properties. Magnetic after-effects (MAEs), which are due to the rearrangement of atomic defect structures, can be treated theoretically within the framework of the thermodynamics of two-level systems (TLS)[1,2], which for long time-scale isothermal processes predicts a change from the predominance of reversible relaxation to the predominance of irreversible relaxation. This phenomenon has recently been observed in the nearly non-magnetostrictive amorphous alloy $Co_{58}Ni_{10}Fe_5Si_{11}B_6$ at 513 K [3,4]. In the present paper we report similar crossover phenomena occurring during long time-scale MAE measurements of the reluctivity r(t,T) of the magnetostrictive amorphous alloy $Fe_{40}Ni_{40}P_{14}B_6$ at 500 K. Our present results, though in qualitative agreement with the predictions of the TLS model, nevertheless suggest that the relaxation mechanisms acting in the amorphous alloy $Fe_{40}Ni_{40}P_{14}B_6$ are more complex than those observed in amorphous cobalt alloys.

2. EXPERIMENTAL TECHNIQUES

The amorphous alloy $Fe_{40}Ni_{40}P_{14}B_6$ was supplied by Allied Chemical Corp. in the form of ribbons with a width of 2mm and a thickness of 50 μ m. Samples were studied both in the as-quenched state and following various thermal and magnetic pre-treatments. Thermal pre-treatment was performed by heating the original as-quenched samples in an evacuated quartz tube at 5×10^{-5} Torr to temperatures in the range $500 \leq T_{a,p}(K) \leq 635$ and maintaining them at these temperatures for times $1 \leq t_{a,p}(h) \leq 4$. Both as-quenched and pre-heated specimens were additionally annealed in magnetic fields. Magnetic annealing was performed by tempering the samples at $T_{a,m}$=500 K during periods of $0 \leq t_{a,m}(h) \leq 64$ in magnetic extra-fields of up to 15 Oe, which were produced in the MAE-specimen-holder along the axis of the ribbon.

Immediately after the magnetic extra-field was switched off, the specimens were demagnetized in an AC field of decreasing amplitude. Subsequently the magnetic relaxation $\Delta r/r_1 = [r(t_2,T) - r(t_1,T)] / r(t_1,T)$, was measured for $t_1=1$ s and $2 \leq t_2(s) \lesssim 10^5$ by means of an automated LC-oscillator technique[5]. The measuring temperature of T=500 K was chosen as being well below the crystallization temperature ($T_{cryst} \simeq 650$ K), so as to allow the alloy to relax towards an amorphous state of lower energy.

3. EXPERIMENTAL RESULTS AND DISCUSSION

Figs. 1-3 show typical isothermal relaxation curves for a series of differently treated specimens. In all cases, the amplitudes, $\Delta r/r_1$, are found to decrease considerably after measuring times greater than 10^3-10^4 s. In terms of the TLS model, the steep rise of $\Delta r/r_1$ observed for shorter times $t < 10^3$ s reflects the predominance of reversible relaxation processes, which are

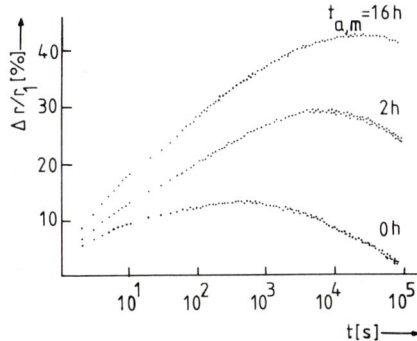

FIGURE 1
Influence of increasing magnetic annealing (Ta,m=500 K ; 0h≤ta,m≤16h) on the MAE of as-quenched specimens.

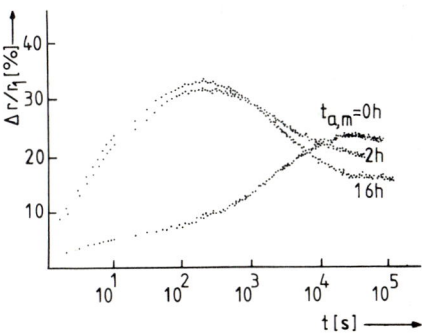

FIGURE 2
Influence of increasing magnetic annealing (Ta,m= 500 K ; 0h≤ta,m≤16h) on the MAE of identically pre-heated (Ta,p=560 K; ta,p=1h) specimens.

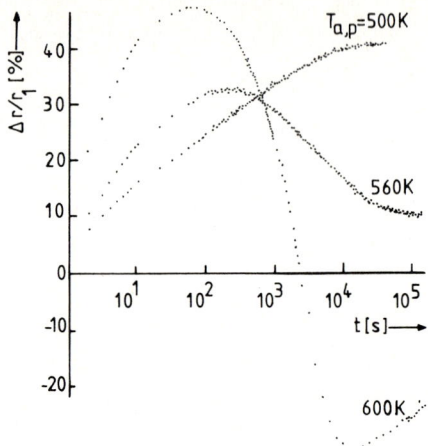

FIGURE 3
Influence of increasing pre-heating
(500 K ≤ Ta,p ≤ 600 K; ta,p= 1h) on
identically magnetically annealed
(Ta,m= 500 K ; ta,m= 16h) specimens.

attributed to the reorientation of mobile atom pairs within magnetic domain walls. The decrease of $\Delta r/r_1$ occurring after longer times $t > 10^4$ s is associated with irreversible relaxation processes, which are assumed to involve the blocking of mobile atom pairs by structural rearrangements. The growth of $\Delta r/r_1$ with increasing magnetic annealing time (Figs. 1,2) is associated with a corresponding increase in the induced magnetic anisotropy. The fact that the various isothermal curves do not tend to the same final quasi-equilibrium value may be explained by assuming a significant rearrangement of the magnetic domain structure to take place in the course of this type of high-temperature relaxation experiment.

REFERENCES

1) H. Kronmüller, Phil. Mag. B 48 (1983) 127.

2) H. Kronmüller, Phys. Stat. Sol. (b) 118 (1983) 661 and 127 (1985) 531.

3) H. Kronmüller, Proc. Intern. Workshop on Amorphous Metals and Semiconductors (Eds. P. Haasen and R.I.Jaffee, Pergamon Press, 1985).

4) H.-Q. Guo, H. Kronmüller, N. Moser and A. Hofmann, Scripta Metallurgica 20 (1986) 185.

5) F. Walz, Phys. Stat. Sol. (a) 8 (1971) 125 and 82 (1984) 179.

MAGNETOSTRICTION OF Fe - BASED AMORPHOUS THIN FILMS [+]

Krzysztof TWARDOWSKI and Jerzy WENDA [x]

Chemistry Department of the Warsaw University, al. Żwirki i Wigury 101, 02 089 Warsaw, Poland

[x]Department of Solid State Physics, Academy of Mining and Metallurgy, al. Mickiewicza 30, 30 059 Kraków, Poland

An influence of metalloid (B+Si, Si) concentration on linear saturation magnetostriction and saturation magnetic field of rf sputtered Fe-based thin films for as-deposited and annealed samples was investigated. It was shown that proper concentration of metalloids in the films can give the high value of saturation magnetostriction and low value of saturation field for ternary FeBSi flims in comparison to binary FeB films. After vacuum annealing of selected FeBSi films further changes of saturation magnetostriction and saturation magnetic field were observed.

1. INTRODUCTION

Transition metal-metalloid amorphous alloys are of interest from both fundamental and technical points of view. Possible applications of these alloys for variable surface acoustic waves devices requires films with a large saturation magnetostriction, low saturation field and proper magnetic anisotropy [1]. In previous paper [2] we have reported a compositional dependence of saturation magnetostriction for FeB films. The maximum value of saturation magnetostriction $(\Delta l/l)_s = 44.3 \times 10^{-6}$ was observed for $Fe_{71.5}B_{18.5}$ film, but at relatively large saturation field 64 kA/m. In this paper we have reported the investigations of saturation magnetostriction and saturation field for FeBSi and FeSi films.

2. EXPERIMENTAL

The investigated films were deposited by rf sputtering technique onto Corning 0211 glass substrates from a "mosaic" target. The measurements of linear magnetostriction were performed by

---

[+]Work supported by Institute of Physics, Polish Academy of Sciences, Warsaw, Poland

cantilever-capacitance method [3]. In order to study an influence of annealing on magnetostriction of these films a process of vacuum annealing was performed. The annealing conditions were chosen as follows : temperature $T = 250°C$, time $t = 2$ h, vacuum of a pressure $p = 1 \times 10^{-5}$ Torr, no magnetic field applied.

3. RESULTS

The values of saturation magnetostriction $(\Delta l/l)_s$ and saturation magnetic field H_s for as-deposited binary FeSi and ternary FeBSi films are shown in Tab. 1. As one can see the value of saturation magnetostriction depends mainly on Fe concentration in the films. The highest values of magnetostriction were obtained for films with the highest Fe concentrations.

Tab. 1 Saturation magnetostriction and field

Composition (at. %)	$(\Delta l/l)_s$ (10^{-6})	H_s (kA/m)
$Fe_{67.5}B_{29.5}Si_3$	29.5	30.4
$Fe_{70}B_{22}Si_8$	33.3	30.4
$Fe_{72.5}B_{12}Si_{15}$	31.4	28.0
$Fe_{73}B_{15}Si_{12}$	44.6	30.4
$Fe_{76}B_{13}Si_{11}$	39.5	36.0
$Fe_{76}B_{14}Si_{10}$	44.1	36.0
$Fe_{77}B_{15}Si_8$	41.3	36.0
$Fe_{77}B_{19}Si_4$	39.1	36.0
$Fe_{79}B_{17}Si_4$	42.9	36.0
$Fe_{85}Si_{15}$	14.6	48.0
$Fe_{82}Si_{18}$	26.0	40.0

For metalloid concentration (B+Si) in the range 21 ÷ 32.5 at. % saturation magnetostriction have changed from 29.5 to 44.6 x 10^{-6}, being higher than magnetostriction of the FeB films with the same Fe concentration [2]. Additionally, the saturation field for FeBSi films was lower than for FeB films. The substitution of B by Si has a effect on value of saturation magnetostriction for FeBSi films, as can be seen from Tab. 1.

It is well known that post-deposition annealing can seriously influence the magnetic properties of amorphous alloys [4,5]. Because the proper annealing can improve magnetostrictive properties of alloys [6] we have investigated the effect of vacuum annealing on some FeBSi and FeSi films. The results of this annealing are shown in Tab. 2.

Tab. 2 Saturation magnetostriction and field after annealing

Composition (at. %)	$(\Delta l/l)_s$ (10^{-6})	H_s (kA/m)
$Fe_{72.5}B_{12}Si_{15.5}$	48.3	16.0
$Fe_{73}B_{15}Si_{12}$	34.0	12.0
$Fe_{82}Si_{18}$	24.4	28.0

As one can see for FeBSi films an increase of saturation magnetostriction about 10% occured accompanied by serious decrease of saturation field.

The performed measurements of magnetostriction for FeBSi and FeSi films proved that it is possible to determine the optimum composition for obtaining films possesing high saturation magnetostriction value and low saturation field. The further improvement of magnetoelastic properties of these films is possible by vacuum annealing.

REFERENCES

1) D. C. Webb, D. W. Forester, A. K. Ganguly, C. Vittoria,
   IEEE Trans. Magn. MAG-15 (1979) 1410.

2) K. Twardowski,
   Acta Physica Polonica Vol. A68 (1985) 383.

3) K. Twardowski, H. K. Lachowicz,
   phys. stat. sol. (a) 53 (1979) 599.

4) H. Hoffmann, M. Takahashi and J. Zweck,
   J. Magn. Magn. Mater. 35 (1983) 211.

5) F. Rottenmeirer, E. Kisdi-Koszo and H. Kronmuller,
   phys. stat. sol. (a)93 (1986) 597.

6) N. Tsuya, K. I. Arai and T. Chsaka,
   IEEE Trans. Magn. Vol. MAG-14 (1978) 946.

FORCED VOLUME MAGNETOSTRICTION OF Zr-CONTAINING Fe-RICH METALLIC GLASSES

L. T. BACZEWSKI[*°], A. WEGRZYN[+]

[*] Institute of Physics, Polish Academy of Sciences, Warsaw (Poland)
[+] Department of Solid State Physics, Academy of Mining and Metallurgy, Cracow (Poland)

1. INTRODUCTION

Magnetic properties of (Fe Ni Co) Zr amorphous alloys have been studied extensively in recent years and particular interest was focused on the Invar type compositions[1]. Also the spin-glass behaviour was found in Fe-Zr metallic glasses[2].

However there are only a few measurements of the forced volume magnetostriction ($\partial\omega/\partial H$) in amorphous systems and even less papers which combine the $\partial\omega/\partial H$ data with pressure effects[3,4].

We have been studying the amorphous $(Fe_{1-x} Ni_x)_{90} Zr_{10}$ system. The temperature dependence of $\partial\omega/\partial H$ and the values of pressure effect on magnetization ($\partial\sigma_0/\partial p$) calculated from $\partial\omega/\partial H$ data are presented in this paper.

2. EXPERIMENTAL

Alloys of $(Fe Ni)_{90} Zr_{10}$ system were prepared by melt spinning technique. Their amorphous structure was confirmed by X-ray diffraction. Three terminal capacitance method was used for magnetostriction measurements.

3. RESULTS AND DISCUSSION

The composition dependence of $\partial\omega/\partial H$ for amorphous $(Fe_{1-x} Ni_x)_{90} Zr_{10}$ alloys, where $0 \leq x \leq 0.2$, is shown in fig. 1. The maximal value of $\partial\omega/\partial H$ at 300 K is observed for x = 0.05, which has T_c only slightly higher than RT. On the other hand, $Fe_{90} Zr_{10}$ alloy has $T_c \simeq 240$ K ; thus the RT value of $\partial\omega/\partial H$ is a forced volume magnetostriction for the paramagnetic state. The values for 0 K were estimated by extrapolation of the data at finite temperature to absolute zero.

The low temperature forced volume magnetostriction is a decreasing function of Ni content. The same tendency was observed for the other amorphous systems like $(Fe Co)_{90} Zr_{10}$[4], (Fe Co) Si B[3,5]. The value at 0 K of $\partial\omega/\partial H = 285 \cdot [10^{-10}$ $Oe^{-1}]$ for $Fe_{90} Zr_{10}$ alloy is the highest ever reported for amorphous systems.

[°] Present address : Physique du Solide - Université de NANCY - I
B. P. 239 - 54506 VANDOEUVRE LES NANCY CEDEX (France)

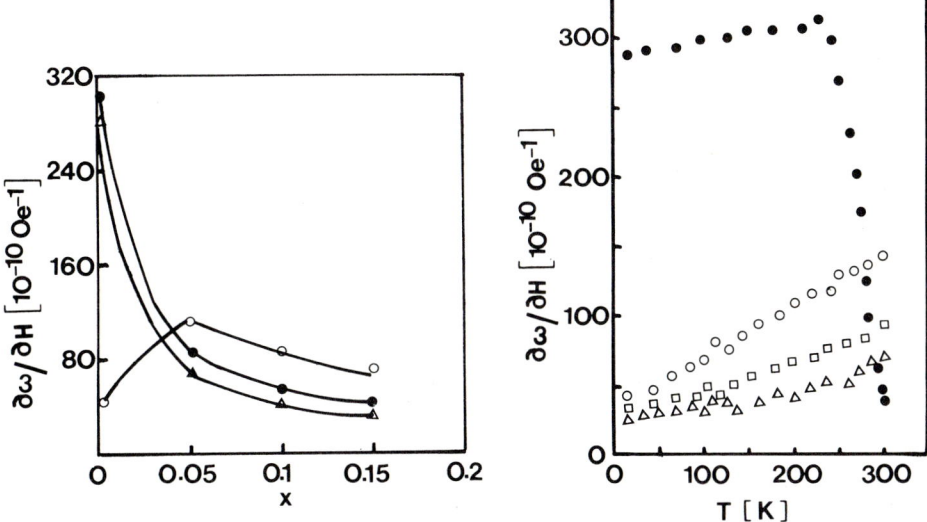

FIGURE 1
Composition dependence of the forced volume magnetostriction $\partial\omega/\partial H$ for $(Fe_{1-x} Ni_x)_{90} Zr_{10}$ amorphous alloys
○ 300 K ; ● 77 K ; △ 0 K

FIGURE 2
Temperature dependence of the forced volume magnetostriction for $(Fe_{1-x} Ni_x)_{90} Zr_{10}$ amorphous alloys
● x = 0 ; ○ x = 0.05 ; □ = 0.1 ; △ x = 0.2

The temperature dependence of $\partial\omega/\partial H$ is presented in fig. 2. The monotonic increase has been observed for $T < T_c$. However for $Fe_{90} Zr_{10}$ alloy no sharp maximum of $\partial\omega/\partial H (T)$ around T_c has been found as is the case for most Fe-based crystalline and amorphous alloys[3].

Pressure dependence of the magnetization σ_o at T = 0 K can be obtained from $\partial\omega/\partial H$ data on the basis of thermodynamic relation :

$$(\partial\omega/\partial H)_{T=0 \ K} = - \rho \ (\partial\sigma_o/\partial p)_{T=0 \ K} \qquad (1)$$

where ρ is the density and σ_o is the magnetization at T = 0 K.

The results of such calculations are shown in fig. 3 where $\partial\sigma_o/\partial p$ is presented as a function of Ni content. The data are similar to those for crystalline Fe-Ni alloys but at the same time one order of magnitude larger than for (Fe Co) Si B amorphous alloys[3]. It can be concluded that generally an increase of iron content in the alloy causes an increase (in the absolute value) of $\partial\sigma_o/\partial p$.

The pressure dependence of magnetization of (Fe Ni) Zr alloys has a large absolute value (fig. 3). As it was proposed for Fe-Ni crystalline alloys there are two possible explanations of this phenomenon. One is a "gradual transition" from the strong ferromagnetism to weak ferromagnetism following Mathon-Wolfarth

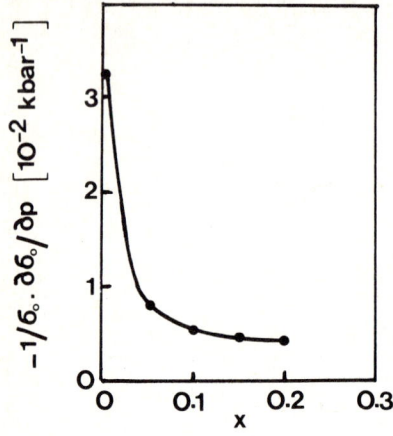

FIGURE 3
Pressure effect on magnetization at T = 0 K as a function of composition for $(Fe_{1-x} Ni_x)_{90} Zr_{10}$ amorphous alloys

theory[6]. The second possible solution is a "rapid" transition from strong ferromagnetism to the spin-glass state in an Invar region[7].

From the last point of view the changes of magnetization with pressure can be explained as a change of the critical composition with pressure. The infinite value of $\partial \sigma_0 / \partial p$ at a critical composition should be expected for homogeneous alloys. However in the amorphous alloys exists a fluctuation of local composition or chemical short range order (CSRO). As was mentioned before the spin-glass properties have been found in Fe Zr alloys. Thus one would expect the existence of spin-glass behaviour in (Fe Ni) Zr alloys for low Ni concentrations. Then the big change of magnetization with pressure may be due to the pressure dependence of the critical concentration for the disappearance of ferromagnetism.

REFERENCES

1) K. Shirakawa, S. Ohnuma, M. Nose, T. Masumoto, IEEE Trans. Mag MAG-16 (1980) 1129.

2) H. Hiroyoshi, K. Fukamichi, J. Appl. Phys. 55 (1984) 1823.

3) H. Tange, M. Goto, S. Yoda, Proc. Conf. RQ 4, eds. T. Masumoto, K. Suzuki Japan Inst. of Metals vol. 2 (1982) 811.

4) T. Jagielinski, A. Wegrzyn, S. Ohnuma, T. Masumoto, Solid. St. Comm. 44 (1982) 225.

5) T. Jagielinski, K. Arai, N. Tsuya, S. Ohnuma, T. Masumoto, IEEE Trans. Mag MAG-13 (1977) 1553.

6) J. Mathon, E. P. Wohlfarth, Phys. Stat. Sol. 30 (1968) K 131.

7) S. Kachi, H. Asano, N. Nakanishi, J. Phys. Soc. Japan 31 (1971) 1278.

STRESS INDUCED ANISOTROPY, MAGNETOSTRICTION AND CREEP OF CURRENT HEATED Co-BASED AMORPHOUS ALLOY

J. KRZYWINSKI, L. ZALUSKI and A. SIEMKO
Institute of Physics, Polish Academy of Sciences, al. Lotników 32/46, 02-668 Warszawa, Poland

The kinetics of recoverable strain and the magnetic anisotropy induced by stress annealing were simultaneously investigated. It is found that only a part of reversible atomic rearrangements characterized by low activation energies is responsible for the recoverable magnetic anisotropy.

1. INTRODUCTION

Stress induced magnetic anisotropy has been widely studied[1,2,3]. This anisotropy may be divided into two contributions: anelastic (recoverable) and plastic[2,3] (unrecoverable). Both of them one can correlate to the anelastic and plastic deformations of the sample[4,5].

In the present work kinetics of stress induced anisotropy and mechanical creep were simultaneously investigated in order to examine the correlation between the anelastic contribution of magnetic anisotropy and the deformation of the sample.

2. EXPERIMENTAL AND DISCUSSION

The investigations of stress induced anisotropy, saturation magnetostriction and creep (i.e. dilatation of length) for the Co-based amorphous alloy - $Co_{57.4}Fe_{5.6}Ni_{10}Si_{11}B_{16}$ ($\lambda_s > 0$) were performed.

Magnetic anisotropy was induced by stress annealing at temperatures higher than the Curie temperature. Before stress annealing, the samples were preannealed. The samples were heated by an electric current flowing through the sample. In order to minimize the plastic component of the deformation, short annealing times were examined.

During annealing, dilatometric measurements were carried out. Annealing temperatures were estimated from dilatometric and magnetic measurements performed during current heating of samples without an external stress.

The easy axis of induced magnetic anisotropy was perpendicular to the ribbon axis. The values of the anisotropy fields H_k was determined from hysteresis loops. Saturation magnetostriction was calculated from stress dependence of anisotropy field.

The anisotropy field and the deformation of the samples resulting from the stress annealing for different times and the temperatures of stress-annealing are shown in Fig.1 (σ = 137MPa). As can be seen in this figure, the anisotropy field H_k and relative elongation $\Delta l/l$ increase with the stress-annealing time. For an annealing time larger than 15s the $H_k/\Delta l_\sigma$ ratio is 0.055±0.005 Oe/µm.

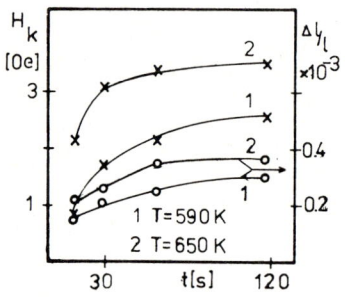
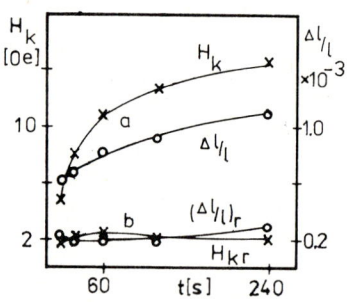

FIGURE 1
Anisotropy field and relative length changes for various annealing times

FIGURE 2
H_k and $\Delta l/l$ vs. annealing time after stress annealing a) and relaxation b)

After stress-annealing the samples were twice annealed without stress for the same time and at the same temperature as for the stress-annealing, and the experiment was repeated for a longer stress-annealing time. The first relaxation process reduced only the anisotropy field to zero while the second one reduced the deformation.

The results obtained for the sample annealed under the stress σ = 550 MPa are shown in Fig.2. In this case after the first relaxation process (parameters like above) the induced anisotropy and strain did not decrease to zero. The strong correlation between H_k and Δl, before and after relaxation, can be seen in this figure. The second relaxation process reduced H_k to zero however the deformation still did not return. This suggested that the relaxation process should be studied in more detail. For this purpose after stress annealing for 240s, relaxation was performed applying a series of increasing length current pulses. The results of this experiment are shown in Fig.3. There is no simple correlation between the time dependence of the anisotropy field and deformation observed during relaxation. As can be seen in this figure the anisotropy field decreases quicker than the induced deformation. This indicates that the mean activation energy connected with the anisotropy is lower than the one connected with recoverable strain. Thus only a part of the reversible atomic rearrangements is responsible for the anelastic magnetic anisotropy induced by stress annealing. From the above one concludes that the strong correlation shown in Fig.1,2 is an artifact because of the deformation remaining after the relaxation.

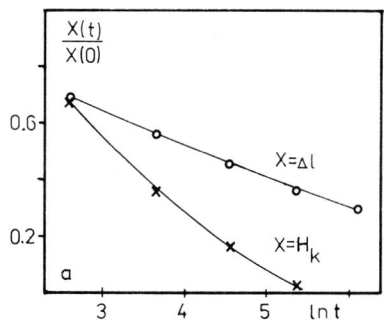 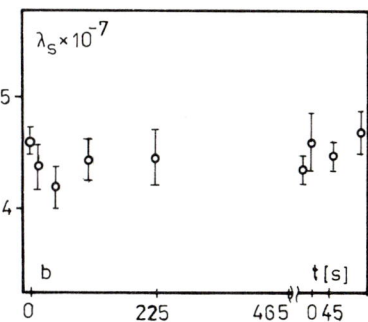

FIGURE 3
Relaxation of H_k and Δl_σ a) and λ_s b) as a function of total time

During relaxation process saturation magnetostriction was measured. The results are presented in Fig. 3b. The measured saturation magnetostriction was $4.4 \pm 0.25 \cdot 10^{-7}$ and did not change during the relaxation process.

REFERENCES

1) H.R. Hilzinger, Proc. 4th Int. Conf. on RQM (Sendai, 1981) 791.

2) O.V. Nielsen, L.K. Hansen, A. Hernando and V. Madurga, J.Magn.Magn.Mat. 24 (1983) 88.

3) O.V. Nielsen, A. Hernando, V. Madurga and J.M. Gonzales, J.Magn.Magn.Mat. 46 (1985) 341.

4) O.V. Nielsen, H.J.V. Nielsen, T. Masumoto and H.M. Kimura, J.Magn.Magn.Mat. 24 (1981) 88.8.

5) Y. Suzuki, J. Haimovich and T. Egami, Phys.Rev. B35 (1987) 2162.

STRUCTURAL RELAXATION AND SURFACE INDUCED AGING EFFECTS OF AMORPHOUS GdTbFe FILMS

S. KLAHN, M. HARTMANN, K. WITTER, H. HEITMANN

Philips GmbH Forschungslaboratorium Hamburg, Vogt-Kölln-Str. 30,
D-2000 Hamburg, FRG

Aging experiments on evaporated amorphous GdTbFe films of different thickness were performed in order to understand the aging kinetics of RE-TM films and to separate between oxidation and structural relaxation processes. The films were coated by Al to prevent oxidation due to exposure to the atmosphere. After aging the films in air at temperatures between 80 °C and 350 °C uniaxial anisotropy constant, K_u, coercivity, H_c, saturation magnetization, M_s, compensation temperature, T_{comp}, and Curie temperature, T_{Curie}, vary linearly with logarithm of aging time. Structural relaxation processes up to 350 °C mainly affect H_c while K_u, M_s, and T_{Curie} remain nearly constant. All magnetic properties are strongly influenced by oxidation effects.

INTRODUCTION

Rare earth transition metal (RE-TM) alloys are the most promising candiadates for MO-recording. However, degradation due to a variety of aging mechanisms, e.g. crystallization[1] and oxidation[2], has been observed. At heating of unprotected GdTbFe-films up to 330 °C Fe-clusters in a RE-oxide matrix has been formed[3]. This has been revealed also by torque measurements[4]. Different amorphous materials show changes of magnetic properties due to structural relaxation processes proceeding linearly with logarithm of aging time[5-8].
In this paper we report on aging treatments of GdTbFe films. The aim of this work has been to separate structural relaxation and oxidation effects.

EXPERIMENTAL

For isothermal aging treatments $Gd_{17.5}Tb_{6.5}Fe_{76}$ films of different thickness, d=18,36,73,146 and 292nm, were evaporated onto glass substrates and covered by 50nm Al in order to reduce surface oxidation. H_c, T_{comp}, and T_{Curie} were determined from substrate incident Kerr hysteresis loops. M_s was obtained from vibrating sample magnetometer measurements. The method of Mijajima et al.[9] for torque measurements provided K_u.

RESULTS AND DISCUSSION

Fig. 1 represents typical magnetization and torque curves of 292nm thick GdTbFe films and their irreversible changes due to aging treatments. The as deposited films exhibit a rectangular hysteresis loop and a uniaxial torque signal. During aging H_c decreases while the torque signal amplitude remains constant. At 250 °C the hysteresis loops deviate from their initial rectangular shape and an additional torque signal (inplane) appears resulting from a second phase[4].

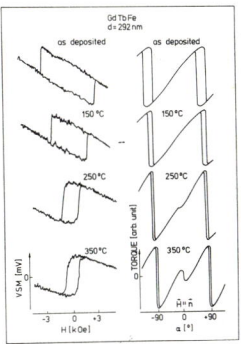

FIGURE 1
Vibrating sample magnetometer hysteresis loops (left) and angular dependence of torque at 10kOe external magnetic field of 292nm thick GdTbFe films as deposited and after aging for 8000 minutes at 150, 250 and 350 °C. The magnetization curves are tilted due to the diamagnetism of the sample holder

Table I summarises the data of 292nm thick samples. The as deposited films exhibit T_{comp} <300K. During aging below 250 °C T_{comp} remains nearly constant but at higher aging temperatures it decreases markedly due to preferential oxidation of the RE-atoms[2] resulting in an increase of M_s. T_{curie} remains constant up to an aging temperature of 350 °C.

TABLE I: Data of 292nm thick evaporated $Gd_{17.5}Tb_{6.5}Fe_{76}$ films on glass substrates coated by 50nm Al as deposited and after aging treatments

		$Gd_{17.5}Tb_{6.5}Fe_{76}$, d = 292 nm					
aging teperature (°C)	aging time (min)	H_c (kOe)	M_s (Oe)	$H_c \cdot M_s$ (10^5 erg/cm³)	K_u (10^5 erg/cm³)	T_{comp} (°K)	T_{curie} (°K)
as deposited		3.00	60	1.80	5.1	210	480
80	800	2.95	59	1.74	5.0	-	-
80	8000	2.76	62	1.71	5.5	-	-
80	80000	2.65	64	1.70	5.5	-	-
150	8000	2.03	65	1.32	5.1	200	490
250	8000	0.81	73	0.59	5.2	115	480
350	8000	0.61	89	0.54	(3.3)	-	470

On aging at 150 °C H_c and K_u change linearly with logarithm of time[8]. In thinner films H_c and K_u are more affected than in thicker ones due to surface effects. SIMS depth profiles[8] show oxygen diffusion from the Al and glass interfaces resulting in a strong decrease of K_u as well as a shift in T_{comp} and thus in a decrease of H_c and an increase of M_s. The average concentration of diffused oxygen in the RE-TM films are expected to be proportional to the ratio $S/V=1/d$, where S is the surface area and V the volume of the film.

Fig. 2 shows the changes in M_s, K_u, and H_c after 8000 minutes aging at 150 °C plotted versus 1/d. Extrapolating the data to pure bulk effects by linear regression (1/d→o) we can determine the influence of strutural relaxation processes. M_s and K_u are nearly unaffected by structural relaxation processes while H_c decreases by 30%. The slopes of curves in Fig. 2 relate to surface reactions. Fig. 3 shows the changes in H_c at different temperatures due to pure bulk effects (1/d→o).

The decrease of H_C due to structural relaxation reveals 'lnt-kinetics'. Thus a broad distribution of activation energies has to be considered[10]. At very high aging temperature (T≥250 °C) even the thickest samples severely oxidize (see table I) and the results of the bulk extrapolation are less accurate.

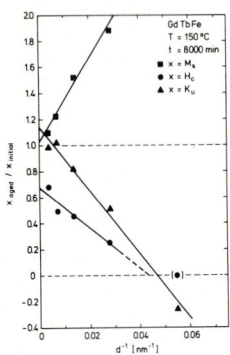 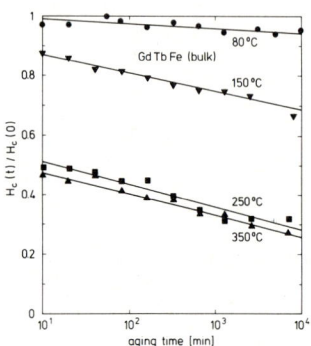

FIGURE 2
M_s, K_u, and H_c data of Al coated GdTbFe films aged at 150 °C for 8000 minutes normalized to the initial values versus the reziprocal value of thickness

FIGURE 3
The degradation of coercivity, H_c, due to pure bulk effects normalized to the initial values (extrapolation 1/d->o)

If oxidation and other surface induced effects can be avoided only H_c will degrade by structural relaxation processes showing 'lnt-kinetics'[8]. This kinetics holds at least over three decades within experimental error. Thus, by extrapolation 17 years can be estimated for the time required to halve H_c by structural relaxation at an aging temperature of 150 °C. At 80 °C this time is by centuries longer. DSC measurements at 1μm thick magnetron sputtered GdTbFe films revealed a crystallization temperature above 600 °C and 2.6eV for the energy of crystallization applying the Kissinger method[11]. Thus, in the temperature range discussed above crystallization is no relevant aging mechanism.

From these experiments we conclude that the lifetime of thin GdTbFe-films will not be limited by structural relaxation phenomena but by surface induced aging mechanisms e.g. oxidation.

REFERENCES
1) F.E. Luborsky, J. Non Cryst. Sol., 61/62, 829 (1984).
2) R.B. v. Dover, E.M. Gyorgy, R.P. Frankenthal, M. Hong, D.J. Siconolfi, J. Appl. Phys. 59, 1291 (1986).
3) N. Saito, Y. Togami, T. Morishita, Jap. J. Appl. Phys. 25, L580 (1986).
4) M.Hartmann,K.Witter,J.Reck,H.J.Tolle, IEEE Trans. Magn. MAG-22, 943 (1986).
5) G.E. Fish, IEEE Trans. Magn. MAG-21, 1996 (1985).
6) T. Egami, J. Magn. Magn. Mat. 31-34, 1571 (1983).
7) T.C.Anthony, J.Brug, S.Naberhuis, H.Birecki, J. Appl. Phys. 59, 213 (1986).
8) M. Hartmann, S. Klahn, K. Witter, INTERMAG '87, Tokio, EC-04.
9) H. Miyajima, K. Sato, T. Mizoguchi, J. Appl. Phys. 47, 4669 (1976).
10) M.R.J. Gibbs, J.E. Evers, J. A. Leake, J. Mat. Sci. 18, 278 (1983).
11) H.E. Kissinger, Anal. Chem. 29, 1902 (1957).

AMORPHOUS IRON-BASE ALLOYS. ELECTROCHEMICAL GROWTH OF OXIDE LAYERS

T.F.OTERO and A.R.PIERNA*
Dpto. de Química-Física. Fac. de Ciencias Químicas.
Apdo. 1072; 20080, San Sebastián. Spain.
* Dpto. de Ingeniería Química E.U.I.T.I.
San Sebastián. Spain

The study of the electrochemical behaviour of amorphous metals has received an increasing interest during the last years[1-3]. In this work we present the results obtained when the amorphous alloy $Fe_{40}Ni_{40}B_{20}$ was subjeted to a potentiodynamic treatment in KOH aqueous solutions.

A clened ribbon, with a surface area of 1 cm^2 was treated to consecutive potential cycles (a constant, previously determined, number of cycles) at high sweep rates. A voltamogram of control was obtained after the treatment recording the i-E electrode answer when a potential cycle was applied to it at low sweep rates.

The iron presence in the solution, after the electrochemical treatment, was checked by visible spectroscopy. Between two consecutive experiments the electrode was cleaned by corrosion in 2N HCl aqueous solution, rinsed with double distilled water and returned into the work solution.

Under those conditions a thick and electrochemically reversible oxide layer was developed on the glass metal. The magnetic properties of samples covered with different thickness of oxide layers, developed potentiodynamically and dried, were mesured.

The potentiodynamic behaviour of the oxide layers, checked at low sweep rate of potential, was quite similar to the obtained with AISI 304 stainless steel[4]. At potentials near to the hydrogen release the oxide layers has a high activity showing a maximun of current density on the anodic voltammogram. At more anodic electrical potentials the layer becomes passivated. Not one of the studied variables give an important iron corrosion in the solution. That is, the corroded metal remains on the glass metal as an electrochemically active oxide layers. In this way it is possible to get an indirect checking of the oxide layers thick by recording of a voltammogram at low sweep tare of electrical potential.

The influence of the different electrochemical or chemical variables on the formation and growth of the oxide layers may be followed by variation of the current density on the voltammogram anodic maximum, or by checking the charge under the anodic branche of the control voltammograms.

In this way, the formation and growth of the oxide layer was studied by consecutive treatment periods, accompanied by its correlative control voltammograms. The oxide layers grews linearly with the first treatment periods.Later it moves toward a stationary thickness, which varies with the studied variables.

When the anodic limit of potential for the treatment periods was shifted, keeping constant the cathodic limit (-1.600 mV. vs. S.C.E.), hte oxide layers was not developed till more anodic limits than -900 mV were reached. Between -300 and -200 mV, depending of the KOH concentration, a maximum of treatment efficiency appears. At higher anodic limits this efficiency decreasses quickly without presence of a significant amount of iron in the solution.

The variation of the cathodic limit of electric potential for the treatment periods, keeping constant the anodic limit on -200 mV, shows no oxide layers growth till more cathodic potential limits than -1.200 mV. At high KOH concentrations the efficiency of the treatment shows a maximum on -1.600 mV. At lower KOH concentrations the treatment efficiency gets a constant oxide layers thickness at more cathodic potential limits.

All those facts point to a great influence of the electric potential on the formation of active or passive oxide layers. The concentration of the electrolyte plays, too, an important rol when we try to obtaining thick oxide layers by an electro-chemical way.

On the other hand, the treatment efficiency decreases when the sweep rate of the potential sweeps increases; an shows a maximum at 40º C. Not presence of iron was detected in the solution.

The oxide layers developed on the metal glass surface shows an oxidiced dark blue colour and a transparent reduced form, which promotes electrochromic display when the electrode potential change. Very thick oxide layers show irreversible gold colour. The blue form is an uniform structure on the metal, as well after the electrochemical treatment as when it was dryed. The presence of pits on the surface may be due to the Cl^- ions remaining there after the electrode cleaning process, nevertheless the washing with double distilled water.

All those results point to an electrochemical behaviour of this metal glass very similar to the obtained with the stainless steel, without a significant influence of the B atoms on the potentiodynamic growth of oxide layers.

The study of the magnetic parameters, obtained form the loop hysteresis when the samples were covered with increasing oxide layers, showed an increase on the internal strees and a decreasing on the magnetostriction constant not very important. A study of the surface properties variation would be more suitable than the bulk mesurements.

ACKNOWLEDGEMENT
   The authors want to thand the Basque Country University for the financial support of this work.

REFERENCES
1) U.Linker, W.Plieth, Werkst. Korrosion. 34 (1983) 391
2) S.Kapusta, K.E.Heusler, Z.Metallkd. 72 (1981) 785
3) U.Unker, W.J.Plieth, Rapidyly Quenched Metals 1465 (1985)
4) T.F.Otero, Y.J.Jiménez, M.T.Ponce, in press.

MAGNETORESISTIVITY OF RANDOM ANISOTROPY a-$Dy_xGd_{1-x}Ni$ ALLOYS*

J.M. MOREIRA, V.S. AMARAL, M.M. AMADO, J.B. SOUSA
Centro de Física da Universidade do Porto - INIC, 4000 Porto, Portugal
B. BARBARA, B. DIENY
Laboratoire Louis Néel, CNRS-USMC 166X, 38042 Cedex, France

High resolution magnetoresistance data, $\Delta\rho(T,H)$, is presented for a series of $Dy_xGd_{1-x}Ni$ amorphous alloys, exhibiting magnetic freezing. In Gd-rich samples, with *large* Imry-Ma domains, $\Delta\rho$ is dominated by spin fluctuations. For Dy-rich samples, with *small* Imry-Ma domains, $\Delta\rho$ is analyzed in terms of electron scattering by antiparallel spin-pairs at Imry-Ma domain boundaries.

INTRODUCTION

The amorphous $Dy_xGd_{1-x}Ni$ alloys exhibit magnetic freezing at characteristic temperatures T_f ranging from about 62 K in GdNi to about 13 K in DyNi[1], due to the competition between predominantly positive exchange interactions (J) and local anisotropy fields (D). The magnetic properties of $Dy_xGd_{1-x}Ni$ change gradually from the low anisotropy ($D/J \sim 10^{-2}$; x=0) to the high anisotropy limit ($D/J \sim 1$, x=1). In the Gd-rich alloys (low anisotropy) the state below T_f consists of a random distribution of spontaneously magnetized domains with characteristic dimensions l_c much bigger than an interatomic distante a (Imry and Ma domains[2]). In Dy-rich alloys, the cost in anisotropy energy prevents the growth of such domains beyond a few interatomic distances. A detailed study in *zero field* has been done recently[3] with high resolution electrical resistivity measurements (ρ, $d\rho/dT$). Whereas in Gd-rich alloys the magnetic resistivity ρ_m decreases below T_f, in the Dy-rich alloys ρ_m *first increases*, before ultimately decreasing at low temperatures. To explain the behaviour of ρ_m, we recently proposed a model[3] based on the competing effects between essentially parallel magnetic moments within the Imry-Ma domains (J>0 in our samples) and essentially antiparallel pairs at the domain boundaries. In order to test this model we now present the results of a detailed investigation of the magnetoresistivity $\Delta\rho(T,H)$ in the $Dy_xGd_{1-x}Ni$ series, from 4.2 up to 100 K and in fields up to 1 Tesla.

---

* Work supported by INIC (Portugal), Reitoria da Universidade do Porto (project 16/85/86) and D.G. Relations Culturelles du Ministère Français des Affairs Etrangères.

RESULTS AND COMPARISON

In the Gd-rich samples (Fig.1a; x=0.1), where the correlation length is large ($\xi \gg a$), $\Delta\rho$ versus T at constant field exhibits a deep negative minimum at T_f. This is attributed to the supression of critical fluctuations by the magnetic field. Also, at these temperatures, no saturation is observed in $\Delta\rho$ at high fields (and T=constant). At low temperatures, where the spontaneous magnetization $\vec{M}_s$ is large, $\Delta\rho$ is positive at low fields (H ≲ 500 Oe), which we associate with the rotation of $\vec{M}_s$ produced by the field (ρ_m depends[4] on the angle between the current $\vec{I}$ and $\vec{M}_s$). However, as the field is increased further, the thermal fluctuations are reduced and a negative $\Delta\rho$ again sets in.

For the Dy-rich samples (Fig.1b; x=1) the magnetoresistance has no singularity at T_f but a smooth decrease towards negative values, suggesting that critical fluctuations play here a secondary role ($\xi \simeq a$). Sufficiently above T_f, an interesting effect is observed in DyNi (Fig.1b): a weak field (H ≲ 700 Oe) enhances the resistivity, whereas in higher fields the reverse occurs.

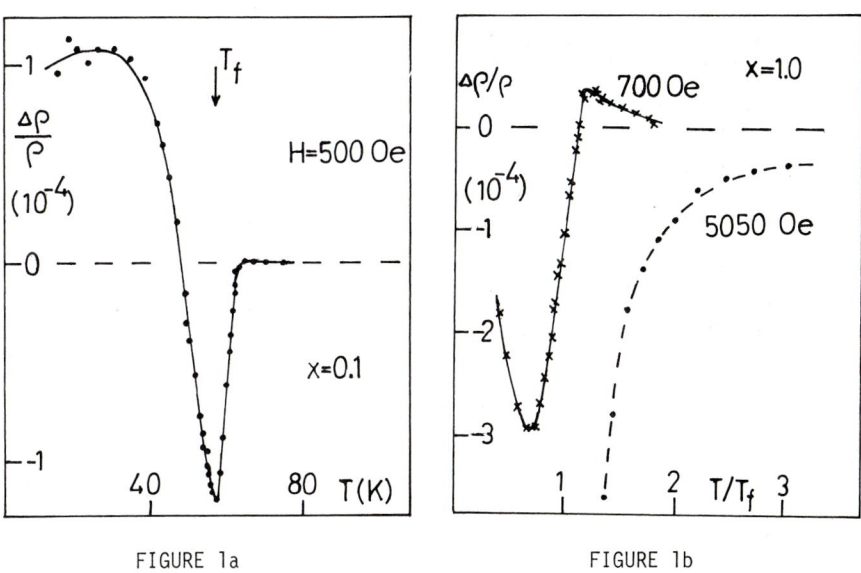

FIGURE 1a   FIGURE 1b

The analysis of the data is better performed with the quantity,

$$\frac{\delta\rho}{\rho} = (\rho_m(T,H) - \rho_{m\infty})/\rho(T,0)$$

as shown in Fig.2 a,b for the samples with x=1 and x=0.1, as a function of temperature. Here $\rho_{m\infty}$ is the full spin disorder resistivity, obtained at $T \gg T_f$. The anomalous positive bump in the curves a (x=.1), attributed to antiferromagnetic pairs, is drastically reduced by the field and ultimately, at high fields, the $\delta\rho/\rho$ curves resemble those in Gd-rich samples (Fig.2b).

CONCLUSION

In the Gd-rich samples ($\xi \gg a$) the magnetoresistance is mainly due to the ordering field effect in the large Imry-Ma domains; evidence also exists for orientational magnetoresistance at low temperatures, where $\vec{M}_s$ is large.

In the Dy-rich samples ($\xi \approx a$) in high fields the magnetoresistance is always negative increasing systematically in magnitude with Dy addition. This effect is attributed to the increasing dominance of the antiparallel pairs between magnetic domains; a strong magnetic field would then destroy progressively such pairs, causing a negative magnetoresistivity. At low fields the situation is more complex, as shown in fig.1b, where $\Delta\rho/\rho$ changes from negative to positive at T sufficiently above T_f.

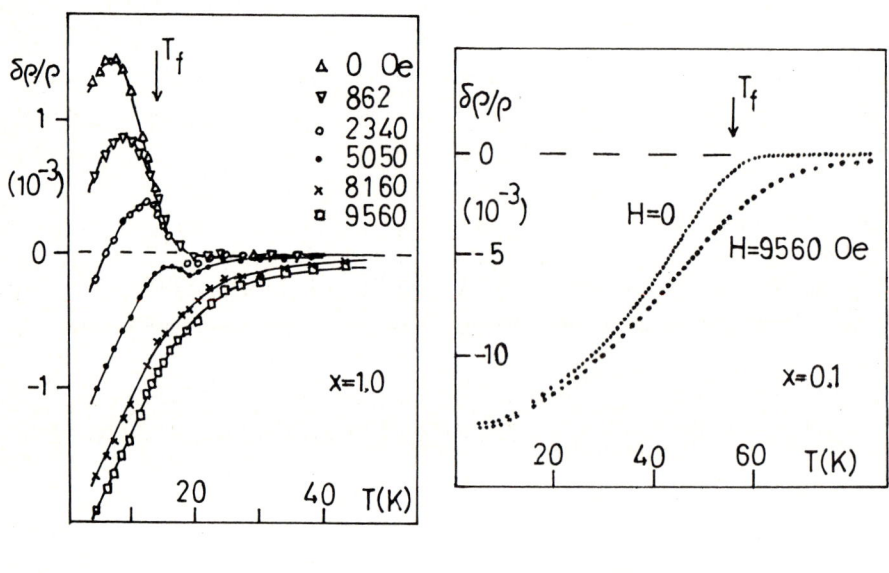

FIGURE 2a   FIGURE 2b

REFERENCES

1) B. Dieny and B. Barbara, Journal de Physique 46 (1985) 293

2) Y. Imry and S. Ma, Phys.Rev.Lett. 35 (1975) 1399

3) J.B. Sousa, J.M. Moreira, V.S. Amaral, M.M. Amado, M.E. Braga, B. Barbara, B. Dieny, J. Filippi, J.Physics F:Metal Physics (submitted for publication)

4) L. Berger, Physica 30 (1964) 1141

THERMAL DEPENDENCE OF ELECTRICAL RESISTIVITY AND CRYSTALLIZATION OF $(Co_{1-x}Ni_x)_{75}Si_{15}B_{10}$ AMORPHOUS RIBBONS

J.C. GOMEZ SAL°, J. RODRIGUEZ FERNANDEZ°, L. FERNANDEZ BARQUIN°
J.M. BARANDIARAN*, F. PLAZAOLA*

° Universidad de Cantabria, Santander, Spain.
* Universidad del Pais Vasco, Lejona, Vizcaya, Spain.

By means of electrical resistivity measurements in $(Co_{1-x}Ni_x)_{75}Si_{15}B_{10}$ for x<0.41 glasses we have pointed out the importance of magnetic contribution to the resistivity as well as the increasing short range order with the Ni amount.

1. INTRODUCTION

In a previous paper [1] we have reported a complet study of the thermal dependence of the electrical resistivity in $(Co_{1-x}(Fe_{0.5}Ni_{0.5})_x)_{75}Si_{15}B_{10}$ metallic glasses. The most significant result was that the magnetic contribution to the resisitivity was not negligeable, specially a change in the slope was observed at the Curie temperature, Tc, when this value is far from the crystallization temperature Tx. Furthermore the variation of the residual resistivity with the composition seems to be related to the pure chemical disorder as well as to the topological short range order effects.

The magnetic properties of $(Co_{1-x}Ni_x)_{75}Si_{15}B_{10}$ have been recently studied[2] by means of the analysis of the magnetostriccion and determining Tc, which are lower than in $(Fe_{0.5}Ni_{0.5}) Co_{1-x}$ glasses. Then (CoNi) based series is a good candidate to deepen in our previous results.

2. EXPERIMENTAL

We have performed resistivity measurements in $(Co_{1-x}Ni_x)_{75}Si_{15}B_{10}$ with x=0, 0.08, 0.15, 0.22, 0.28, 0.34 and 0.41. The samples were kindly supplied by Dr. M. Vazquez of the U.C. (Madrid), and were produced by the simple roller technique (typical cross section was 0.4 mm x 20 μm). The resistivity was determined by the a.c. four probe method [3]. D.S.C. measurements have been also performed in order to determine the crystallization temperature.

3. RESULTS AND DISCUSSION

In figure 1, we present the thermal variation of the reduced resistivity $\rho/\rho(300)$, for all the samples studied between 10 K and 900 K. In all the compounds a change in the slope of the resistivity is found, corresponding to the Curie temperature Tc. For temperatures above Tc an almost linear behaviour is

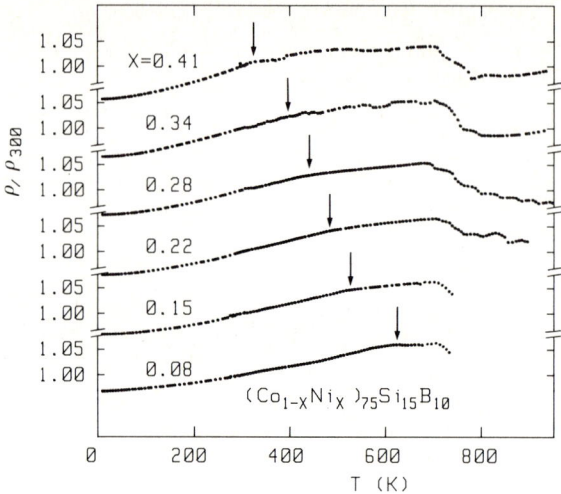

FIGURE 1

Thermal variation of the reduced resistivity between 10 and 900 K. Arrows indicate Curie temperatures.

observed. The crystallization temperatures, Tx, can be also determined, the several drops observed show different crystallization steps according to the D.S.C. measurements. In figure 2 we compare the Tx obtained by the different methods and the Curie temperatures determined by resistivity and magnetic measurements [2].

Our results confirm the importance of magnetic contribution to the resistivity in these metallic glasses. The change in the slope at Tc, shows that for T>Tc exists a constant spin disorder resistivity (as occurs in crystalline magnetic materials). The linear increase observed between Tc and Tx is only due to the non magnetic effects. In this way according to the Ziman framework [4] the θ^*_D (Debye temperature) calculated, taking the slope values in the paramagnetic range are more acurate than those obtained with the slope of the posible linear behaviour at 300 K, θ_D (see table 1) which have not physical meaning.

The relative differences between the resistivities at 10 K and room temperatures (presented in table 1) increase with the Ni content. This fact can be due to the lowering of the Curie temperature, then magnetic effects are more significant in this temperature range. Other posibility to explain this fact is the increase of short range order with the increase of Ni amount, in other words the variation of the structure factor $S(2K_F)$ with temperature[5] is more important in rich Ni compounds, having positive temperature coeficient.

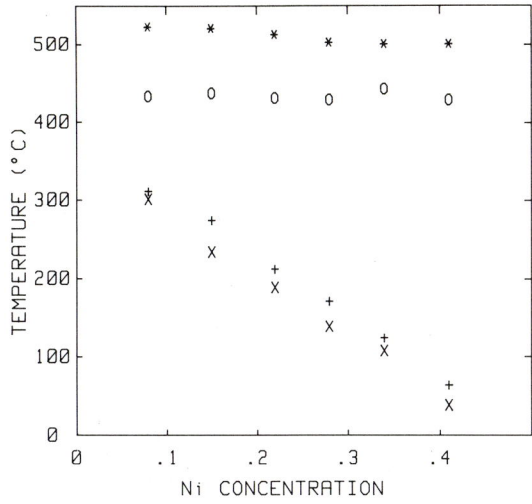

FIGURE 2
Cristalization temperature Tx
(*) from D.S.C.
(°) from Resistivity
Curie temperature Tc
(x) from (Ref. 2)
(+) from Resistivity
as a function of the composition.

The variation of the residual resistivity (table 1) with composition is in oposition with the Nordheim rule (the lowest is at x=0.41), this fact also indicates the increase a short range order which overcomes the chemical disorder.

In order to ascertain these conclusions, it will be interesting to continue our measurements in compounds with x>0.41, with Tc lower than 300 K.

TABLE 1
Resistivity and Debye temperatures

x	$\rho(10K)$ ($\mu\Omega \times cm$)	$\dfrac{\rho(300K)-\rho(10K)}{\rho(300)}$	θ_D	θ_D^*
0	105.1	2.94×10^{-2}	898	---
0.08	82.5	3.19 "	739	350
0.15	92.0	4.11 "	828	346
0.22	62.6	4.52 "	855	358
0.28	76.5	4.92 "	769	313
0.34	71.8	5.54 "	691	256
0.41	62.0	6.68 "	798	247

REFERENCES

1) J.M. Barandiaran, J.C. Gomez Sal, J. Rodriguez, R.J. Lopez and O.V. Nielsen, Phys. St. Sol. (a) 99 (1987) 243
2) M. Vazquez, A. Hernando and O.V. Nielsen, J. Mag. Magn. Mat. 61 (1986) 390
3) J. Rodriguez, R.J. Lopez and J.C. Gomez Sal, An. Fis. B. 82 (1986) 203
4) N. Banerjee, R. Roy, A. Majumdar and R. Hasegawa, Phys. Rev. B 24 (1981) 6861
5) S.R. Nagel, Phys. Rev. B. 16 (1977) 1964.

ELECTRICAL AND MAGNETIC PROPERTIES OF AMORPHOUS FeZr AND CoZr FILMS

T. STOBIECKI*, G. BAYREUTHER AND H. HOFFMANN

Institut für Angewandte Physik, Universität Regensburg, D-8400 Regensburg, FRG

Results of magnetic moment and Hall effect of amorphous $Fe_{1-x}Zr_x$ and $Co_{1-x}Zr_x$ films ($0.10 \leq x \leq 0.84$) are presented. The samples are ferromagnetic for $x \leq 0.62$ and $x \leq 0.42$, respectively. The ferromagnetic films show a large anomalous Hall effect which originates from the quantum side jump mechanism. The normal Hall effect in paramagnetic FeZr and CoZr films shows a strong influence of the band structure in these amorphous alloys.

1. INTRODUCTION

It is known that liquid-quenched ribbons of FeZr and CoZr can be obtained as stable amorphous alloys only in a limited range of concentration. Sputtered films of these alloys, however, are stable in the amorphous phase in a wider range of composition. The electrical and magnetic properties of amorphous $Fe_{1-x}Zr_x$ and $Co_{1-x}Zr_x$ films have been reported only for $x \leq 0.45$ [1-3]. In this paper we present and discuss magnetic moment and Hall effect on rf sputtered FeZr and CoZr films over a wide range of composition ($0.10 \leq x \leq 0.84$).

2. MAGNETIC MOMENT

Figure 1 shows the concentration dependence of the magnetic moment per Fe and Co atom in amorphous FeZr and CoZr films and ribbons[4]. Our results (denoted as open circles, fig.1) show that μ_{Fe} decreases monotonically with increasing x and disappears around $x=0.62$. Fe-rich amorphous samples ($0.08 \leq x \leq 0.11$) (ribbons[4] and films) show a sharp decrease of magnetic moment (from $1.72\mu_b$ to $1.38\mu_b$) associated with transition from crystalline to amorphous state. The extrapolated moment of pure amorphous Fe is about $1.6\mu_B$, i.e. lower than $2.2\mu_B$, the crystalline value of bcc iron. The reduced value has been explained by band calculations in amorphous FeB[5]. The magnetic moment of Co decreases also monotonically with x and disappears around $x=0.42$ (Fig.1). The earlier vanishing of μ_{Co} than μ_{Fe} is due to the fact that Co is a strong ferromagnet whereas Fe is a weak ferromagnet. In the case of CoZr amorphous films the pure Co amorphous moment is about

* Academy of Mining and Metallurgy, Solid State Physics Dept. PL-30-059 Krakow, Poland

$1.7\mu_B$ as for crystalline Co.

3. HALL EFFECT

Two types of scattering mechanism have been discussed in the anomalous Hall effect (AHE)[6], both are described by

$$R_s = a\rho + b\rho^2 \qquad (1)$$

where R_s is the spontaneous Hall coefficient, ρ is the resistivity, a and b are constants rougly independent of temperature[6]. The first term in eq. (1) describes the classical asymmetric scattering due to magnetic impurities. The second term in eq.(1) is associated with the quantum mechanical side jump Δy of the charge carrier trajectory at the point of scattering. In amorphous alloys the side jump is expected to occur because of a high resistivity and a very small mean free path of the conduction electrons. In the Fe- and Co-based amorphous alloys the side jump mechanism is responsible for the AHE because R_s versus ρ very closely follows a square law as seen in fig.2. Our present measurements and earlier calculations of the side jump parameter Δy of FeB and FeZr films[3] have shown the correlation between Δy and μ_{Fe}. The side jump parameter Δy for amorphous Fe-based alloys lies between 0 to $2*10^{-11}$ m.

In the case of paramagnetic FeZr and CoZr films the Hall coefficient R_H can be written as

$$R_H = 1/\mu_0 (\partial \rho_H / \partial H) = R_0 + \chi R_s \qquad (3)$$

where χ is the paramagnetic susceptibility and R_0 is the normal Hall coefficient. The concentration dependence of R_H (fig.3) in FeZr films shows a very sharp increase of R_H at the transition from para- to ferromagnetic state ($x \approx 0.62$). For CoZr films we observed a change of the sign of R_H around $x=0.5$. The sign reversal is directly associated with the sign of the derivative of the density of states at the Fermi level[7] ($R_0 \approx -g'(E_F)/g^2(E_F)$). In Zr-rich Zr-TM alloys (TM=Fe,Co,Ni,Cu), E_F is located at the Zr 4d-band[9] where $g'(E_F)<0$ hence $R_0>0$. With increasing TM concentration, E_F shifts slightly to the position where $g'(E_F)>0$ hence $R_0<0$. The sign reversal takes place in CuZr, NiZr and CoZr amorphous alloys, where the TM 3d-band and the Zr 4d-band are well separated from each other. In the case of FeZr the Zr 4-band nearly completely overlaps with the Fe 3d-band[9] and therefore we do not observe any sign reversal of R_0 in FeZr.

The strong concentration dependence of R_H (fig.3) in paramagnetic FeZr films results from a large contribution of the spontaneous Hall effect to the total coefficient R_H.

AKNOWLEDGEMENTS

We would like to thank the Deutsche Forschungsgemeinschaft for financial support.

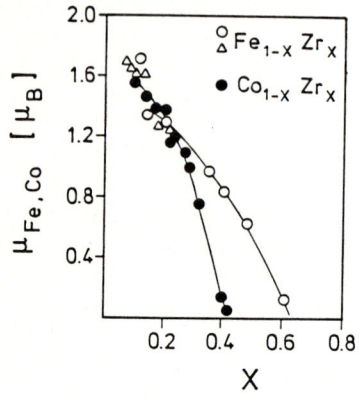

FIGURE 1
Magnetic moment vs. x in amorphous FeZr and CoZr films and ribbons[4] (△) at T=4K.

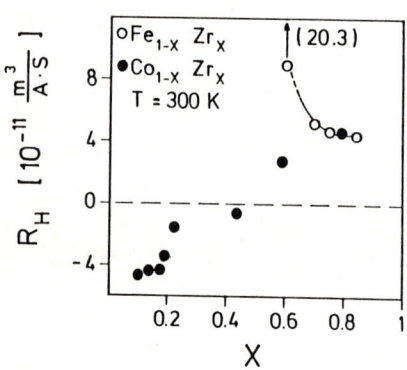

FIGURE 3
The Hall coefficient R_H of FeZr and CoZr films vs. x.

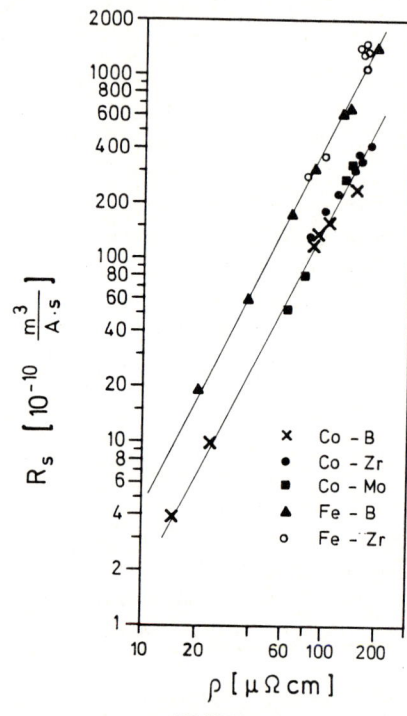

FIGURE 2
LogR_S vs. logρ for amorphous Fe- and Co- based alloys. Slope 1.97 ± 0.04

REFERENCES

1) K. Fukamichi, et al., J. Appl. Phys. 53 (1982) 2310
2) T. Yamagata and S. Ito, J. Magn. Mag. Mat. 31-34 (1983) 1475
3) T. Stobiecki and M. Przybylski, phys. stat. sol. (b) 134 (1986) 131
4) H. Hiroyoshi, et al., Sci. Rep. Ritu, A-33 (1986) 68
5) S. Krompiewski, et al., J. Magn. Mag. Mat. in print
6) L. Berger, Phys. Rev. B2 (1970) 4559
7) D. Nguyen Manh, et al., Phys. Rev. B33 (1986) 5920
8) R. W. Cochrane, et al., Phys. Rev. B27 (1983) 5955
9) V. L. Moruzzi, et al., Phys. Rev. B27 (1983) 2049

ANISOTROPIC MAGNETORESISTANCE WITHIN THE NFE MODEL

E. PILIPCZUK, H. MATYJA

Institute of Materials Science and Engineering
Warsaw University of Technology
ul. Narbutta 85, 02-524 Warsaw, Poland

Anisotropic magnetoresistance defined as:

$$\frac{\Delta\varrho}{\varrho_o} = \frac{\varrho_\| - \varrho_\perp}{\varrho_o} \qquad (1)$$

was a subject of many papers concerning the physical properties of the transition metal systems [1]. In nickel crystalline alloys the magnetorisistance can be estimated using the CFJ theory[2]. The extension made by Malozemoff[3] allows to determine the magnetoresistance of nickel amorphous alloys containing metalloids, but these theories are still no valid for others transition metal systems.

In this paper we consider the cobalt-based alloys, The classical NFE resistivity formula[4] do not provide anisotropic properties of the system. For this reason we applied the Boltzmann equation scheme with the relaxation time approximation to put out the conductivity tensor of the system. By evaluating it we made the general improvement to the NFE theory[5], namely in ferromagnetic metallic glasses the Fermi surface (FS) i.e. the sphere was replaced by the ellipsoid having the symmetry axis parallel to the direction of magnetic field (z-axis):

$$E(\bar{k}) = \frac{\hbar^2}{2m_\perp} \cdot (k_x^2 + k_y^2) + \frac{\hbar^2}{2m_\|} \cdot k_z^2 \qquad (2)$$

The fig. 1 presents qualitatively the $\Delta\varrho/\varrho_o$ plot versus the FS-parameter $\xi = m_\perp/m_\|$ and q parameter; q reflects the rate of the relaxation times for the up- and down-spin electrons: $q = \tau_\uparrow/\tau_\downarrow$. The positive value of the anisotropic magnetoresistance observed experimentally in considered alloys is noticed for $\xi < 1$, i.e. when FS is flatted along the z-direction.

This theory permits the prediction of the $\Delta\varrho/\varrho_o$ value for among others $Co_{80-x}Tm_xB_{20}$ glasses (Tm being an early transition metal like Mn, Cr, V). The FS parameter we assumed to be associated with the average magnetic moment $\langle\mu\rangle$. The higher value of $\langle\mu\rangle$ forces the greater effect of FS deformation. The q parameter can be estimated using Mott theory[6]: $q = \tau_\uparrow/\tau_\downarrow = N_{d\downarrow}(E_F)/N_{d\uparrow}(E_F)$. If the electronic 3d-band structure of the cobalt-based alloys is considered using the SBM model[7] and the ellipsoidal deformation of FS is taken into account, this theory predicts maximum of $\Delta\varrho/\varrho_o$ only in the alloys containing Mn[5]. This maximum is connected with crossing zero value by the linear magnetostriction constant λ_s. The alloys containing Cr and V do not exhibit maximum of $\Delta\varrho/\varrho_o$, even if the λ_s reaches its zero value.

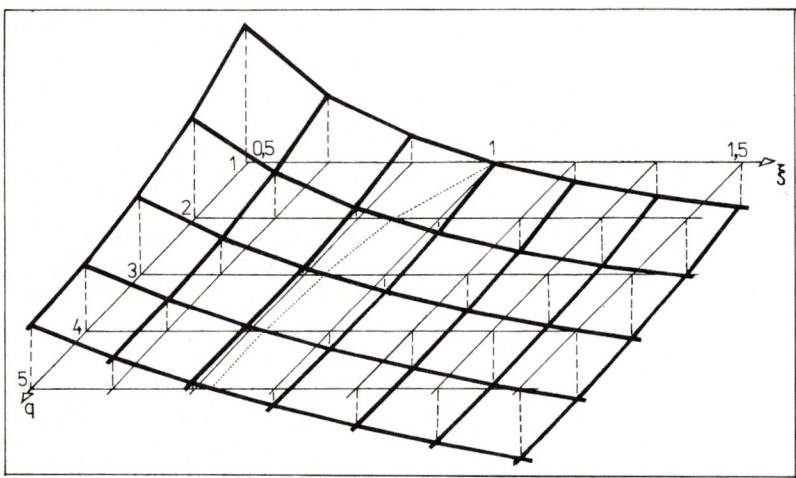

Fig. 1. The anisotropic magnetoresistance versus the parameters ξ and q (calculated for small magnetic field $\omega_c\tau = 0.01$).

The theoretical consideration was confirmed by an experiment. Under investigation the following alloys was taken: $Co_{80-x}Mn_xB_{20}$, $Co_{78-x}Mn_xB_{20}$, $Co_{80-x}Cr_xB_{20}$ and $Co_{80-x}V_xB_{20}$. Magnetoresistance was measured using specially build system. In order to have high sensitive experimental data (of an order of at least 10^{-3}) we applied AC compensational method. This method is intendent for the measurement of the resistivity changes of samples caused by different external conditions. The main idea of this method consist in a compensation of the difference voltage taken from a sample by the reference signal being not phase shifted, The compensated signal being measured by the LOCKIN-nanovoltmeter is separated from the background noise. The application of LOCKIN additionally allows to record experimental data thanks to analog output. The fig. 2 shows the measurement system which is controlled by the microcomputer. The linear magnetostriction for the investigated alloys was checked by the "three terminal" capacitance method[8]. The fig. 3 presents anisotropic magnetoresistance for the investigated alloys compared with linear magnetostriction constant λ_s.

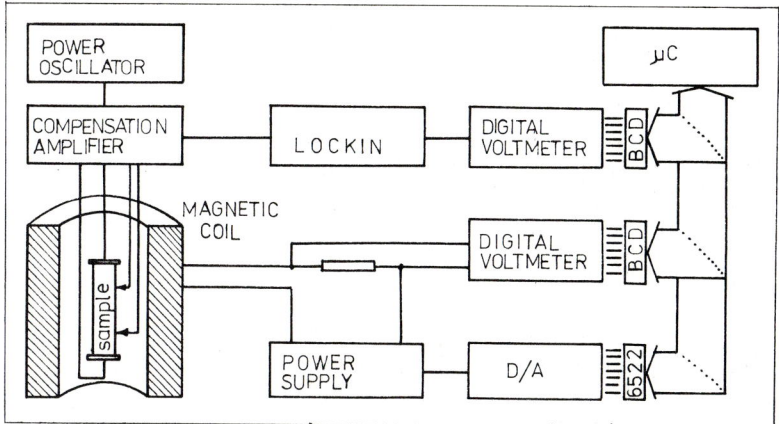

Fig. 2. The system for magnetoresistance measurement.

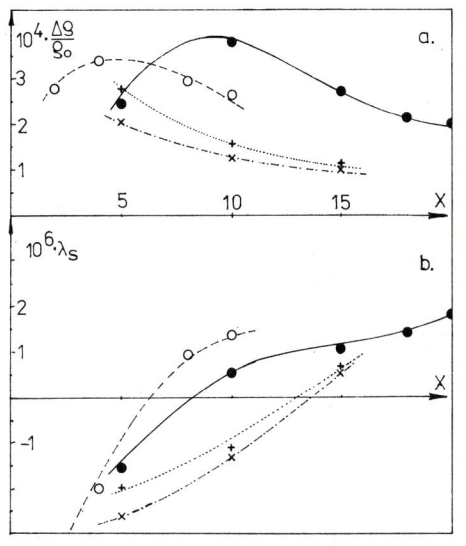

Fig. 3. Anisotropic magnetoresistance (a) and linear magnetostriction constant (b) versus the contents of early transition metal.

—●— $Co_{80-x}Mn_xB_{20}$

---o--- $Co_{78-x}Mn_xB_{22}$

········+········ $Co_{80-x}Cr_xB_{20}$

—·×—·— $Co_{80-x}V_xB_{20}$

REFERENCES

1) H.C. van Elst, Physica 25(1959)708; I.A.Campbell,J.Phys.F4(1974)L181
2) I.A.Campbell,A.Fert,O.Jaoul,J.Phys.C Suppl.1(1970)S95
3) A.P.Malozenoff,Phys.Rev.B Rap.Comm.Vol.32,No.9(1985)6080
4) T.E.Faber,J.Ziman,Phil.Mag. 11(1965)153
5) to be published
6) N.F.Mott, Adv.Phys. 13(1964)325
7) R.C.O'Handley in "Amorphous Metallic Alloys",Butterwords Monographs,p.257
8) N.Tsuya et al., Phys.Stat.Sol.31A(1976)557.

LOCALIZATION IN AMORPHOUS Nb/Ni MULTILAYERS

M.T. PÉREZ FRIAS and J.L. VICENT
Dept. Física Materia Condensada, Facultad de Ciencias Físicas,
Universidad Complutense, 28040 Madrid, Spain

1. INTRODUCTION

Advances in thin-film deposition technology have motivated interest in artificially prepared materials exhibiting different physical properties than those that occur naturally[1]. We present data on an artificially layered metallic system Nb/Ni. Small angle X-ray diffraction patterns and resistivity measurements show the clear layered character of the samples, with small angles peaks[2] and anomalies in the resistivity[3], strongly depended on layer thickness.

2. EXPERIMENTAL

Amorphous Nb/Ni multilayers have been obtained by vacuum deposition technique using a triode sputtering system. Corning glasses were used as substrates. The individual thickness of each layer, λ_{Nb} and λ_{Ni} varies from 20 to 250 Å with total film thickness of about 2000 Å. The system is initially pumped down to a vacuum of about 10^{-7} torr and typically films are deposited at Argon pressures of 0.5 mtorr. The deposition rate is about 1 Å / sec for both metals.

The alternative deposition of two metals by sputtering produces sometimes the appearance of superstructure in the growing direction, perpendicular to the film plane. These metallic superlattices can be characterized with X-Ray diffraction. Under certain conditions, the interface between the two materials is so perfect that diffraction peaks appear for small angle, namely for $D = \lambda_{Nb} + \lambda_{Ni}$ being D the corresponding to the periodicity of the superstructure, as it can be easily deduced from the Bragg's law: $2 D \sin \theta = n \lambda$.

Standard D.C. four probe technique was used for resistance measurements at low temperatures down to 5 K. The samples were photolithographed with the appropiate pattern and the electrical contacts were made using indium.

3. RESULTS

Fig. 1 and 2 show the X-Ray diffraction patterns $\theta \neq 2\theta$ for a superlattice of 22 Å Nb/ 22 Å Ni at high and small angle respectively. In figure 1, the only structures are due to the glass substrate and an incipient (1,1,1) texture for the Ni layers. Samples grown under the same conditions with layer thickness higher than 70 Å exhibit (1,1,1) texture for Ni and (1,1,0) for Nb layers. In figure 2, it can be seen a sharp small angle peak which corresponds to D = 44 Å and reflects the good layering of this sample. The coherence length is 530 Å estimated from the Scherrer expression. Rocking curve in the inset of the same figure gives mosaic spread of Δω ≃ 0.1º.

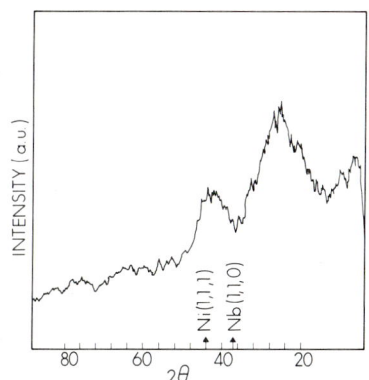
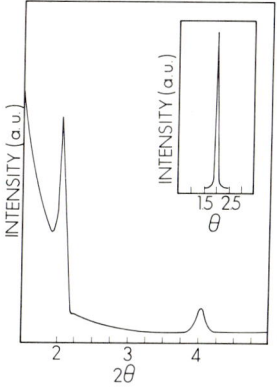

FIGURE 1
X-Ray diffraction pattern (high angle) for 22Å Nb/22Å Ni superlattice.

FIGURE 2
X-Ray diffraction pattern (small angle) for a 22Å Nb/22Å Ni superlattice. Rocking curve (inset)

Resistivity measurements are shown in figure 3. The influence of the individual thickness of the layers is clearly shown. The results for five different samples are plotted up to 50 K. For the samples with the thinner layer thicknesses, the resistivity shows a minimum around 20 K. Samples with layers thicker than 70 Å have a normal metallic behaviour. The temperature dependence of the resistivity in the nonmetallic regime for samples with thinner layer thicknesses can be due to incipient weak localization mechanism[4] or to the so-called interaction effects[5]. Both mechanisms predict a logarithmic behaviour of the temperature dependence of the resistivity at low enough temperatures. Such a behaviour is shown in figure 4 for temperatures below the minima.

Hall effect and magnetization SQUID measurements in these samples show ferromagnetism down to λ_{Ni} = 20 Å. So far we know, there are some controversies about the minima in disorder ferromagnetic materials[6,7]. Nb/Ni multilayers correspond to such a case: they are disorder ferromagnetic and present minima below 70 Å. Therefore an analysis based in usual theories[4,5]

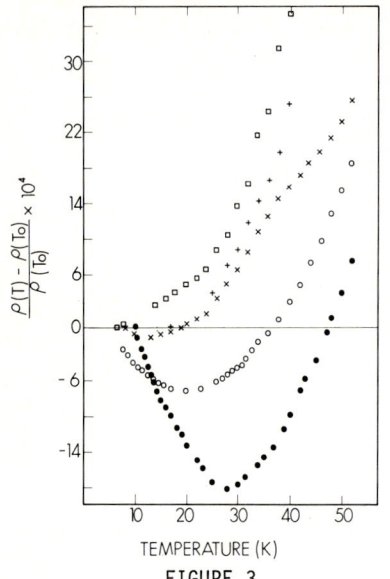
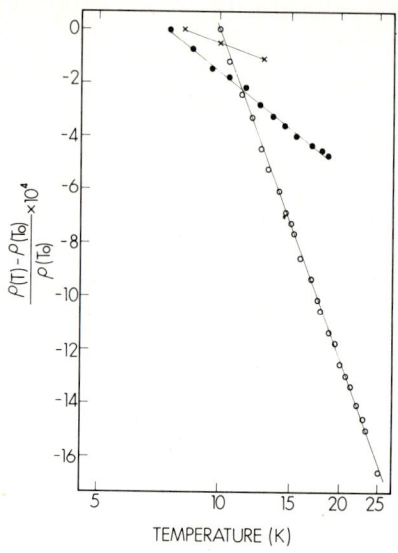

FIGURE 3
Reduced resistivity versus temperature for the following Nb/Ni superlattices: ● 20/20, ○ 30/30, ✕ 50/50, + 100/100, □ 250/250.

FIGURE 4
Reduced resistivity versus temperature in logarithmic scale for the ○ 20/20, ● 30/30, ✕ 50/50 Nb/Ni superlattices.

could not be done. Magnetoresistance and Hall effect allow to distinguish between Coulomb correlations[8] and Anderson localization[9], but the spin-disorder scattering in ferromagnetic samples obscures the analysis of galvanomagnetic effects.

REFERENCES

1) L.L. Chang, B.C. Giessen, Synthetic Modulated Structures (Academic Press, 1985).

2) E.M. Gyorgy, D.B. McWhan, J.R. Dillon, L.R. Walker, J.V. Waszczak, Phys. Rev. B 25 (1982) 6739.

3) T.R. Werner, I. Banarjee, Q.S. Yang, C.M. Falco, I.K. Schuller, Phys. Rev. B 26 (1982) 2224.

4) P.W.Anderson, E.Abrahams, T.V.Ramakrishnan, Phys.Rev.Lett.43 (1979) 718.

5) B.L. Altshuler, A.C. Aranov, P.A. Lee, Phys. Rev. Lett. 44 (1980) 1288.

6) R.W. Cochrane et al. Phys. Rev. Lett. 35 (1975) 676.

7) G.S. Grest, S.R. Nagel, Phys. Rev. B 19 (1979) 3571.

8) P.A. Lee, T.V. Ramakrishnan, Phys. Rev. B 26 (1982) 4009

9) A. Kawebata, Sol. St. Comm. 34 (1980) 431.

MAGNETORESISTANCE IN AMORPHOUS $Fe_{81.5}B_{14.5}Si_4$

J. FLORES and J.L VICENT*

Departamento de Física (Fac. C. Químicas), Universidad de Castilla la Mancha, 13071 Ciudad Real, Spain

* Laboratorio de Magnetismo (Fac. C. Físicas), Universidad Complutense, 28040 Madrid, Spain

INTRODUCTION

This study is a completion of other studies carried out, from the standpoint of the properties of magnetic anisotropy and magneto-elastic properties of this metal glass[1,2]. On the other hand, it provides fresh data for the studies carried out on other amorphous ferromagnetic substances.

EXPERIMENTAL METHOD

For temperatures between 4,2 and 77 K a superconducting solenoid of 7 Teslas was used as the magnetic field source. For temperatures between 77 and 300 K, we used equipment based on a copper coil cooled by liquid nitrogen which produces field impulses controlled by a micro-computer[3].

The magneto-resistance was measured using equipment designed by us based essentially on a differential amplifier connected to an adding device compensating the signal provided by the resistivity, and only the alterations produced by the activation of the magnetic field remained at the ouput. In the copper coil equipment the measurement was controlled by a micro-processor. The signal is mesured immediately prior to the field impulse and during the field impulse, with a time difference of 0.1 sec. This procedure allows the magnetic field effect to be separated from other alterations in the signal, which are slower.

RESULTS

Both variables taken into consideration for the magneto-resistance study are temperature and the angle of direction of the intensity with regard to the magnetic field. At all temperatures the transversal magneto-resistance $\Delta\rho\perp/\rho$ (90º angle) was measured and the longitudinal magneto-resistance $\Delta\rho"/\rho$ (0º angle). From these measurements, the anisotropy of the resistance $(\rho" - \rho\perp)/\rho$ for low fields was obtained.

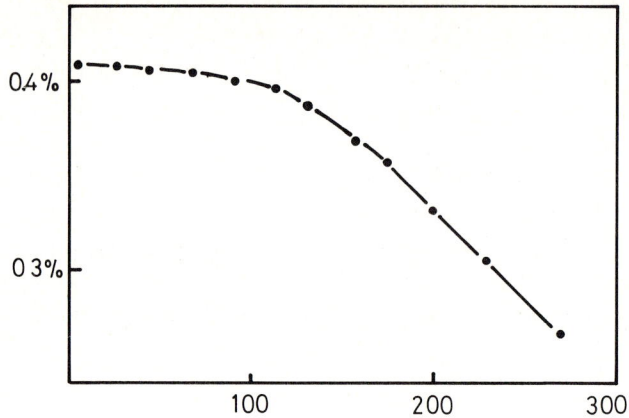

Fig. 1 Anisotropy of the resistivity versus temperature
X axis: Temperature (K)
Y axis: Anisotropy of the resistivity (%)

Figure 1 shows the alteration in the anisotropy with the temperature. For low temperature 50 K the anisotropy has a constant value (as indicated for these cases by R.W. Cochrane)[4]. For higher temperatures it was reduced in a similar manner to that observed in other amorphous ferromagnetic substances[5,6].

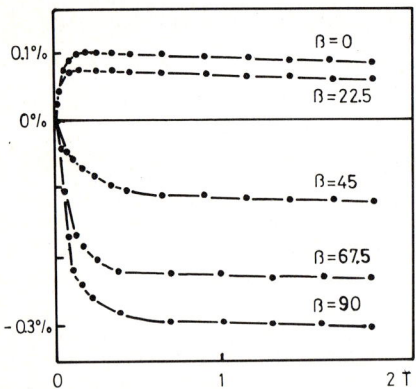

Fig. 2 Magneto-resistance grafs for different angles (ß) at 77 K
X axis: Magnetic field (Teslas)
Y axis: Magneto-resistance (%)

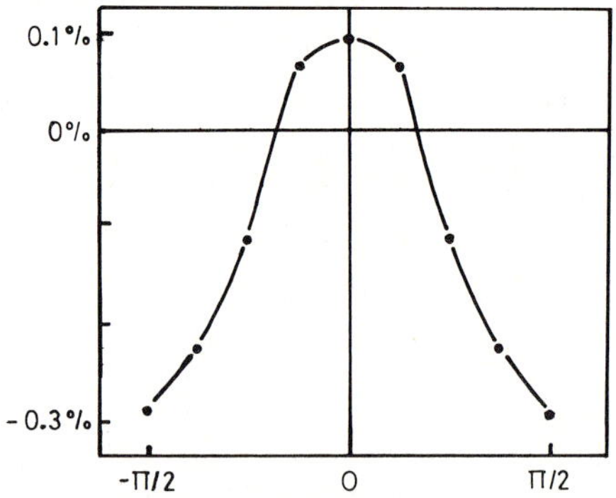

Fig. 3. Magneto-resistance for 77 K and 0.5 T versus the electric current-magnetic field angle.
X axis: Angle (Radian)
Y axis: Magneto-resistivity (%)

Figure 2 shows an example of the magneto-resistance graphs obtained for different angles, and in particular for a temperature of 77 K. In these graphs two types of slopes can be observed. For high fields a small negative slope can be seen, which does not depend on the angle. In amorphous ferromagnetics, positive and negative slopes habe been observed. The origin of positive is not clear[7]. Negative slopes such as the ones found in our case, are due to less electron-magnon scattering with increasing applied mangetic field[6,7].

Figure 3 shows the magneto-resistance based on the angle for 0.5 T and 77 K. The behaviour with regard to the angle for low fields corresponds to the normal anisotropy of magneto-resistance observed in ferromagnetic materials, both crystalline and amorphous. Recently, A.P. Malozemoff[8] has modified the theory of Campbell-Fert-Jaoul[9,10] for dilute crystalline alloys to be applied in the type of materials we are discussing here. The lack of specific heat data in a-Fe 81.5 B 14.5 Si 4 does not permit us to study our data from the point of view of this theory.

REFERENCES

1. J.M.Gónzalez, J.L. Vicent. J. Appl. Phys., 57 (1985) 5400
2. M. Liniers. Tesis Universidad Complutense 1987
3. J. Flores, J.L. Vicent. Rev. Sci. Inst. (J. Phys. E.) to be published
4. R.W. Cochrane, J.O. Strom-Olsen. J. Phys. F 7 (1977) 1799
5. K.V. Rao. "Amorphous Metallic Glasses" Ed. F.E. Luborsky (Butterworths) 1983
6. A.K. Majundar, A.K. Nigan. J. Appl. Phys. 51 (1980) 4218
7. A.K. Nigan, A.K. Majundar. Physica 95B (1978) 385
8. A.P. Malozemoff. Phys. Rev. B33 (1985) 6080
9. A. Campbell, A. Fert, O. Jaoul. J. Phys. C. Suppl. 1, S95 (1970)
10. O. Jaoul, I.A. Campbell and A. Fert. J. Magn. Magn. Mater. S (1977) 23

AMORPHOUS MELT SPUN Co-BASED ALLOYS WITH HIGH METALLOID CONTENT:
THERMAL STABILITY, MAGNETIC AND ELECTRICAL PROPERTIES STUDIES

M. PONT[+], K.V. RAO[+] and A. INOUE[§]
[+] Department of SSP, Royal Institute of Technology, Stockholm, Sweden
[§] Visiting scientist from Toyoku University, Sendai, Japan

Co-based amorphous alloys with metalloid content (B and C) up to 40 % have been produced by melt spinning. The metalloids are found to enhance the stability of these glassy alloys. The magnetic phase diagram as well as the electronic properties are discussed.

1. INTRODUCTION

The role of metalloids in enhancing the glass formation and thermal stability of amorphous alloys is a topic of considerable importance. We have studied this phenomenon in Co-based alloys containing as much as 40 at % B and C. These concentrations of metalloid extend well over the eutectics for this system.

In this paper we present our studies on the thermal stability, Curie temperatures as well as Hall and electrical resistivities for the Co-B-C system.

Special attention has been paid to understand the role metalloids play in the physical properties of these alloys.

2. EXPERIMENTAL

Two series of amorphous Co-based alloys were produced by rapid quenching from the melt using the melt spinning technique. The two series, $Co_{76-x}B_{24}C_x$ with x=4,8,12,16 and 20, and $Co_{88-y}B_yC_{12}$ with y=12,16,20,28 and 32 were examined and confirmed to be amorphous by x-ray diffraction.

Ribbons in general are about 0.7 mm wide and thickness, estimated from the density-weight-lenght method, ranges from 7 to 11 µm.

The thermal stability of the alloys were studied by differential scanning calorimeter using Perkin-Elmer DSC-4 equipment.

The Curie temperature was determined using a Thermogravimetry Analizer (TGS-Perkin-Elmer). If the sample is subjected to a low magnetic field then the temperature dependence of the magnetization can be determined accurately.

Hall and electrical resistivities were measured using the double ac-technique (1), i.e. a method in which adopting two different low frequency dependencies for the applied ac-field and ac-current provides a possibility to separate the Hall and Ohmic voltages unambiguously.

3. RESULTS AND DISCUSSION

a) Crystallization temperatures

The crystallization temperatures, defined as the onset of the crystallization peaks in the DSC scans, are show in Fig 1 for all the alloys. In both the series, for the alloy with 56 at % Co the structural transformation occurs at temperatures above the upper limit (873 K) of our DSC. As seen in Fig 1, Tx increases with increasing the metalloid content. It is clear that increasing metalloid content stabilizes the glassy phase even more.

The effective activation energy defined as $Q \equiv E/n$ for the crystallization process has been calculated using the Kissinger method (2) in which,

$$- Q/R = d(Ln(\dot{T}/Tp^2))/d(1/Tp)$$

Here Tp is the peak temperature and $\dot{T}$ (10, 20, 40 and 80 K/min) the heating rates.

Following the same trend as Fig 1, it is found that initially Q decreases with Co content for both the series. The evaluated activation energies are given in Table I. Even though some of the alloys crystallize in more than one phase we only present in this paper results for the first stage of the Co-B-C system.

b) Curie temperatures

Fig 2 shows the Tc determined for the different alloys as a function of the Co concentration. Tc in both series decreases monotonically with decreasing Co.

The concentration dependence of Tc is found to obey a $(x-x_0)^{1/2}$ behaviour in both the series, where x_0 is the critical concentration at which ferromagnetism disappears. x_0 is found to be ≈54 at % Co for both the series. Such a concentration dependence has been also reported for Co-B alloys (3). Although the exponent (1/2) can be indicative of a dilution process, this simple model can not explain the loss of ferromagnetism at concentrations of Co as high as 54 at %.

From a magnetic point of view it seems that B and C play a similar role when alloying with Co.

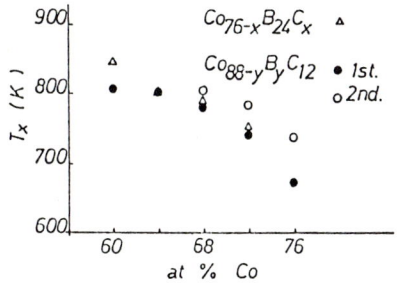

Fig 1. Temperature of the onset of crystallization peaks at heating rate of 20 K/min for Co-B-C alloys

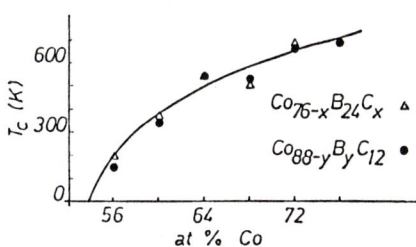

Fig 2. Magnetic phase diagram for the Co-B-C system

c) Hall effect and electrical resistivity measurements

Our preliminary results in Hall and electrical resistivities are presented in table I. The monotonic concentration dependence of the Hall coefficient R_H for all the alloys is shown in Fig 3. The positive values for R_H arises from the magnetic contribution to the Hall coefficient. The resistivity data, measured at room temperature show a large increase with decreasing the Co content. This increase is found to be much more pronounced in the $Co_{88-y}B_yC_{12}$ series.

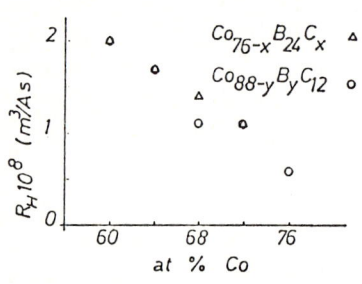

Alloys with 56 at % Co are found to be non-magnetic at room temperature. For these alloys we have also measured the temperature dependence of the Hall and Ohmic resistivity. Hall resistivity shows the expected sharp drop at the ferromagnetic transition temperature. These results will be discussed in detail elsewhere.(4). The temperature coefficient for the electrical resistivity is found to be positive for both the alloys. It is $\simeq 0.2 \cdot 10^{-4}$ ΩmK^{-1} for $Co_{56}B_{24}C_{20}$ and $\simeq 0.4 \cdot 10^{-4}$ ΩmK^{-1} for $Co_{56}B_{32}C_{12}$.

This work is supported by the Swedish Technical Board for Research and Development.

x	Tx (K)	E/n (eV)	Tc (K)	ρ (μΩcm)	R_H (10^8 m³/As)
4	751	3.25	680	117	1.1
8	791	3.00	509	117	1.4
12	803	4.50	539	130	1.7
16	845	5.20	366	152	2.0
20	—	—	193	162	—

x	Tx (K)	E/n (eV)	Tc (K)	ρ (μΩcm)	R_H (10^8 m³/As)
12	671	2.80	681	162	0.6
16	740	3.50	665	157	1.1
20	782	4.20	527	143	1.1
24	803	4.50	539	130	1.7
28	809	4.00	377	263	2.0
32	—	—	143	438	—

Table I. Crystallization temperatures, activation energy, Curie temperatures, electrical resistivity and Hall coefficient for the Co-B-C system
a) $Co_{76-x}B_{24}C_x$
b) $Co_{88-y}B_yC_{12}$

REFERENCES
1) R.Malmhäll, PhD Thesis, University of Umeå, (Sweden), 1980
2) H.E.Kissinger, Annal. Chem. 29 (1957) 1702
3) N.S.Kazama, Y.Masumoto and M.Mitera, J.Magn.Magn.Mat. 15-18 (1980) 1331
4) M.Pont and K.V.Rao, to be published

STUDY OF THE STRESS DEPENDENCE OF THE HYSTERESIS LOOP AND MAGNETOSTRICTION OF DIFFERENT "ZERO-MAGNETOSTRICTIVE" AMORPHOUS RIBBONS

R. GRÖSSINGER, A. PÖNNINGER and G. HERZER*
Institut f. Experimentalphysik, Techn. University Vienna, Austria
* Vacuumschmelze, Hanau, Germany

1. INTRODUCTION

"Zero-magnetostrictive" amorphous ribbons are ideally suited for studying short range order effects by means of magnetostriction measurements. The magnetostriction was found to consist of a negative single ion contribution and a positive two ion contribution. In the present paper the effect of substituting Mn in such ribbons in the as cast state but also applying a heat treatment is studied.

2. EXPERIMENTAL

Ribbons of the composition $Co_{75-x}Fe_1Mn_xMo_1Si_{14}B_9$ ($2 \leq x \leq 7$) were prepared by the single roller technique by Vacuumschmelze. All samples were studied in the as cast state. The ribbon with $x = 4$ was heat treated between 200°C and 450°C for 1 h in air. All measurements were performed at room temperature. The magnetostriction was determined using the SAMR (Small Angle Magnetization Rotation) method (1). Additionally the stress dependence of the hysteresis loop was measured in order to determine λ_s with the Becker-Kersten method (2).

3. RESULTS AND DISCUSSION

Table 1: Comparison of λ_s values determined with the SAMR method and with the Becker-Kersten method of the ribbons $Co_{75-x}Fe_1Mn_xMo_1Si_{14}B_9$.

x	2	4	5	7
$\lambda_s(0)$E-6	-1.4	-0.2	-0.15	+1
λ_s(hyst)	-1.6	-0.2	-0.15	-
J_s(T)	0.84	0.82	0.82	0.77
$d\lambda_s/d\sigma$ (MPa)$^{-1}$	-	-3.5E-10	-3.2E-10	

$\lambda_s(0)$ and $d\lambda_s/d\sigma$ were obtained by using the relation $\lambda_s = \lambda_s(0)+(d\lambda_s/d\sigma)\sigma$ for SAMR measurements. λ_s(hyst) was determined from the stress dependence of the hysteresis applying the Becker Kersten method. J_s(T) is the saturation polarization in Tesla.

The ribbons with x = 4 and x = 5 (showing the smallest λ_s values) exhibit a stress dependence of λ_s. A similar behaviour was found recently (3,4). It is believed that this stress dependence of λ_s is a consequence of a stress induced anisotropy which can change the measuring values (from which λ_s is calculated) if an indirect method based on the linear dependence of the magnetoelastic anisotropy on the applied stress is used. Therefore the extrapolated λ_s value at zero stress ($\lambda_s(0)$) is the correct one (5).

The ribbon with x = 4 was annealed at various temperatures and then measured again.

Table 2: Results as obtained on a heat treated ribbon of the composition $Co_{71}Fe_1Mn_4Mo_1Si_{14}B_9$.

T (C)	--	200	250	300	350	400	425	450
$-\lambda_s(0)$E-8+	38	30	26	24	43	55	62	68
$-\lambda_s$(VAC)	51	37	27	--	38	56	63	72
$d\lambda_s/d\sigma$ E-10+	-3	-3	-4	-4	0	0	0	0
$d\lambda_s/d\sigma$ E-10	-2.2	-3	-3.8	--	-2.5	0	0	0
H_c(A/m)	20	17	15	11	13	21	30	31

T.... annealing temperature. +.....values obtained by the SAMR method. The (VAC) data are results determined from the stress dependence of the hysteresis loop. The ribbons used for both methods were not identical! H_c...coercivity field obtained at σ = 200 MPa.

Up to an annealing of 300°C λ_s decreases (and is stress dependent) then it increases again. It is well known that in soft magnetic material where the magnetoelastic energy is dominant the coercivity is strongly influenced by the magnetostriction. In a mathematical sense this can be expressed as (6):

$$H_c(\sigma) = M_r(\infty)(-3\lambda_s/\mu_0 M_s)\sigma$$

This fact becomes visible in fig. 1 where the dependence of the coercivity and the magnetostriction on the annealing temperature is shown. The character of both curves is similar.

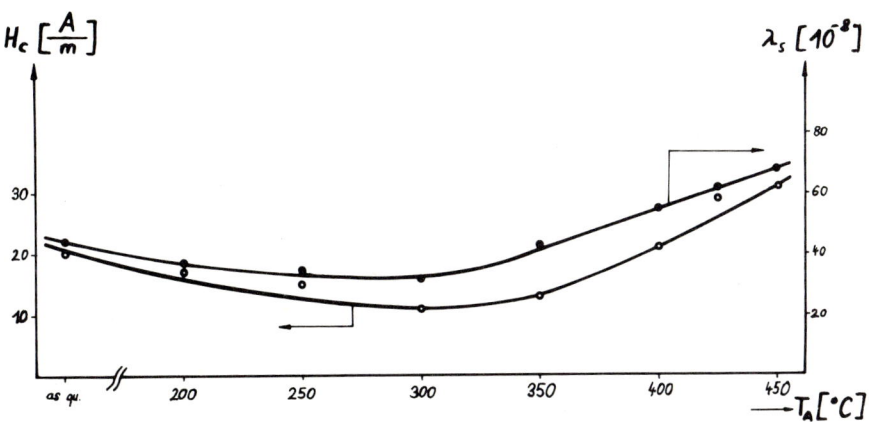

Fig. 1: Dependence of the magnetostriction λ_s and the coercivity H_c on the annealing temperature T_A.

ACKNOWLEDGEMENT:

This work was supported by the Austrian "Fonds zur Förderung der wissenschaftlichen Forschung", proj. number: P-5020.

REFERENCES

1) K. Narita, J. Yanasaki, H. Fukunaga, IEEE Trans. on Magn. MAG-16 (1980) 435

2) R. Becker, K. Kersten, Z. Phys. 64 (1930) 660

3) G. Herzer, Proc. Conf. on Stoft Magnetic Mat. 7, Blackpool (1985); Ed. Wolfson Centre, Cardiff, England

4) J.M. Baradiaran, A. Hernando, V. Madurga, O.V. Nielson, M. Vazquez I, M. M. Vazquez II, Phys. Rev. B 35 (1987) 5066

5) A. Siemko, H.K. Lachowicz, JMMM 66 (1987) 31

6) M. Vazquez, M. Fernengel, H. Kronmüller, phys. stat. sol. (a) 80 (1983) 195

CREEP-INDUCED MAGNETIC ANISOTROPY OF METALLIC GLASSES

Luděk KRAUS, Niva ZÁRUBOVÁ, Karel ZÁVĚTA and Pavol DUHAJ[*]

Institute of Physics, Czechoslovak Academy of Sciences, Na Slovance 2, CS - 180 40 Praha 8, Czechoslovakia
[*]Institute of Physics, Slovak Academy of Sciences, Dúbravská cesta 9, CS - 842 28 Bratislava, Czechoslovakia

Creep-induced anisotropy of amorphous $Fe_{80}Cr_2B_{14}Si_4$ and $Co_{55}Fe_5Ni_{14}B_{10}Si_{16}$ alloys was studied. The anisotropy constants of the two alloys are of the opposite signs. The kinetics of induction and recovery of K_u shows that only anelastic strain is responsible for the anisotropy of the Fe-rich alloy.

1. INTRODUCTION

Mechanical creep deformation is known to induce magnetic anisotropy. Nielsen[1] proposed that the creep-induced anisotropy can be divided into two parts: (1) K_{an} which saturates at large stress-annealing times t_a and is recoverable by subsequent annealing without stress, and (2) K_{pl} which increases steadily with t_a and is non-recoverable. Haimovich et al.[2] have demonstrated that anelastic polarization produces structural anisotropy which results in the magnetic anisotropy K_{an}. The question about the existence and the origin of the plastic contribution K_{pl} is still open. We report here the first systematic study of the creep-induced anisotropy of an Fe-rich amorphous alloy.

2. EXPERIMENTAL

The amorphous ribbons of $Fe_{80}Cr_2B_{14}Si_4$ and $Co_{55}Fe_5Ni_{14}B_{10}Si_{16}$ were prepared by the single roller liquid quenching technique. The annealing treatments performed in air atmosphere were divided in three steps: (1) pre-annealing for the time t_p at the temperature T_p followed by (2) stress-annealing for t_a at T_a under applied stress σ, finally some samples were (3) stress--relaxed for t_r at T_r. The induced anisotropy was measured on circular discs spark-cut from the stress-annealed ribbons by the method of biased transverse susceptibility. Domain structures were observed by means of JEOL 733 Superprobe SEM on shiny sides of the ribbons.

3. RESULTS AND DISCUSSION

In both alloys a rather large anisotropy can be induced by stress-annealing at temperatures well above T_c (see Figs. 3 and 4). This proves that the magnetoelastic coupling can not be the main cause of the anisotropy.

The anisotropy constants of the two alloys are of the opposite signs. While the easy axis of the Co-rich alloy is perpendicular to the tensile stress applied during the stress-annealing, the easy axis of the Fe-rich alloy lies parallel to it. The domains are perpendicular or parallel to the ribbon axis in these two cases as is shown in Figs. 1 and 2.

K_u of the Co-rich alloy (Fig. 3) steeply increases with the stress-annealing temperature T_a between 200 and 300°C and then becomes proportional to T_a (dotted line). If the anisotropy is anelastic in nature[2], then

$$K_u(t_a) = K_u(\infty)\, g(\sigma, T_a, t_a), \qquad (1)$$

where $g(\sigma,T,t)$ is the relaxation function $(0 < g < 1)$. At low temperatures the change of K_u is controlled mainly by the kinetics of the process $(g < 1)$, while at higher T_a within $t_a = 1$ h the saturation value $K_u(\infty)$ proportional to T_a is nearly reached[3].

The dependence of K_u on the temperature T_a, applied stress σ, the pre-annealing conditions and the kinetics of its induction and recovery were investigated in more details for the Fe-rich alloy. The creep-induced anisotropy shows the typical anelastic behaviour: (1) K_u approaches some saturation value $K_u(\infty)$ at large stress-annealing times and (2) it is fully recoverable by subsequent annealing without stress. The temperature dependence of K_u (Fig. 4) is, however, quite different from that of the Co-rich alloy. The wider region of temperatures where the value of K_u is controlled by the kinetics of the process (from 200 up to about 400°C) implies that the relaxation times are higher and more spread in this alloy. The sharp decrease of K_u at high temperatures T_a may be caused rather by the stabilization of the amorphous structure during the pre-annealing at high $T_p \,(= T_a)$ than by the stress-annealing itself. It has been observed that

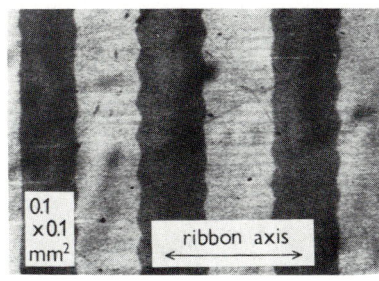

FIGURE 1
Domain structure of a stress-annealed $Co_{55}Fe_5Ni_{14}B_{10}Si_{16}$ ($T_p = T_a = 250°C$, $t = t_p = t_a = 1$ h, $\sigma = 860$ MPa)

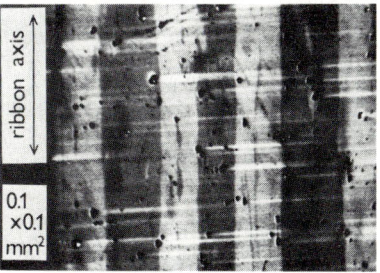

FIGURE 2
Domain structure of a stress-annealed $Fe_{80}Cr_2B_{14}Si_4$ ($T_p = T_a = 375°C$, $t_p = t_a = 1$ h, $\sigma = 720$ MPa)

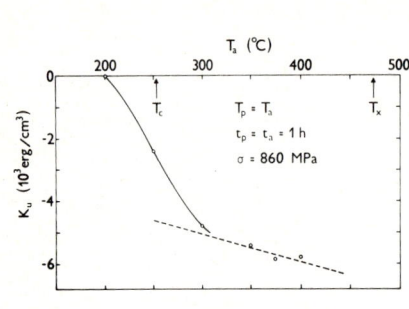

FIGURE 3
Temperature dependence of the creep-induced anisotropy of amorphous $Co_{55}Fe_5Ni_{14}B_{10}Si_{16}$

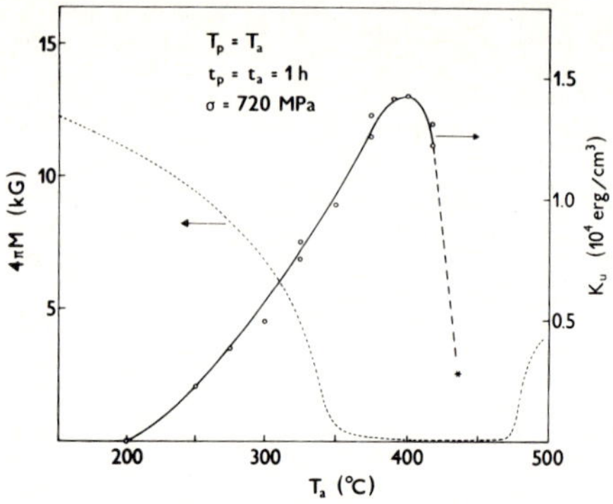

FIGURE 4
Temperature dependence of the creep-induced anisotropy of amorphous $Fe_{80}Cr_2B_{14}Si_4$

K_u considerably decreases after increasing the pre-annealing time t_p or the temperature T_p near the glass transition. At the present time we are not able to decide whether this decrease of K_u is caused by the decrease of $K_u(\infty)$ or by an extreme slowing down of the kinetics of induction.

The kinetics of induction and recovery of K_u are both linear in log t indicating a broad distribution of activation energies with the average value about 2 eV. The induction is more rapid than the recovery at the same temperature. The recovery cannot be described by a unique spectrum of activation energies because an appreciable dependence of its kinetics on the thermal history of the sample has been observed. We can thus conclude that the reversible relaxation processes responsible for the creep-induced anisotropy are influenced by some internal structural relaxation of the alloy.

The dependence of K_u on the applied stress σ is linear up to 1.1 GPa at the annealing temperature $T_a = 275°C$ but at higher T_a a deviation from linearity is observed. The stress dependence at $T_a = 375°C$ can be well fitted by a sinh-dependence theoretically predicted by Argon[4] for anelastic strain. From his model this fit corresponds to an activation volume of 87 $Å^3$.

REFERENCES

1) O.V. Nielsen, IEEE Trans. Mag. MAG-21 (1985) 2008.

2) J. Haimovich, T. Jagielinski and T. Egami, J. Appl. Phys. 57 (1985) 3581.

3) T. Jagielinski and T. Egami, IEEE Trans. Mag. MAG-21 (1985) 2005.

4) A.S. Argon, J. Appl. Phys. 39 (1968) 4080.

MAGNETIC AFTER-EFFECTS IN BINARY AMORPHOUS ALLOYS

Pavol VOJTANÍK

Dept. of Exp. Physics, Faculty of Sciences, P.J.Šafárik University
nám. Február. víťazstva 9, CS- 041 54 Košice, Czechoslovakia

The time dependence of reluctivity, $r = r_1 + \dot{r} \cdot \ln t$, has been investigated in Co-B and Fe-B alloys in the temperature range 150-450 K. It has been found that at higher temperatures and longer measuring times $\dot{r} = \frac{dr}{dt}$ makes change of its value in cobalt alloys as well as its sign in iron alloys. Amplitude dependences of AC permeability show perminvar behaviour. Dependences of stabilization field on magnetic polarization derived from them show the different presence of 180°- and 90°-domain walls and the difference in order of stabilization energy value in Co-B and Fe-B alloys.

1. INTRODUCTION

In all so far known amorphous soft magnetic alloys, rather intensive magnetic after-effects, related to the rearrangement of their quasistable amorphous structure, have been observed[1-3]. These magnetic after-effects are caused by magnetic relaxation of uniaxial point defects and their clusters in accordance with the principle of free energy minimum. During relaxation the actual positions of magnetic moments are stabilized. This causes the change in all magnetization characteristics of ferromagnetic bodies. In this work we study two typical manifestations of magnetic relaxations, reluctivity after-effect and perminvar effect.

2. EXPERIMENTAL

Time and amplitude dependences of AC susceptibility (reluctivity) have been investigated by means of a mutual inductance bridge with frequency $f = 970$ Hz in the temperature range 150-450 K. Amorphous samples having compositions $Co_{100-x}B_x$ (x= 18-30 at.% B) and $Fe_{100-x}B_x$ (x= 14-23 at.% B) were prepared by rapid melt quenching on one rotating disc in KFKI Budapest and in IFTT Chernogolovka, respectively. They looked like a bunch of 10 cm long and 1 mm wide amorphous ribbons.

3. RESULTS AND DISCUSSION

In the work we are dealing with the analysis of time dependences of the reluctivity $r = r_1 + \tilde{r} \ln t$, where $r_1 = r(t=1s)$, $\tilde{r}$ is a curve slope, t - time (Fig. 1). In case of $\tilde{r}$ = const., activation parameters of reversible magnetic relaxation were easily calculated[1,2], in agreement with later works[3]. In some cases, especially at higher temperatures and longer times, $\tilde{r}$ changes with respect to time. This fact indicates the presence of irreversible relaxation processes, which differ qualitatively in the Co-B and Fe-B alloys.

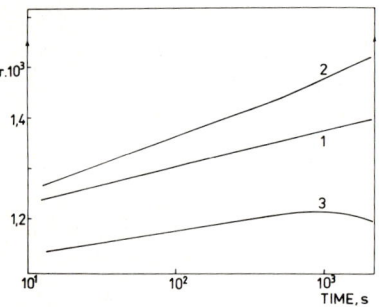 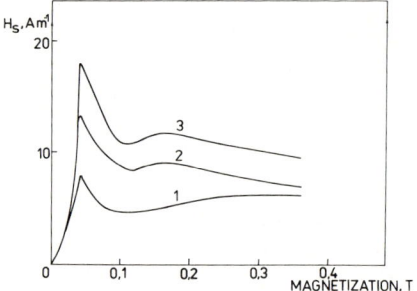

FIGURE 1
$Fe_{86}B_{14}$ - time dependences of reluctivity. Curve 1 - temperature T = 240 K, 2- 340 K, 3- 450 K.

FIGURE 2
$Co_{75}B_{25}$ - stabilization field versus magnetization, T=450 K. Curve 1- stabilization time t_s = 30 min., 2- 6 h, 3- 12 h.

In Co-B alloys $\tilde{r}$ was a constant only well below room temperature, above RT $\tilde{r}$ increased with respect to time. Repeated demagnetization has shown that the reluctivity increase is permanent, which proves the existence of irreversible magnetic relaxation. Boron concentration has influenced this process to a small extent. In Fe-B alloys $\tilde{r}$ behaves in some cases similarly to Co-B alloys, however, at higher temperatures there is a prevailing $\tilde{r}$ decrease as well as its gradual sign change (Fig.1). The lower boron concentration, the lower stability of samples. These results show that the relief of inner stresses, which is connected with free volume homogenization, becomes a dominant process in Fe-B alloys at high enough temperatures.

In the second part of the work, we have studied the AC susceptibility dependence on the amplitude of measuring magnetic field. Measured dependences show typical perminvar behaviours in both Co-B and Fe-B materials[4,5]. The susceptibility remains a constant up to a certain critical field value, connected with reversible domain walls motion throughout the potential well bottoms of parabolic shape. Having reached its critical value, the outer field pressure on domain walls becomes equal to the reversing pressure due to magnetic inhomogenities and magnetic relaxations. Domain walls motion becomes irreversible, as indicated by a rapid increase of susceptibility and magnetization.

Different results have been achieved in different alloy series. Critical fields in Co-B alloys reach the values up to 50 Am^{-1}, which are much higher values than their coercive force. In Fe-B alloys these values are of order 10 Am^{-1}, comparable with their coercive force.

From perminvar curves the stabilization field dependences on magnetization $H_s(I)^6$ were derived (Fig.2). Their shape is considerably different for different alloy types. On one hand, in Co-B alloys the H_s-field changes only slightly after reaching its maximum. In Fe-B alloys, on the other hand, H_s abruptly decreases to zero. This behaviour of H_s is related to the stabilization of different domain wall types^7. Accordingly, in Co-B alloys, mainly 90^0-domain walls are present, while in Fe-B alloys, the magnetic domais are separated mainly by 180^0- domain walls.

The area under the curves $H_s(I)$ determines the density of the partial stabilization energy E_s. This energy increases with logarithm of time, approximately linearly. As it has been found, there a big difference in the value of stabilization energies in different alloy types. While in Co-B alloys E_s reaches the values of the order 1 Jm^{-1}, in Fe-B alloys E_s is only of the order 0.1 Jm^{-1}. In Co-B alloys, the value of stabilization energy approx-imates the magnetization work for one hysteresis loop.

ACKNOWLEDGEMENT

The author is thankful to D.Macko and J.Petrík for their assistance at measurings.

REFERENCES

1) P. Vojtaník and I.B. Kekalo, Phys. stat. sol. (a) 60 (1980) K 45.

2) P. Vojtaník et al., J. Magn. Magn. Mat. 41 (1984) 385.

3) F. Rettenmeier and H. Kronmueller, Phys. stat.sol. (a) 93 (1986) 221.

4) P. Vojtaník and I.B. Kekalo, Acta phys. slovaca 31 (1981) 113.

5) P. Vojtaník et al., Acta phys. polonica   in print.

6) P. Allia and F. Vinai, IEEE Trans.Magn. MAG-17 (1981) 1481.

7) L. Néel, J. Phys. Rad. 13 (1952) 429.

ELASTIC PROPERTIES OF AMORPHOUS FERROMAGNETS

Kazuo KAMIGAKI, Shunya ABE and Hiroyasu FUJIMORI

Institute for Materials Research, Tohoku University,
Katahira, Sendai 980, Japan

Elastic properties of iron-base amorphous alloys are analyzed on the basis of the process of magnetization. A large change of elasticity at the intermediate state of magnetization is caused by a rotation process of the magnetic moment.

1. INTRODUCTION

A large change of elastic constant has been observed in the course of the magnetization process in some amorphous ferromagnetic compounds[1,2]. An analysis is given here on the elastic properties of amorphous ferromagnets in connection with the magnetization process in the external field[3,4].

2. ELASTIC CONSTANT

In ferromagnetic materials an additional change of elasticity is caused by a deformation through a change of magnetization when the external force is applied. The relative change of elasticity with magnetic field is defined here as $(\Delta E) = (E_s - E)/E_s$, where subscription s denotes the saturation magnetization.

The magnetization curve and the corresponding domain structure were observed in $Fe_{80}P_{13}C_7$ [5]: in thin ribbons, three dimensional magnetic domain structure was observed at the demagnetized state, a rotation of magnetization was observed at the intermediate state to the saturation magnetization. Hence, the magnetization process is analyzed in a similar way to the ordinary crystalline ferromagnets. In a simple case where the rotation of magnetic moment is

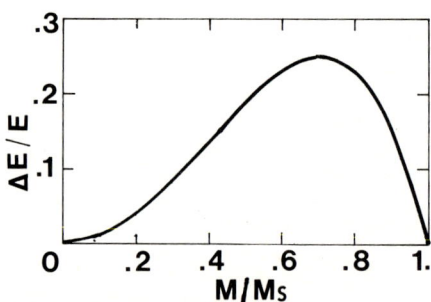

FIGURE 1   Change of elastic constant with relative magnetization.

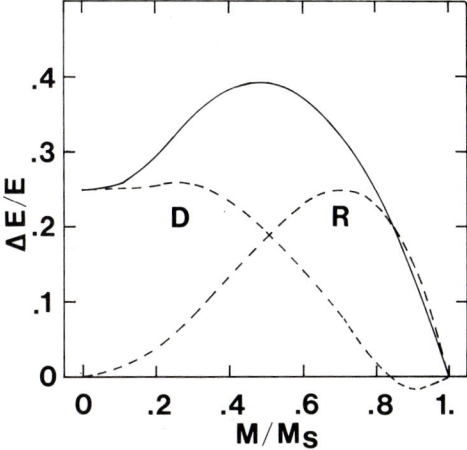

FIGURE 2   Change of elastic constant in $Fe_{75}P_{15}C_{10}$

a major process of magnetization, the elastic constant is expressed as a function of the relative magnetization M/M_s. The magnetization dependent part of (ΔE) is expressed as $(M/M_s)^2(1 - (M/M_s)^2)$. The value is zero for $M = 0$ and $M = M_s$, and has maximum at around $M/M_s = 0.7$, the feature is shown in Fig. 1.

The elastic constant measured in a ribbon of $Fe_{75}P_{15}C_{10}$ is shown in Fig. 2 as a function of the relative magnetization. (ΔE) has a maximum at around $M/M_s = 0.5$, and has a finite value at the initial state of the magnetization process. The evaluated $(\Delta E/E)$ in Fig. 1 is shown as a curve R and a difference to the observed curve is illustrated by a curve D. The result is interpreted as: at the initial stage of magnetization, 90° wall displacement is prefered as characterized by the curve D, then the rotation process takes place as shown by the curve R.

3. EFFECT OF MAGNETIC FIELD-COOLING

By annealing and cooling the ribbon in the magnetic field, a prefered distribution of the magnetic domains is expected. $Fe_{78}Si_{10}B_{12}$ -ribbon was cooled in the magnetic field transverse to the ribbon direction. The magnetization curve measured along the ribbon direction showed a typical character of the rotation of the moment. The elastic constant measured longitudinal to the ribbon is shown in Fig. 3. In the figure, $\Delta E/E$ has a maximum around $M/M_s = 0.7$ in good correspondence with the curve in Fig. 1 as shown by the curve R, remaining part D is smaller in comparison with R. The rotation of the magnetic moment occupies a overwhelming part in the magnetization process in this case.

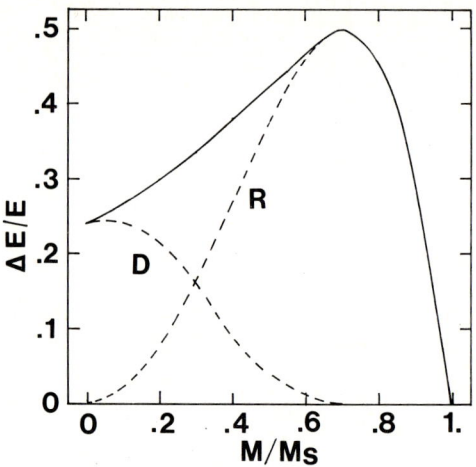

FIGURE 3 Change of elastic constant in $Fe_{78}Si_{10}B_{12}$.

The internal friction has been measured on the amorphous ferromagnets as a function of the relative magnetization[1]. The results were analyzed on the basis of the magnetization process[3], and a good accordance was obtained in the similar way.

In view of these results, it is concluded that the magneto-elastic properties of amorphous ferromagnets examined here are similar in principle with those of polycrystalline ferromagnets.

REFERENCES

!) B. S. Berry and W. C. Pretchet, J. Appl. Phys. 47 (1976) 3295.
2) K. I. Arai, N. Tsuya, M. Yamada and T. Masumoto, IEEE Trans. Mag. M-12 (1976) 936.
3) K. Kamigaki and S. Abe, Internal Friction and Ultrasonic Attenuation in Solids, ed. Hashiguchi (Univ. Tokyo Press., Tokyo 1977) 273.
4) B. S. Berry, Elastic and anelastic behavior, in Metallic Glasses, eds. Gilman and Leany (ASM, Ohio, 1978) pp. 161 - 189.
5) Y. Obi, H. Fujimori and H. Saito, Japan J. Appl. Phys. 15 (1976) 611.

ELECTRONIC STRUCTURE, BONDING AND MAGNETISM IN AMORPHOUS ALLOYS

R.C. O'HANDLEY, A. COLLINS, and M.E. McHENRY

Massachusetts Institute of Technology, Cambridge, MA 02139 USA

This paper combines a somewhat tutorial review of simple concepts of electronic structure, bonding and magnetism with a summary of some of the general aspects of those subjects observed experimentally and theoretically in amorphous alloys. We identify both covalent and polar components in the bonding of amorphous alloys and note the conditions for which bonding decreases the strength of magnetic interactions.

I. INTRODUCTION

Amorphous materials lack long range order but otherwise have appreciable chemical short-range order (CSRO) that often resembles that of a related equilibrium or metastable crystalline phase. In this paper we consider the effects on magnetism of the <u>chemistry</u> necessary for the stability of the glassy state. Specifically, we ask how does the presence of glass formers in transition metal-metalloid (T-M) alloys - and how does the mixing of two <u>different</u> transition metal species in T_1-T_2 alloys - affect magnetism. It is already well established that <u>technical</u> magnetic properties of amorphous and crystalline materials differ dramatically because microstructural inhomogeneities and magnetocrystalline anisotropy, factors that hinder the magnetization process, are absent in the former. The aspects of magnetism focussed on here are those determined by the electronic structure. Electronic interactions set the conditions for moment formation and thus determine the fundamental magnetic parameters, primarily magnetic moment μ_i per atom of species i and Curie temperature T_C, and secondarily magnetic anisotropy energy K and its strain derivative, magnetostriction λ.

The fundamental magnetic properties are determined by the electronic structure. In a metal, electronic structure is a very local phenomenon. Hence the fundamental magnetic properties of amorphous alloys do not reflect the absence of long-range order. Instead they generally resemble those of a phase with similar chemical and topological short-range order. So the magnetic properties of most amorphous alloys should be comparable to those of the corresponding crystalline phase or phases. Where no corresponding crystalline phase can be found, the magnetic properties may give clues to the nature of the short-range order.

II. Magnetism and Bonding

To provide a common background for the body of the paper, we outline briefly the principal interactions responsible for magnetism and then describe in simple terms the consequences of alloying and bonding on electronic structure and hence on magnetic interactions.

Four aspects of chemical short-range order can be identified as important to magnetism: the number, type, distance and symmetry of the nearest neighbors about a given site. We can illustrate this with the molecular field expression for the Curie temperature

$$T_C = J(r) \, Z \, S(S+1)/3k_B \qquad (1)$$

or the Stoner criterion for the existence of ferromagnetism

$$I(E_F)D(E_F) > 1 \qquad (2)$$

where $J(r)$ is the distance-dependent interatomic exchange integral, Z is the magnetic coordination, S is the spin quantum number, k_B is Boltzmann's constant $I(E_F)$ is the Stoner (intra-atomic) exchange parameter and $D(E_F)$ is the paramagnetic density of states (DOS) at the Fermi energy. The number, type and distance of nearest neighbors enter these expressions explicitly through Z, S and $J(r)$ and implicitly through $I(E_F)$ and $D(E_F)$. The symmetry of the nearest neighbor arrangement clearly affects $D(E_F)$ by changing the degeneracy of the electronic states. The nature of the bonding interactions among nearest neighbors strongly influences both $I(E_F)$ and $D(E_F)$. Local symmetry will also determine local anisotropy and magnetostriction.

The chemical bonding interactions that affect the electronic states and determine the physical properties can be most simply reduced to two types: polar bonds and covalent bonds.[1] Covalent bonds are formed between orbitals on two atoms A and C (anion and cation) that have similar electronegativities (i.e. similar electronic energies $E_A = E_C$), as well as satisfying symmetry and overlap conditions. In a covalent bond, charge is delocalized from each of the atomic sites and builds up between the atoms. Bonding and antibonding hybrid orbitals are created. Covalent bonds formed between partially occupied (i.e. near E_F) valence orbitals will often involve magnetic states. The d states become more delocalized as a result of covalent bonding. This delocalization results in a decrease in d-character (hence weaker intra-atomic exchange $I(E_F)$) and in a suppression of $D(E_F)$, both of which weaken moment formation (Eq. 2).

Polar bonds are formed between orbitals that differ significantly in their electronegativity ($E_A \neq E_C$). The orbitals must also satisfy symmetry and overlap conditions. In the formation of a polar bond, charge is transferred from the orbital of higher energy (lower electronegativity) to that of lower energy. As a result of this charge transfer the bond is biased toward the more electronegative species A. If a polar bond is formed in a metal, the conduction electrons will redistribute themselves to screen the charge transfer and maintain some degree of local charge neutrality. The screened polar bond still contributes to the chemical stability of the alloy. However it will only affect the magnetic properties if one of the orbitals involved contributes to the magnetism (e.g. a 3d orbital).

We will now illustrate these competing bonding and magnetic interactions in a number of amorphous alloys, keeping in mind that the effects are unique to amorphous alloys only in the sense that the chemical additions are necessary to stabilize the glassy state.

III. BONDING AND MAGNETISM

A. Cluster Calculations

The oldest and most important class of amorphous alloys are of the transition metal-metalloid type T-M. We begin by illustrating the basic principles of bonding and magnetism in simple Fe-M clusters (M = B, C, N, Si). We then compare the calculated and experimental state densities.

Messmer[2] applied the self-consistent-field, X-alpha, scattered-wave, molecular-orbital (SCF-Xα-SW-MO) method to tetrahedral clusters of Fe and Ni atoms adding central B and P atoms in order to study chemical bonding effects important in T-M metallic glasses. He found that the formation of metal-metalloid bonding states decreases the amount of d character about the T atoms and hence decreases the magnetic moment. These occupied d-(sp) hybrid states appear at lower energy than the corresponding states in the pure metal clusters, accounting for the stabilization upon alloying.

Recently Collins et al[3] have extended Messmer's work to gain a clearer insight into the difference among various metalloids in bonding with iron. Eigenvalues and wave functions in M-centered tetrahedral clusters Fe_4M (M = B, Si, C, N) are compared and shed new light on a number of aspects of electronic structure and magnetic properties of T-M alloys. The results are summarized in Fig. 1.

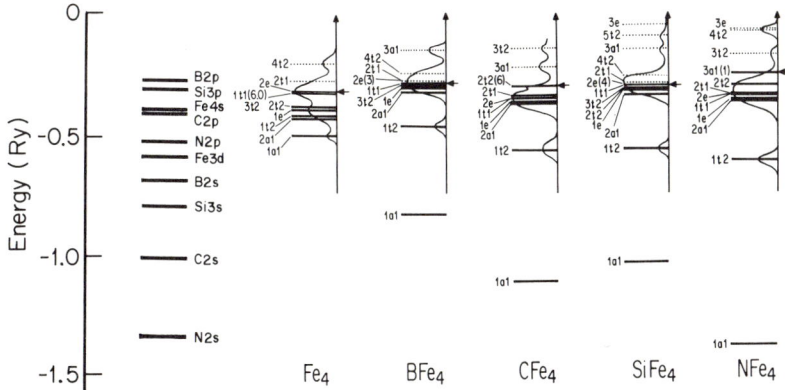

Fig. 1 Molecular orbital energy levels (electronegativities) and their Gaussian broadened cluster DOS for Fe$_4$ and Fe$_4$M clusters (M = B,C,N, and Si). At the far left the relevant free atom valence levels are shown (Ref. 3).

At the far left the free atom eigenvalues for the various unbonded species are shown. The extent of <u>covalent</u> mixing of states, when simulating an alloy by formation of an Fe$_4$M cluster, will increase as the difference in energy between states decreases. The degree of <u>polarity</u> in a bond will increase as the difference in electronegativity between the initial states increases. We thus expect that the four metalloid atomic p states immediately above the Fe 3d state will form a predominantly p-d covalent orbital, with the greatest covalent mixing taking place between Fe and N or C. The bonding between the Fe 3d level and the lower lying metalloid s states will be largely polar in character and again strongest for N and C whose s orbitals are most electronegative relative to Fe 3d.

The second column of Fig. 1 shows the molecular orbital eigenvalues for an empty Fe$_4$ cluster. States are labeled acording to the irreducible representations of the tetrahedral group. Solid lines indicate occupied states and dashed lines, empty states. Partial occupancy (near E$_F$ located by an arrow) is indicated by numbers in parentheses. The cluster DOS obtained by Gauussian broadening of cluster eigenvalues (which are really measures of electronegativity[4]) are drawn in lighter lines over the discrete states for comparison. The extent to which Fe$_4$ and larger iron cluster results are acurate and useful in understanding physical properties of surfaces and bulk solids are discussed elsewhere.[5] Allowing the spin degree of freedom to the cluster states results in a large moment for the Fe$_4$ cluster, calculated to be 3 μ_B/Fe, because the conditions in Eq. 2 are satisfied. Yang et al [5] have shown the importance of antibonding states at E$_F$ in obtaining large values of the Stoner exchange integral because of their greater localization relative to bonding states.

The effects of the various metalloids on bonding, paramagnetic DOS and D(E$_F$) are shown in the next four columns. Note that B and Si atomic p states lie well above the atomic iron 3d states and should therefore show only weak covalent hybridization. Calculated charge distribution contours show that the d and p electrons at E$_F$ tend to stay localized within their atomic spheres. (This electron localization holds less for the larger Si atom than for the smaller B atom). Consequently D(E$_F$) remains high for these clusters where appreciable d character remains and thus the spin-unrestricted Fe$_4$B and Fe$_4$Si cluster calculations give large moments, 2.6 and 2.4 μ_B, respectively. These calculated results are consistent with experimental observations. Amorphous Fe$_{80}$B$_{20}$

shows a saturation moment of approximately 2 μ_B/Fe atom[6] and Si additions are observed to yield a slightly larger moment.[7] Amorphous $Fe_{1-x}Si_x$ thin films[8] show a significantly smaller magnetic moment than is observed in amorphous $Fe_{1-x}B_x$.

For C, and for N especially, the covalent pd hybridization is large due to small energy differences between the initial atomic states. Calculated charge distributions for the molecular orbitals near E_F do indeed show more delocalization and (sp)-d mixing for C and N than for B and Si. This results in increased interstitial charge indicative of delocalized character and leaves the product $I(E_F) D(E_F)$ low. Calculated moments for Fe_4C and Fe_4N clusters are, therefore, found to be low, 0.2 and 0.0 μ_B/Fe, respectively. Amorphous $Fe_{1-x}N_x$ has not been fabricated but N additions to Fe base glasses are reported to have mixed effects.[9] Carbon additions to $Fe_{80}B_{20}$ glasses weakly suppress the saturation moment.[7]

There is a degree of polarity in these Fe-M bonds that is most clearly seen in the position of the 1a1 level which marks the bottom (bonding state Fig. 1) of the valence band. The degree to which this level is stabilized relative to its position in the Fe_4 cluster is porportional to the difference between the atomic Fe-4s level and the M-s levels. Charge density contours also reveal this bond polarity. While these large polar components to bonding are a major factor in alloy stability they do not affect the magnetic properties as much as the covalent interactions because here they involve states that are largely s like.

Fujiwara[10] has calculated electronic states for clusters of 1500 atoms arranged as determined by dense random packing. The locations of spectral features arising mainly from boron s states (-12.5eV), from iron-spd bonded with boron p states (-6 to -8eV), and from iron-d states can be identified. These features also appear in electronic DOS calculated by tight-binding methods[11] from large random clusters generated by an algorithm different from Fujiwara's and in Moruzzi's augmented spherical wave band structure calculations[12] for Fe_3B. Thus, the essentials of the metal-metalloid bonding as indicated by the presence and position of the sp-d bonding state, are the same in a variety of calculations. Smaller clusters show a narrower valence band and thus the bonding state is not found at as low an energy as it is in larger clusters.

B. Experimental Results for T-M Alloys.

Ammamou and Krill have compared the experimental XPS valence band spectrum of pure iron with those of two amorphous iron-base alloys.[13] The valence band of the pure iron is mainly of 3d character. Alloying results in significant covalent mixing of 3d with metalloid p character, broadening the XPS valence band. The broadened valence band has a correspondingly reduced $D(E_F)$ which suppresses moment formation (Eq. 2). The experimental XPS spectrum for Fe_3B shows the Fe-B bonding state 9.5 eV below E_F. Thus all calculational approaches agree quite well with experiment on the basic chemical physics of bonding and electronic structure. These results are summarized in Table I.

C. Split-bands

Metallic glass compositions often contain two or more transition metal species with or without the metalloids, T_1-T_2-M or T_1-T_2. The d-d bonding between the T species supplements or replaces the d-(sp) bonding between T and M. Typically the T species further to the right in the Periodic Table has the greater electronegativity and hence is considered the anion in the formation of a polar d-d bond.

The first direct observation of split bands in amorphous alloys was made by Guntherodt's group[14] using ultraviolet photoelectron spectroscopy on Pd-Zr and Cu-Zr T_1-T_2 alloys. The spectra showed two well-resolved

Table I Binding energy (in eV below E_F) for boron s states and Fe-d - B-(sp) hybrid bonding states as determined by four different calculations and by XPS studies.

Method	Ref.	B s states	Fe-d-B-(sp) hybrid
Fe_4B Cluster	3	10.4	7
Fe_2Ni_2B Cluster	2	-	6.5
1500 atom $Fe_{80}B_{20}$	10	12.5	6 - 8
ASW Fe_3B band struc.	12	-	8 - 10
Experimental Fe_3B XPS	13	11.5	9.5

peaks due to the d bands of the two species, Zr states near E_F and Cu or Pd states several eV below E_F. The energy shift of the more electronegative species d states from their pure metal positions is a measure of the bonding responsible for stabilization of the alloy. Oelhafen et al[15] measured these d-band shifts ΔE in a number of T_1-T_2 alloys and found them to correlate with the valence difference between the two T species. In addition to this polar component to the d-d bonding, there is a degree of covalent d-d mixing. This covalent mixing decreases as the electronegativity difference increases. In effect, the wave function at higher energy sees mainly the core potentials of the cations and that at lower energy sees mainly the core potential of the anions. As a result, the widths of the two components of the split band spectrum are less than their widths in the pure metallic state because in the alloy they have fewer like neighbors with whom they interact strongly. Thus the d bands of T_1-T_2 alloys are <u>narrower</u> than those of the pure metallic constituents in contrast to the d bands of T-M alloys which are <u>broader</u> (due to d-sp covalent mixing) than those of the pure T metal.

Because the bonds in T_1-T_2 alloys involve d electrons, the greater the bonding the less the d-character remaining for magnetism. While it is difficult to compare the magnetic properties of the 3d alloys in Fig. 5 because of their small late transition metal content, Buschow[16] has compared the effective Co and Ni moments and effective Fe hyperfine fields in metal-base amorhpous alloys. His results confirm that the stronger the bonding (more negative heat of formation) the weaker is the magnetic moment.

Several simple models exist which incorporate these concepts into formulas useful for comparison with experimental data.[17]

D. Electronic Structure and Magnetostriction

Magnetic anisotropy results when the atomic spin moment couples to a local atomic environment of less-than-spherical symmetry. This coupling can occur either through anisotropic exchange between local spins, $\Sigma_j J_{ij}(S_i^z \cdot S_j^z)$, or, more commonly, by the coupling of the spins and orbits on a given site, $\xi L \cdot S$, and an electrostatic interaction of the magnetic orbital with the crystal field of its environment. For either mechanism to produce significant anisotropy, the crystal field must have a low symmetry. But the latter mechanism also requires that the orbitals of the atom probing that field must be asymmetric (i.e. $L \neq 0$). Hence the importance of the symmetry of the orbitals near E_F in determining local anisotropy.

If the spin-orbit interaction or some anisotropic exchange mechanism were sufficiently large to induce even a small distortion so as to break the local symmetry, then energy would be gained (providing the d band is not completely filled) by populating the lower lying states of the manifold split by the lowered symmetry. This strain is called magnetostriction.

The condition for zero magnetostriction (L·S = 0) can be met by a filled or empty d band. These cases are not interesting because they give no magnetic moment. Spin-orbit energy is also zero when the d bands split into two or more energetically distinct components (each of which satisfies $\Sigma_i \ell_i = 0$) and E_F lies between two of these sub-bands. Split bands can result from a polar bonding between d bands of sufficient electronegativity difference as described in Section C above. The effect of this split-band condition on magnetostriction was modeled successfully by Berger[18] in crystalline alloys and by O'Handley and Berger[19] in amorphous alloys.

The split-band, zero magnetostriction condition can be reached in amorphous Fe-Ni-M glasses for compositions to the Ni rich side of the permalloy ratio Ni:Fe::4:1.[19] It should also be achieved in certain T_1-T_2 amorphous alloys (e.g. Ni-Zr, Co-Zr) when the charge displaced from the Zr d states just fills the Ni or Co d-bands. Unfortunately, at this point the Zr bands are not spin split (because while they are weakly antibonding with respect to Ni or Co, they are delocalized and bonding rather than localized and antibonding with respect to Zr) and no moment exists. It remains a challenge for this split-band model and for alloy design to discover systems other than cobalt-base glasses for which $\lambda = 0$ and $\mu \neq 0$.

References

1. W.A. Harrison, Electronic Structure and the Properties of Solids, (W.H. Freeman and Co., San Francisco, 1980).

2. R.P. Messmer, Phys. Rev. B23, 1616 (1981).

3. A. Collins and R.C. O'Handley, submitted for publication.

4. M.E. McHenry, K.H. Johnson, and R.C. O'Handley, Phys. Rev. B 35, 3555 (1987).

5. C.Y. Yang, K.H. Johnson, D.R. Salahub, J. Kaspar and R.P. Messmer, Phys. Rev. B 24, 5673 (1981).

6. R.C. O'Handley, R. Hasegawa, R. Ray and C.-P. Chou, Appl. Phys. Lett. 29, 330 (1976).

7. M. Mitera, M. Naka, T. Masumoto, N. Watanabe, phys. stat. sol. (a) 49 K 163 (1978) and N.S. Kazama, M. Mitera and T. Masumoto in Rapidly Quenched Metals III, vol 2, op cit. p. 164 (1978)

8. G. Marchal, P. Mangin, M. Piecuch, C. Janot and J. Hubsch, J. Phys. Met. Phys. 7, L 165 (1977).

9. M. Nagakubo and M. Naoe, paper to be presented at Rapidly Quenched Metals VI, Montreal, August 1987.

10. T. Fujiwara, J. Phys. Metal Phys. 12, 661 (1982) and T. Fujiwara, J. Non-cryst. Sol. 61-62, 1039 (1984).

11. S. Krompiewski, U. Krey and H. Ostermeier, J. Mag. Mag. Mat. (1987).

12. V.L. Moruzzi, unpublished results.

13. A. Amamou and G. Krill, Sol. St. Comm. 33, 1087 (1980).

14. P. Oelhafen, E. Hauser, H.J. Guntherodt, and K. Benneman, Phys. Rev. Lett. 43, 1134 (1979).

15. P. Oelhafen, E. Hauser, and H.J. Guntherodt, Sol. St. Comm. 35, 1017 (1980).

16. K.H.J. Buschow, J. de Physique, colloq. (1984), K.H.J. Buschow and P.G. Van Engen, Mat. Res. Bull. 16, 1177 (1981).

17. B.W. Corb, R.C. O'Handley, and N.J. Grant, J. Appl. Phys. 53, 7728 (1982) and B.W. Corb, R.C. O'Handley, and N.J. Grant, Phys. Rev. B 27, 636 (1983), A.P. Malozemoff, A.R. Williams and V.L. Moruzzi, Phys. Rev. B 29, 1620 (1984), R.C. O'Handley and Y. Hara, Phys. Rev. B 35, 5276 (1987).

18. L. Berger, Physica B91, 31 (1977).

19. R.C. O'Handley and L. Berger, in Physics of Transition Metals, ed. M.J.G. Lee, J.M. Perz and E. Fawcett, (Inst. of Physics, Bristol, England, 1978) p. 477. R.C. O'Handley, Phys. Rev. B18, 930 (1978).

INVESTIGATION OF DOMAINS WALLS IN AMORPHOUS MATERIALS USING
SCANNING ELECTRON MICROSCOPY WITH SPIN POLARIZATION ANALYSIS

J. UNGURIS, G. HEMBREE, C. AROCA (*), R.J. CELOTTA, and D.T. PIERCE
National Bureau of Standards, Gaithersburg, MD 20899, USA
(*) Dpt. Materia Condensada, Fac. Física, Universidad Complutense
    28040, Madrid, Spain

   Domain walls in amorphous ferromagnetic materials have been observed with
a field emission scanning electron microscpoe with electron spin
polarization analysis. The domain wall thickness of these materials can
be the order of microns, and are therefore suitable for internal structure
studies. In this work we have studied domain walls in different amorphous
samples and obtained a profile of the magnetization distribution in the
domain wall.

1. INTRODUCTION

   Scanning electron microscopy with polarization analysis, SEMPA, is used to image magnetic microstructure with high spatial resolution (1-5). It is based on the result that the low energy secondary electrons generated when a ferromagnetic material is probed by an electron beam are spin polarized (6-8). The secondary electron polarization is directly related to the magnitude and direction of the magnetization in the area probed by the electron beam. Scanning the sample with the incident electron beam, an image of the magnetic microstructure and a standard SEM topographic image are independently obtained.

   We have used this system to obtain images of the magnetic microstructure of the domain walls in amorphous ferromagnetic samples with different anisotropy and magnetostriction. By processing the resulting digital image, we have measured the profile of the magnetization direction of the domain wall at the sample surface. The two in-plane magnetization components were directly obtained and the perpendicular one was deduced from the other two.

2. EXPERIMENTAL

   A schematic of the apparatus is shown in fig. 1. The SEM has an electron emission source that can produce beam diameters of 10 Secondary electrons from the sample are accelerated to 1500 eV by a collection lens and focused into an hemispherical energy analyzer to filter out the elastically scattered electrons. All three components of the spin polarization are measured by electrostatically switching betwee two orthogonal analyzers. Since each detector only measures the transverse component of the polarization, orthogonal detectors are required to measure the three projections of the magnetization vector.

The resolution of the microscope was limited by beam vibrations caused by external sources of hundred of amstrongs. To get information of the domain wall structure we decided to measured samples with large domain wall thickness.

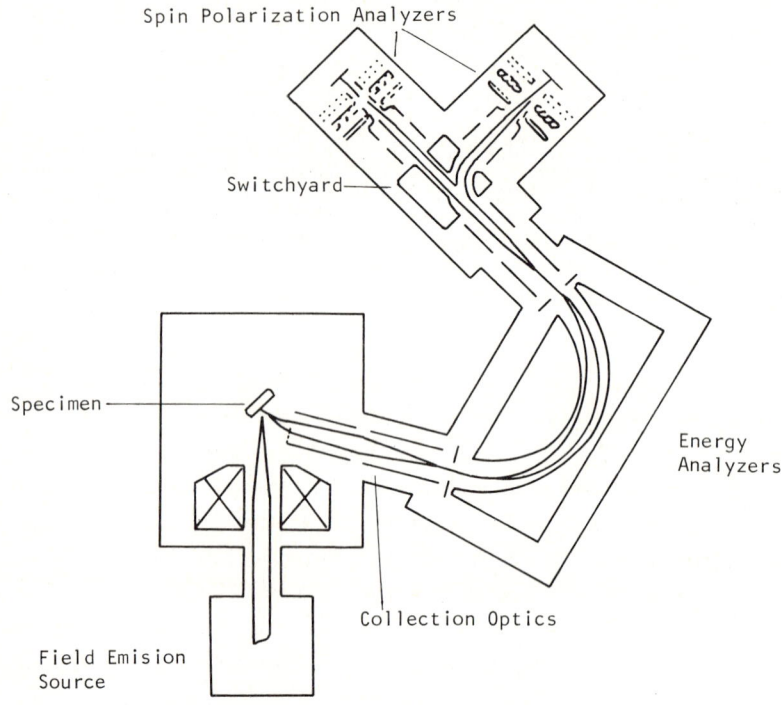

Fig. 1

The samples studied were low-magnetostriction Metglass 2750 Mn low and Metglass SASC3 and high magnetostriction Metglass 2605 Co. The anisotropy of these samples was below the 500 j/m^3. The anisotropy is not known accuratedly because induced stresses caused by the sample holder, the necessary ion bombardment and the posterior annealing needed to clean the sample.

In the process of changing from one analyser to the other there were a change of the focused area. Thefore it not was possible to get directly the three projections of the magnetization. We decided to obtained directly the two in plane projections and from them to calculate the perpendicular one, $M_z = M - (M_x^2 + M_y^2)^{.5}$. To perform this calculation it was necessary to evaluate M. We therefore found an area of the sample with all the magnetization in the plane and parallel to the domain wall. This is an area far away from the domain wall. It was not possible to apply this technique to image with the high resolution because of the domain wall almost fills the total image.

3. RESULTS

Fig. 2 shows two images of the same wall in Metglas SASC3. The analyzers are approximately oriented along the diagonals of the imagen direction. That

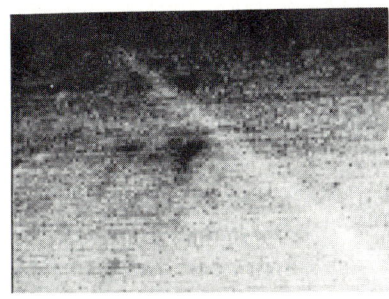
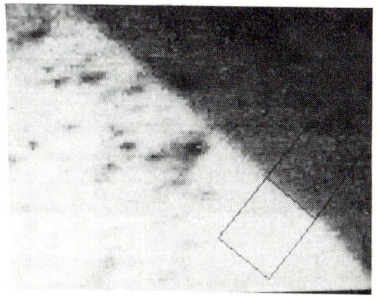

Polarización x
a
Fig. 2
Polarización y
b

is, the P_y analyzer is almost parallel to the domain wall an the P_x analyzer almost perpendicular. Then the level of gray color in fig. 2b is proportional to the projection of the magnetization parallel to the domain wall and in fig. 2a to the magnetization perpendicular to the domain wall, then the magnetization is parallel to the domain wall in both images, except in the domain wall itself were it is perpendicular to the domain wall. If all the magnetization were parallel to the surface the $(M_x^2 + M_y^2)^{\frac{1}{2}}$ value will be constant.

The same sequence is shown in Figs. 4,5. As can be observed there is also a component of magnetization perpendicular to the domain wall, but there is a change in its sign at both sides of the domain wall.

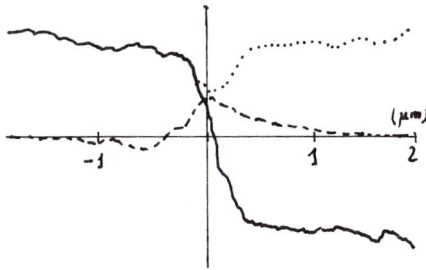

Fig. 3

Fig. 6 shows an image with higher magnification of the SASC3 sample along

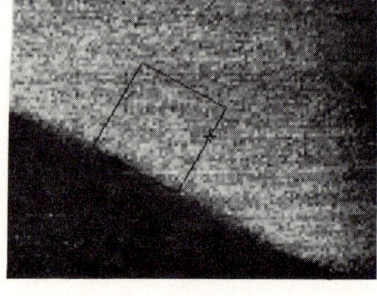

Polarization x
a

Polarization y
b

Fig. 4

with profiles of the magnetization parallel and perpendicular to the domain wall (fig. 7).

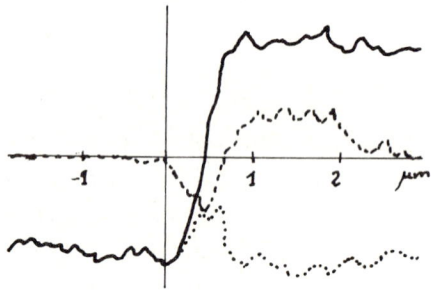

Fig. 5

The domain width obtained for these samples was .5 Mn for SASC3 and 2605 Co, and 1 Mn for 2705 Mn. But we found a variation in these measurements, probably due to the stress induced by the sample holder an to the ion bombardment. There was also a variability in profiles of different walls in the same sample. Here we show the expremes of the different profiles obtained, but normally the profiles were a combination of both.

If we compare these results with the theoretical models 9,13, they correspond to an asymmetrical and a symmetrical Bloch domain wall (fig. 8). In the mode asymmetrical wall there is no magnetization perpendicular to the sample surface. We have observed a perpendicular component in all of the profiles but one. This is in agreement with previous work 14. However, the secondary electrons, due to the depth of penetration of the electron beam, can come from the inner portion or the sample where the magnetization of the domain wall is perpendicular to the surface.

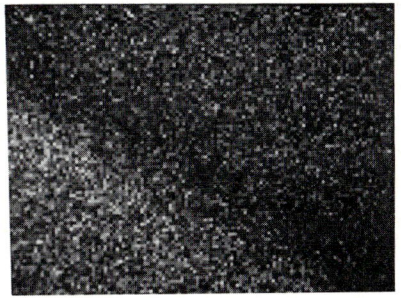

Polarization x
a

Polarization y
b

Fig. 6

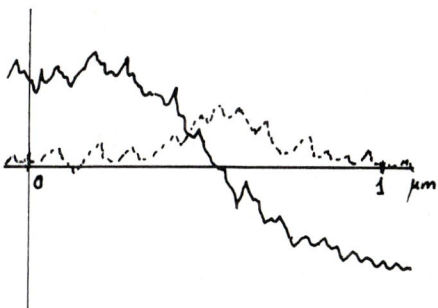

Fig. 7

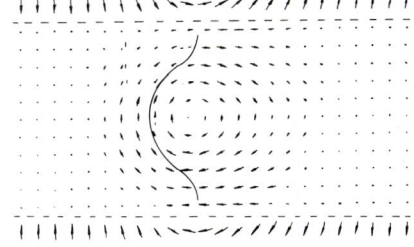

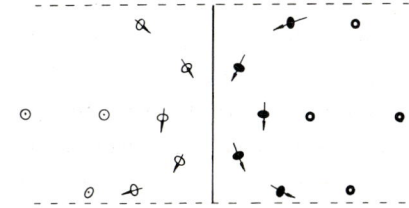

Fig. 8

a

b

The observed symmetrical wall in the zero magnetostriction sample is surprising from a theoretical point of view, such a wall appears occasionally appears even not in all the obtained profiles in the same sample.

4. CONCLUSION

It has been proved that the SEMPA is an instrument useful for the study of domain wall structure. It will be informative to continue this work with a systematic study in samples with different anisotropy and magnetostriction. We plan to modify the sample holder to make possible the application of -stresses to the sample, in order to study the effect of the anisotropy on domain wall structure.

ACKNOWLEDGEMENT

We are grateful with Dr. H.T. Savage and Dr. K. Hathaway for their encouragement, discussions, and to supply us the samples. This work has been supported by the Comité conjunto Hispano Americano.

REFERENCES

1) K. Koike and K. Hayakawa. J. Appl. Phys. 57 (1985) 4244
2) R.J. Celotta and D.T. Pierce in: Microbeam analysis. ed. K.F.J. Heinrich (San Francisco Press, California 1982), p. 4212
3) T.H. DiStefano. IBM Tech. Disc. Bull. 20 (1978) 4212
4) J. Kirschner, Appl. Phys. A 36 (1985) 121
5) J. Unguris, G. Hembree, R.J. Celotta and D.T. Pierce. J. Mag. Mag. Mat. 55-57 1986
6) J. Ungris, D.T. Pierce, A. Galejs and R.J. Celotta. Phys. Rev. Lett. 49 (1982) 72
7) M. Landolt and D. Mauri. Phys. Rev. Lett. 49 (1982) 1783
8) H. Hopster, R. Raue, E. Kisker, G. Guntherodt and M. Campagna. Phys. Rev. Let. 50 (1983) 70
9) A. Hubert, IEEE Trans. on Magn. MAG-11 (1975) 1285
10) A. Hubert. Phys. Stat. Sol. 32 (1969) 519
11) A. Hubert. Phys. Stat. Sol. 38 (1970) 699
12) A. Aharoni. Phys. Stat. Sol. 18 (1973) 661
13) A.E. LaBonte. J. Appl. Phys. 40 (1969) 2450
14) C. Aroca, E. López and P. Sánchez. J. Magn. Magn. Mat. 23 (1981) 193

EVALUATION OF PERPENDICULAR ANISOTROPY OF MAGNETIC THIN FILM
USING THE SPONTANEOUS HALL EFFECT

Kensho OKAMOTO*

Universität Regensburg, Institut für Physik-Angewandte Physik
Universitätsstrasse 31, Postfach 397, F. R. Germany

This paper introduces a new method to investigate the perpendicular magnetic anisotropy in RE-TM amorphous films. In this method, the perpendicular anisotropy is evaluated quantitatively or qualitatively using their spontaneous Hall effect.

1. INTRODUCTION

There are several methods to investigate the magnetic anisotropy (H_k, K_u) of perpendicular anisotropy films. Typical ones are the torque curve method and the measurement of M-H curve for in-plane magnetic field. However, it is often difficult or troublesome to use such methods for rare earth-transition metal amorphous alloy films (e.g. GdCo, GdFe, TbFe, etc.). Because these films are ferrimagnetic substance and they show very small value of magnetization at room temperature. The author has developed a new technique to evaluate the perpendicular magnetic anisotropy field using the spontaneous Hall effect. This technique have already been published to a certain extent in Ref.1. In this paper, more detailed explanation of this method are described.

2. THEORY

Soohoo and Lee proposed a method to evaluate the uniaxial magnetic anisotropy and its dispersion in a bubble film from the transverse magnetization curve measured with additional in-plane magnetic field [2]. The new technique described below is partly based on their method.

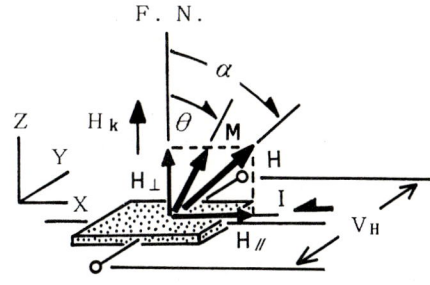

FIGURE 1
Application of magnetic field **H**

The publication of this paper was assisted by the Alexander von Humboldt Foundation in the Federal Republic of Germany.
*On leave from Kagawa University, Faculty of Education, Saiwaicho 1-1, Takamatsu, Kagawa 760, Japan.

Let us consider a Hall sample made of magnetic thin film with perpendicular magnetic anisotropy as illustrated in Fig.1. Here, we assume that the film forms a single magnetic domain in the initial state (remanence state).

Now, if magnetic field H, whose in-plane and perpendicular components are respectively $H_{\parallel}$ and $H_{\perp}$, is applied obliquely in the XZ plane then magnetization M tilts by angle θ from Z axis(film normal) towards X axis. In this case, the free energy E of the film is given by

$$E = K_u \sin^2\theta + 2\pi M^2 \cos^2\theta - MH_{\parallel}\sin\theta - MH_{\perp}\cos\theta, \quad (1)$$

where K_u is the uniaxial anisotropy constant.

In order to find the minimum energy condition, let $dE/d\theta = 0$, then we get

$$(K_u - 2\pi M^2)\sin 2\theta = MH_{\parallel}\cos\theta - MH_{\perp}\sin\theta. \quad (2)$$

Substituting the relationship $K_u = MH_k/2$ (H_k is the anisotropy field) into Eq.(2) and rearranging Eq.(2), we find

$$H_k - 4\pi M = H_{\parallel}/\sin\theta - H_{\perp}/\cos\theta. \quad (3)$$

If $H_{\perp}=0$ in Eq.(3), that is, magnetic field H is applied in the direction of the X axis along the film plane, Eq.(3) reduces to

$$H_k' = H_k - 4\pi M = H_{\parallel}/\sin\theta, \quad (4)$$

where H_k' is the effective anisotropy field.

In the case of Fig.1, the Hall voltage V_H produces in proportion to the perpendicular component of the magnetization, M_s. That is,

$$V_H \propto M_s \cos\theta. \quad (5)$$

In a film which has an ideal rectangular M-H loop (e.g. GdFe, GdCo, etc., also MnBi[3]), the Hall loop also shows a rectangular shape like Fig.2 and the remanence Hall voltage V_{Hr}, which arises from the remanence magnetization $M_{r\perp}(\simeq M_s)$, is given by

$$V_{Hr} \propto M_s. \quad (6)$$

From Eqs.(4), (5) and (6), we get

$$H_k' = H_{\parallel}/\sqrt{1-(V_H/V_{Hr})^2} \quad (7)$$

or

$$H_{\parallel}^2/H_k'^2 + V_H^2/V_{Hr}^2 = 1. \quad (8)$$

Thus, effective magnetic anisotropy field of the film can be determined using Eqs.(7) or (8). Figures 2 and 3 are the graphic

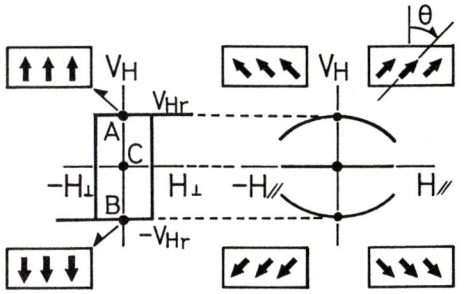

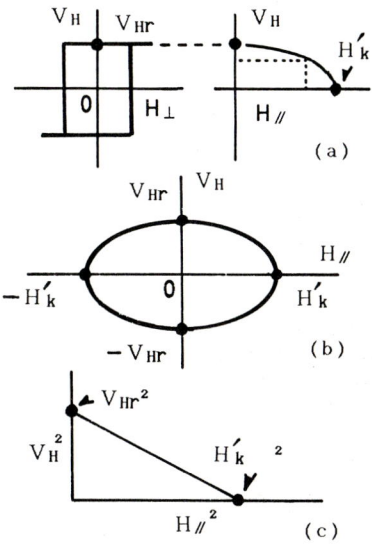

(a) H⊥film plane (b) H∥film plane

FIGURE 2 (Left)
Change of remanence Hall voltage V_{Hr} with in-plane field $H_{\parallel}$

FIGURE 3 (Right)
Graphic expressions of Eqs.(7) and (8)

expressions of Eqs.(7) and (8), respectively. In Fig.3(b), the curve forms an ellipse, and the value of H_k' is known from that of the applied field $H_{\parallel}$ where V_H becomes to zero.

However, in most cases, RE-TM films have anisotropy dispersion, and that the magnitude of H_k' is very large (in the order of from several kilooersteds to more than tens of kilooersteds), so that it is more practical to use the graph shown in Fig.3(c). In Fig.3(c), H_k' can be known from the point on the abscissa where the extrapolated $V_H^2 - H_{\parallel}^2$ line intersects the abscissa.

We can also evaluate the quality of the perpendicular anisotropy qualitatively from the completeness and/or the symmetry of the form of ellipse in Fig.3(b).

Figure 4 shows an example of $V_H - H_{\parallel}$ characteristic measured in a rf-sputtered (Vb= -50V) GdCo amorphous film.
From Fig.4, the following facts are known. That is, this film has an an-

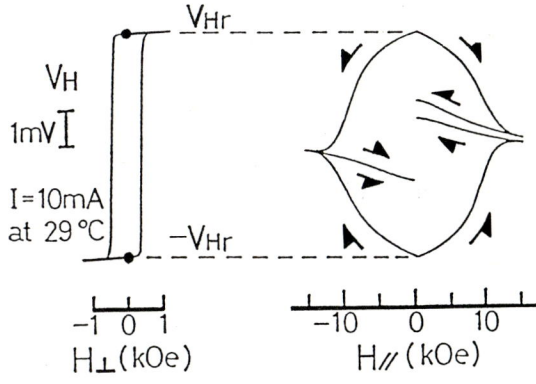

FIGURE 4
$V_H - H_{\parallel}$ characteristic in a GdCo film

isotropy dispersion, because the V_H-$H_{\parallel}$ curve is skirt-like in shape, and that this film includes the in-plane magnetic anisotropy component, because the polarity of V_H is changeable by the in-plane magnetic field $H_{\parallel}$ as shown by the arrows.

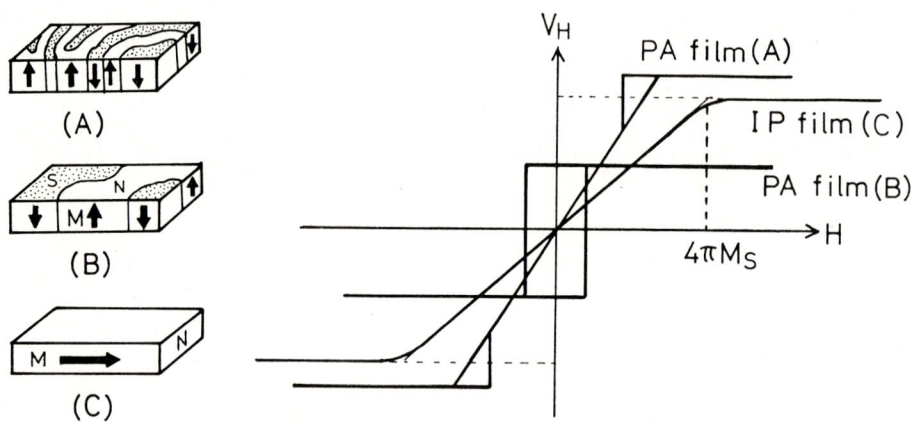

FIGURE 5
Relationship between magnetic domain pattern and Hall loop (A) and (B): perpendicular anisotropy(PA) film
 (A); linear M-H hysteresis curve film (non-remanence film)
 (B); rectangular M-H hysteresis curve film (remanence film)
(C): in-plane(IP) film

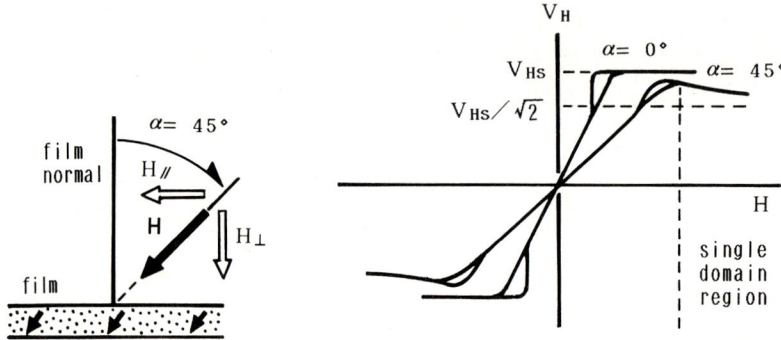

FIGURE 6
Measurement of oblique Hall effect with $\alpha = 45°$

FIGURE 7
Transverse($\alpha = 0°$) and oblique($\alpha = 45°$) Hall characteristics for linear M-H film

However, this approach can not be applied to the case of perpendicular anisotropy film which possesses a linear M-H characteristic and accordingly shows a linear V_H-H loop as illustrated in Fig.5 (PA film-A). Because such a type of film exhibits a fine maze or stripe magnetic domain pattern at the remanence state as illustrated in Fig.5(A) ($M_{r\perp}= 0$), so that always $V_{Hr}= 0$.

In such a case, we apply a large magnetic field sufficient to make the whole film single domain in the oblique direction $\alpha = 45°$ (i.e. $H_{\parallel} = H_{\perp}$) as shown in Fig.6, then H_k' can be calculated by the following equation;

$$H_k' = H\{1/\sqrt{1-(V_H/V_{Hs})^2} - 1/(V_H/V_{Hs})\}/\sqrt{2}, \qquad (9)$$

where V_{Hs} is the saturation transeverse Hall voltage in the case of $\alpha = 0°$ (see Fig.7).
Equation (9) is obtained from Eq.(3) by letting

$$\cos\theta = V_H/V_{Hs}, \qquad (10)$$

$$\sin\theta = \sqrt{1-\cos^2\theta} = \sqrt{1-(V_H/V_{Hs})^2}, \qquad (11)$$

and

$$H_{\parallel} = H_{\perp} = H/\sqrt{2}. \qquad (12)$$

The method described above is especially effective for rare earth-transition metal amorphous alloy films for magnetooptical recording, because they show very large spontaneous Hall voltage independently of their very small value of magnetization M_S due to the ferrimagnetism. Also, this technique may be useful for some perpendicular magnetization films other than RE-TM films, for example CoCr-system films.

ACKNOWLEDGMENTS

The author wishes to thank Dr. Yuji Togami of the NHK Science and Technical Research Laboratories for providing the GdCo films. He also would like to thank the Alexander von Humboldt Foundation in F.R. Germany for the grant for the publication of this paper.

REFERENCES

1) K. Okamoto, Digests of 3rd Conf. on Magnetics in Japan 3 (1979), 32.

2) R.F. Soohoo and Kenneth Lee, AIP Conference Proceedings No.24, (1974), 590

3) K. Okamoto et al., AIP Conf. Proc. 34 (1976) 55.

STRESS DEPENDENCE OF SATURATION MAGNETOSTRICTION IN METALLIC GLASSES

Henryk K. LACHOWICZ and Henryk SZYMCZAK
Institute of Physics, Polish Academy of Sciences,
al. Lotnikow 32/46, 02-668 Warszawa, Poland

The effect of stress-dependent saturation magnetostriction, λ_s, observed in metallic glasses is discussed. The experimental stress dependence of λ_s, obtained at different temperatures for low-magnetostrictive metallic glasses is presented. These results are discussed in terms of the pair-ordering model and also of the model of preferred distribution of atomic bonds, phenomenon occurring in metallic glasses subjected to stress. This analysis shows that stress-dependent λ_s can be satisfactorily interpreted in terms of the models mentioned above.

1. INTRODUCTION

Magneticians who have for a long time been dealing with amorphous magnets, and in particular, with metallic glasses, know that they still surprise us by showing new and unexpected phenomena. From the very beginning of their history, while we were strongly convinced that the magnetic order should not occur in structurally disordered media, these materials have continued to surprise us. For example, metallic glasses have shown us that large magnetic anisotropy can be created in these materials by subjecting them to a mechanical strain caused by a tensile stress applied along the ribbon axis at an elevated temperature[1]. This anisotropy reaches a value which is an order of magnitude larger than the one induced by field-annealing[2], the latter procedure applied successfully to crystalline magnets for a very long time.

Since magnetic anisotropy and magnetostriction are of the same origin, one would expect that stress should also affect saturation magnetostriction in metallic glasses, though our thinking, developed in dealing with materials exhibiting translational symmetry, hesitates to accept such an effect.

However, metallic glasses showed once more that this seemingly unbelievable phenomenon does indeed occur. It has been experimentally shown, independently in our[3] and also in other laboratories[4,5], that saturation magnetostriction does depend on stress and that λ_s changes linearly with the tensile stress applied along the metallic glass ribbon's axis. Since these changes are relatively small, being of the order of 10^{-10} MPa^{-1}, the effect can be experimentally detected only in low-magnetostrictive metallic glasses, e.g.: Co-based alloys.

2. THE PHENOMENON

When the magnetoelastic anisotropy field H_K^σ of a conventional magnetic material is measured as a function of the applied stress σ, the dependence obtained is linear according to the well-known equation:

$$H_K^\sigma = (3\lambda_s/\mu_0 \cdot M_s)\,\sigma , \qquad (1)$$

since

$$K_u^\sigma = (\mu_0 H_K^\sigma M_s/2) = (3/2)\lambda_s\sigma , \qquad (2)$$

where μ_0 is the magnetic constant, M_s is the saturation magnetization.

For highly-magnetostrictive metallic glasses (e.g. the Fe-based alloys which show $\lambda_s \approx 30 \times 10^{-6}$) the stress-dependent contribution to saturation magnetostriction is five orders of magnitude smaller, and the effect is not detectable.

However, if one measures a low-magnetostrictive metallic glass, such as the one shown in Fig.1, the dependence of H_K^σ on stress becomes non-linear, as is seen in this figure. This dependence (the dependences obtained for other Co-based metallic glasses of different compositions, exhibiting both positive and negative magnetostrictions, are of the same character) is best fitted by a second order polynomial in stress, of the form:

$$H_K^\sigma = a\sigma^2 + b\sigma , \qquad (3)$$

where a and b are constants, the values of which are obtained by numerical fitting.

Since a derivative of the anisotropy energy gives a measure of magnetostriction, one can differentiate eq.(1) with respect to stress, under the assumption that λ_s is stress-dependent. This gives:

$$dH_K^\sigma/d\sigma = (3/\mu_0 M_s)[(d\lambda_s/d\sigma)\sigma + \lambda_s] . \qquad (4)$$

Differentiating now eq.(3) one can easily find, by comparing the obtained derivative with eq.(4), that stress dependent saturation magnetostriction changes linearly with stress according to:

$$\lambda_s(\sigma) = (d\lambda_s/d\sigma)_{\sigma=0} \cdot \sigma + \lambda_s(0) , \qquad (5)$$

where the coefficients in the above equation can be expressed in terms of the calculated polynomial constants:

$$(d\lambda_s/d\sigma)_{\sigma=0} = 2a\mu_0 M_s/3 , \qquad (6a)$$

$$\lambda_s(0) = b\mu_0 M_s/3 , \qquad (6b)$$

giving us a possibility of calculating the coefficients of stress-dependent saturation magnetostriction.

At room temperature the slope, calculated according to eq.(6a) from the experimental data obtained for low-magnetostrictive metallic glasses, is always negative, independent of the sign of the initial saturation magnetostriction $\lambda_s(0)$ (for zero stress), with the absolute value of the order of 10^{-10} MPa^{-1}.

Therefore, for samples exhibiting very low positive initial magnetostriction a change in sign of $\lambda_s(\sigma)$ is observed at a certain value of stress, as is seen in Fig. 1.

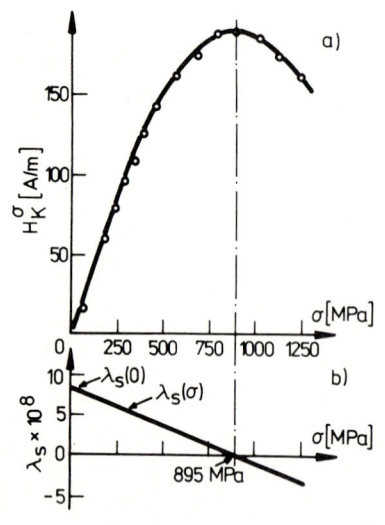

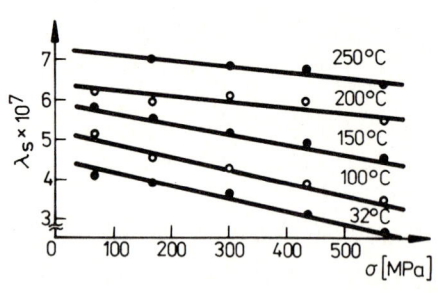

FIGURE 2
Stress dependence of saturation magnetostriction for metallic glass sample of the $Co_{74}Fe_6B_{20}$ alloy measured at different temperatures (after ref.12).

FIGURE 1
Room temperature stress dependences obtained for the $Co_{67}Fe_4Cr_7Si_8B_{14}$ metallic glass ribbon: (a) magnetoelastic anisotropy vs. tensile stress (the SAMR method); (b) stress-dependent saturation magnetostriction obtained by numerical fitting of the experimental data given in (a) to eq.(3) - (after ref.7).

The latter seems to be important not only from the point of view of fundamental research, but also for applications since in some devices utilizing metallic glasses the sign of magnetostriction should be strictly controlled. Upon subjecting the metallic glass to stress, introduced for example, by technological processes, the performance of the device can deteriorate dramatically.

Furthermore, the effect of stress-dependent λ_s has serious implications on the measurement of this quantity when λ_s of low-magnetostrictive metallic glasses is measured by means of indirect methods, in which the sample is subjected to mechanical stress (for the details see[6,7]).

3. THEORETICAL MODELS

The models proposed to explain the experimental dependence of saturation magnetostriction on stress were derived by utilizing the previously described phenomenological approaches to magnetoelastic interactions in amorphous alloys, namely: the pair-ordering (two-ion)[8] and random anisotropy (one-ion)[9,10] models. Because of limited space, only the final results of the calculations will be presented here (for the details see[11]).

In the case of the pair-ordering model, which is based on the assumption that the interaction between atoms in a pair is of pseudo-dipolar character, the final result is given by the following inequality:

$$(d\lambda_\sigma/d\sigma) \, \text{sgn}(p) < 0 , \qquad (7)$$

where λ_σ is the stress-dependent contribution to saturation magnetostriction and p is the structural anisotropy parameter, which is defined by the probability function $P(\beta)$ determining the probability that the line joining two atoms makes an angle β with the stress direction:

$$P(\beta) = (1/4\pi)[1 + p(3\cos^2\beta - 1) + ...] . \qquad (8)$$

The function above is normalized on a unit sphere:

$$2\pi \int_0^\pi P(\beta) \sin\beta d\beta = 1 . \qquad (9)$$

The random anisotropy model is based on the cluster theory of net anisotropy induced by preferential ordering of the local anisotropy axes. An analysis of this model has shown that, in the isotropic approximation, the stress-dependent contribution to saturation magnetostriction is equal to zero, $\lambda_\sigma = 0$.*

4. EXPERIMENTAL AND INTERPRETATION

The stress dependence of saturation magnetostriction was measured[12] using the small angle magnetization rotation technique[13] (SAMR). The device was equipped with a furnace located coaxially with the d.c.-field solenoid, making it possible to heat the sample *in situ*. Fig.2 shows this dependence measured at different temperatures for a metallic glass ribbon exhibiting low positive magnetostriction (the dependences obtained for negative $\lambda_s(0)$ are of similar character). As has been the case in the previous experiments[3,4,5], these dependences are linear. However, the slope of the straight lines shown in Fig.2 does not change its sign for higher temperatures, as has been shown by Barandiaran et al.,[5] who have heated the sample by a current flowing through the ribbon.

In the case discussed, the slope is negative in the whole measured temperature range (as is seen in Fig.2). Moreover, the observations performed have shown that even for higher temperatures, up to 350 °C, there is no tendency to

---

*Due to an error in calculations presented in the previous publication[12], the contribution originating from the random anisotropy model was found to be non-zero.

change the sign of the slope. It is worth noting that in the higher temperature range, the measurements are difficult to perform because of creep development, which should significantly disfigure the measured quantities.*

To explain the results obtained in terms of the pair-ordering model developed, one could compare inequality (7) with the experimental data. By performing such an analysis, one can easily find that the pair-ordering model is consistent with the experimental data but only when the structural anisotropy parameter p is positive.

It can be shown that this condition is fulfilled if one considers the recently developed model of the prefered distribution of atomic bonds[14,15]. In the lower temperature range under stress, the original isotropic distribution of atomic bonds becomes anisotropic in the axial direction (then, $p > 0$) due to elastic deformation. The system tends to recover isotropy under stress by severing the bonds in the axial direction and forming new ones in the perpendicular direction (in metallic glasses the bonds are relatively easily cut and formed[15]). Such a rearrangement needs thermal excitations and, therefore, it occurs in a particular temperature range. For temperatures above this range, the rearrangement process should tend to saturate. The temperature dependence of the slope of the straight lines in Fig.2 shows a behavior consistent with this picture.

The experimental data shown in Fig.2 can also be satisfactorily interpreted in terms of the model of two-level systems[11], most frequently used to explain the phenomena observed in amorphous magnets (see e.g.[16]).

For this model, one can find that the structural anisotropy parameter tends to reach zero for a sufficiently high temperature[11]. This result is, at least qualitatively, consistent with the experimental data, since for an increase of temperature the slope of the straight lines decreases as is seen in Fig.2. The decrease of this slope is equivalent to an decrease of the structural anisotropy parameter.

5. CONCLUSIONS

In conclusion, one can say that stress-dependent saturation magnetostriction can be satisfactorily interpreted in terms of existing phenomenological models. The observed effect is shown to be related to microstructural parameters of metallic glasses. It is worth noting that the structural anisotropy parameter is strongly temperature dependent for a sample under stress and tends to zero for a sufficiently high temperature increase.

---

* In order to minimize the influence of time-dependent effects (e.g.creep) on the measurements, these should be carried out while keeping the sample under load for as short a time as possible. The best way of fulfilling this requirement seems to be to use a fully self-operated measuring set with automatic loading.

ACKNOWLEDGEMENT

We would like to express our thanks to Mr. A.Siemko from the authors' Institute, who has performed the experiment, and with whom we have fruitfully discussed the problems presented in the present paper. Discussions with Dr. R. Żuberek are also gratefully acknowledged.

REFERENCES

1) H.-R. Hilzinger, in: Rapidly Quenched Metals, vol.II, eds. T.Masumoto and K.Suzuki, (Japan Institute of Metals, Sendai, 1982) p.791.

2) T.Egami, Atomic short-range ordering in amorphous metal alloys, in: Amorphous Metallic Alloys, ed. F.E.Luborsky (Butterworths, London, 1983) p.100.

3) A.Siemko, H.K.Lachowicz and B.Lisowski, paper presented at the 3rd Intern. Conf. on Physics of Magnetic Materials, Szczyrk (Poland) Sept.9-14, 1986, Acta Phys.Polon., in print.

4) G.Herzer, in: Proc.Intern.Conf.Soft Magnetic Materials 7, (Wolfson Centre for Magnetic Technology, Cardiff, 1986) p.355.

5) J.M.Barandiaran, A.Hernando, V.Madurga, O.V.Nielsen, M.Vazquez and M.Vazquez-Lopez, Phys.Rev., B35 (1987) 5066.

6) A.Siemko and H.K.Lachowicz, J.Magn.Magn.Mat., 66 (1987) 31.

7) A.Siemko and H.K.Lachowicz, IEEE Trans.Magn., in print.

8) H.Szymczak, J.Magn.Magn.Mat., in print.

9) H.Szymczak and R.Żuberek, IEEE Trans.Magn., MAG-17 (1981) 2843.

10) E.du Tremolet de Lacheisserie, J.Magn.Magn.Mat., in print.

11) H.Szymczak and R.Żuberek, IEEE Trans.Magn., in print.

12) H.K.Lachowicz, A.Siemko and H.Szymczak, paper presented at the Intern.Symposium on Physics of Magnetic Materials, Sendai (Japan) April 8-11, 1987, in print in the Conf.Proc. (World Scientific Publ.Co., Singapore, 1987).

13) K.Narita, J.Yamasaki and H.Fukunaga, IEEE Trans.Magn., MAG-16 (1980) 327.

14) J.Haimovich, T.Jagielinski and T.Egami, J.Appl.Phys., 57 (1985) 3581.

15) Y.Suzuki, J.Haimovich and T.Egami, Phys.Rev., B35 (1987) 2162.

16) H.Kronmüller and N.Moser, Magnetic after-effects and the hysteresis loop, as in ref.2, p.341.

MAGNETIC AND STRUCTURAL RELAXATION IN AMORPHOUS METALS

J.A. LEAKE

Department of Materials Science and Metallurgy, University of Cambridge, Cambridge, CB2 3QZ, U.K.

Many properties of amorphous metals change with time because of structural relaxation. Magnetic properties can be particularly sensitive. A summary of the principal experimental results is presented. Care is taken to distinguish between reversible and irreversible structural relaxation and stress relaxation. Important observations to be explained include isothermal 'log time' kinetics and the temperature dependence of those kinetics, 'crossover' phenomena, and the different kinetics of reversible and irreversible processes. The correlation between the kinetics of change of different properties provides useful additional information about the nature of the relaxation processes, as do variations between glasses of related compositions. Microstructural changes that can mask relaxation effects are described. Of the various theoretical models that have been advanced the Activation Energy Spectrum model is shown to be particularly useful. Broad Gaussian spectra are proposed for the reversible and irreversible processes. The problem of relating possible atomic rearrangements to activation energy spectra is discussed. The concept of thermally-activated two-level systems (TLS) provides useful insights.

1. INTRODUCTION

Amorphous metals in the as-prepared condition are not only metastable with respect to the appropriate crystalline phases(s) but are also unstable with respect to a differently configured, denser amorphous state. Atomic rearrangements within the glass lead gradually towards this 'ideal' amorphous structure. The kinetics of these rearrangements are of interest and importance because the changes in structure may give rise to significant changes in properties. For most practical amorphous metals these changes are so slow as to be negligible at room temperature but they become increasingly rapid at moderately enhanced temperatures, well below the crystallisation temperature. An appreciation of the nature and kinetics of relaxation is important if practical amorphous metals are to be used in a condition in which their desirable properties have been optimised and stabilised for use over long periods at their operating temperature. Several detailed review papers have been published, e.g.[1,2].

Throughout the discussion true **structural** relaxation is considered. This involves **atomic-scale** rearrangements of the structure and of the internal stresses but these internal stresses are assumed to average to zero over distances much greater than a typical atomic diameter. Similar rearrangement

processes will be involved in the relaxation of macroscopic stresses but such
'stress relaxation' is not dealt with in detail. The term 'reversible'
structural relaxation is applied to changes which are reversed by reversal of
a temperature change.

2.   EXPERIMENTAL OBSERVATIONS OF RELAXATION

Changes due to structural relaxation have been measured in many
properties[1,2]. For ferromagnetic amorphous metals these have included Curie
temperature, coercive field, initial susceptibility, field induced anisotropy
and magnetic after-effect. Because crystallisation usually occurs close to
the glass transition temperature nearly all observations of relaxation in
amorphous metals have been made below that temperature. Experiments are
normally conducted by annealing at constant temperature or with successive
anneals at alternate, constant temperatures or using a constant heating rate.
Use may be made of pre-annealing at a different constant temperature for a
suitable time interval before beginning the sequence of measurements.
Important, frequently-reported observations include: a linear dependence of
property change on the logarithm of the isothermal annealing time ('log-time
kinetics') with a rate of change that depends on the annealing temperature;
'crossover' phenomena when a switch in annealing temperature is made; and a
combination of reversible and irreversible changes.

The magnitudes of the changes vary greatly from property to property and
from glass to glass. Ferromagnetic metal-metalloid glasses containing more
than one magnetic ion generally show large, substantially reversible changes
of Curie temperature. Relative changes in electrical resistivity are usually
rather smaller and are predominantly but not entirely irreversible. Unless
measured with very great precision, changes in length (or density) appear to
be wholly irreversible below the glass transition temperature. Particularly
instructive are measurements of different properties of identical specimens
subjected to identical treatments.

It is usually tacitly assumed that the specimen has remained amorphous
and statistically uniform. There is increasing evidence that this is not
always so. A small amount of crystallisation in the bulk or at the surface
can be very difficult to detect but may lead to changes in the composition of
some of the glass which can produce changes in properties comparable with
those due to relaxation[3]. Even without crystallisation, incipient phase
separation or changes in the composition near the surface, because of
evaporation of one species or oxidation, can also lead to measurable changes
of property. Anelastic changes associated with the relaxation of macroscopic
stresses which had not been eliminated before the experiment can also be

troublesome.

3. THEORETICAL MODELS

The principal approaches to a mathematical description of the kinetics of relaxation involve the use of a spectrum of relaxation times, a spectrum of activation energies, an 'extended' exponential or the concept of free volume. These have been compared in more detail elsewhere[4].

The activation energy spectrum model[5] describes the kinetics in a system in which there is a spectrum of processes available to yield relaxational changes. The change in a property, P which will have occurred after isothermal annealing for time, t at temperature T is given by the equation:

$$\Delta P(t) = \int_0^\infty c(E) q(E,T) [1-\exp[(-\nu_0 t\{\exp(-E/kT)\})]] dE \quad (1)$$

$c(E)$ is a coupling function which relates the occurrence of a process of activation energy E to the consequential change in measured property. $q(E,T)$ is the (possibly temperature-dependent) spectrum (number density) of available processes. The expression in square brackets is the characteristic annealing function. The frequency factor, ν_0 is commonly taken to be of the order of a Debye frequency. A more thorough analysis leads to a slightly modified equation[6], which is not explored further here.

In simplest applications of this model a constant coupling function independent of energy, a box spectrum and a step function approximation to the annealing function are used. This step, which in reality is blurred over a range of a few kT, implies that no processes with activation energies greater than $E_0 (= kT\ln\nu_0 t)$ and all those with activation energies less than E_0 will have taken place by time t at temperature T. In this form the model has been used successfully to explain log-time kinetics[5], the interdependence of certain annealing times in crossover experiments[7], and the behaviour of field-induced magnetic anisotropy on annealing in a field and in (effectively) zero field[8].

More thorough applications of the model make use of a more realistic shape for the initial spectrum of available processes, e.g. a Gaussian[9], and the proper analytical form of the characteristic annealing function. From this it is straightforward to derive a 'master curve' representing the relative change in property as a function of the parameter E_0 in an isothermal anneal. For an initial spectrum which is independent of temperature, this master curve is also independent of temperature to a good approximation over the applicable range of temperatures because the characteristic annealing function does not change rapidly with temperature. Comparison of such a master curve with experimental data shows quite good agreement but also indicates that the initial spectrum is not wholly independent of temperature. From the same calculated master curve it is also possible to predict the results of isothermal annealing experiments at a series of temperatures. This shows that there will be a range of annealing temperatures for which linear log-time kinetics would be observable over several decades of time with a linear rate of change of the property that depends on temperature. This is in good agreement with experimental observations.

To proceed further it is necessary to incorporate a temperature dependence of the spectrum[9]. From consideration of a two-level system model, it can be shown that the equilibrium spectrum of processes should depend quadratically on temperature. Equation (1) can be modified to extend the calculation to non-isothermal situations. Thus it is possible to simulate constant heating-rate experiments in differential scanning calorimetry.

To use the activation energy spectrum model to analyse relaxational changes it is necessary to identify the parameter to which equation (1) applies. Thus in viscous flow, it is the accumulated shear deformation at a given stress which corresponds to $\Delta P(t)$[10]. If the viscosity changes linearly with time, as is often observed, the shear deformation varies linearly with log time.

4. REVERSIBILITY

As indicated above, most properties display a combination of reversible and irreversible structural relaxation. For a given material some properties may show a large reversible component and others a negligible one. The sign of the reversible change in electrical resistivity is not always the same as the sign of the irreversible change but in enthalpy changes the signs are the same[4]. These differences in behaviour indicate that fundamentally different structural changes must be involved. Separate activation energy spectra must be considered for each. These spectra may overlap in energy but their means need not coincide. For most properties the irreversible processes are generally slower suggesting that the spectrum for those processes is centred on a higher energy. However for some magnetic effects it appears that the irreversible processes occur at generally lower energies[11]. Until more is known about the details of atomic rearrangements occurring in structural relaxation it may not be possible to explain this difference.

An interaction between reversible and irreversible processes has been demonstrated in careful measurements of field-induced anisotropy[12]. These show a significant reduction in the **rate** of reversible relaxation as irreversible relaxation proceeds.

Reversible processes are often said to be associated with chemical short-range order (i.e. the statistical distribution of atomic species over a fixed pattern of sites); irreversible processes with topological short-range order (i.e the statistically defined pattern of atomic sites without regard to the distribution of atomic species)[1,2]. There is good evidence that some reversible processes do correspond to chemical short-range order-disorder processes[13]. However it is not clear that all reversible processes belong to this category.

5. STRUCTURAL INTERPRETATIONS

Although structural relaxation leads to measurable changes in the X-ray and neutron structure factors and in Mössbauer spectra, convincing interpretations in terms of specific atomic rearrangements have not been presented so far. It seems likely that many of the rearrangements involve the coordinated motion of groups of atoms rather than the jumping of a single atom to a vacant site. Computer modelling has shown how it is possible to distinguish rearrangements on the basis of atomic-scale stress distributions[14].

So far, the most fruitful approach in the search for a connection between activation energy spectra and actual atomic rearrangements has been the use of the concept of **thermally-activated** two-level systems (TLS)[5]. This leads to the idea that reversible processes correspond to TLS with relatively low splittings between the two states whereas irreversible processes correspond to TLS with relatively large splittings[6]. A careful analysis of measurements of enthalpy changes indicates that there is no correlation between the splitting and the barrier height (activation energy) which has to be surmounted for a transition to occur from the higher to the lower state[15]. TLS of different splittings will contribute differently to different physical properties.

6. CONCLUSIONS

Structural relaxation allows the possibility of optimising and stabilising properties, provided that the kinetics of such relaxation are carefully established. Quantitative modelling of the kinetics is becoming well-established, particularly through the activation energy spectrum model. Much remains to be done to relate the parameters of the chosen model to actual atomic rearrangements in the structure and to understand how the operation of a specific process of rearrangement is related to the magnitudes of the consequential changes in various physical properties. However, even without this deeper understanding, useful quantitative predictions of changes in properties can be made.

ACKNOWLEDGEMENTS

I am grateful to many colleagues for helpful discussions, especially Dr E. Woldt, Dr J.E. Evetts, Dr M.R.J. Gibbs, Dr A.L. Greer and Professor T. Egami, to whom I am also grateful for arranging for me to spend a sabbatical in the Department of Materials Science and Engineering of the University of Pennsylvania.

REFERENCES

1) T. Egami, Proceedings of the Acta-Scripta Workshop on Amorphous Metals and Semiconductors, eds. P. Haasen and R.I. Jaffe (Pergamon, 1986) p.222

2) A. von den Beukel, Proceedings of the ASM International Conference on Rapdily Solidified Materials, (ASM, 1986) p.193

3) Z. Altounian, J.O. Strom-Olsen and M. Olivier, Rapidly Solidified Alloys and Their Mechanical and Magnetic Properties, eds. B.C. Giessen, D.E. Polk and A.I. Taub (MRS, 1986) p.81

4) J.A. Leake, Metallic and Semiconducting Glasses, ed. A.K. Bhatnagar (Trans Tech) to be published

5) M.R.J. Gibbs, J.E. Evetts and J.A. Leake, J. Mat. Sci. 18 (1983) 278

6) G. Hygate and M.R.J. Gibbs, J. Phys. F. 17 (1987) 815

7) M.R.J. Gibbs, Proceedings of the Fifth International Conference on Rapidly Quenched Metals, eds. S. Steeb and H. Warlimont (North Holland, 1985) p.643

8) P.D. Hodson and J.E. Evetts, J. Magn. Magn. Mat. 59 (1986) 81

9) J.A. Leake, E. Woldt and J.E. Evetts, Proceedings of the Sixth International Conference on Rapdily Quenched Metals (1987), to be published in Mat. Sci. Eng.

10) A. Hernando, O.V. Nielsen and V. Madurga, J. Mat. Sci. $\underline{20}$ (1985) 2093

11) H.-Q. Guo, A. Hofmann, W. Fernengel and N. Moser, ref. (7), p.127

12) T. Jagielinski and T. Egami, IEEE Trans. Magn. MAG-21 (1985) 2002

13) E. Woldt and J.A. Leake, Proceedings of the Sixth International Conference on Liquid and Amorphous Metals (1986), to be published

14) T. Egami and V. Vitek, J. Non-Cryst. Solids 61 & 62 (1984) 499

15) E. Woldt, Ph.D. Thesis, University of Cambridge, U.K. (1986)

MAGNETIC PROPERTIES OF AMORPHOUS IRON-RICH BINARY ALLOYS

D.H. RYAN

McGill University, Physics Department, 3600 University Street, Montreal, Quebec, Canada H3A 2T8

Amorphous alloys of the form a-Fe_xM_{100-x} (x>85) exhibit a range of magnetic properties. Alloys with M=B are ferromagnetic, while those with Y and Sc are asperomagnets. A development from ferromagnetic to asperomagnetic ordering is seen as x→100 for alloys with M=Zr and Hf, and a transition due to the freezing of transverse spin components is observed well below T_c. An explanation in terms of ferromagnetic spin glass theory is proposed, and the properties of pure amorphous iron extrapolated.

Amorphous iron-based binary alloys of the form Fe_xM_{100-x} in the iron-rich limit (x ≥ 85) for the four alloying elements : B, Y, Zr and Hf exhibit two distinct forms of magnetic ordering as x→100[1]. Amorphous Fe-B is ferromagnetic, with an ordering temperature (T_c) which falls from 570K with increasing x, while the iron moment is constant at $2.1\mu_B$[2]. By contrast, a-Fe-Y exhibits a random, non-collinear, or asperomagnetic, spin structure[3] and while the iron moment is $2.0\mu_B$ in a-$Fe_{88}Y_{12}$, the ordering temperature is only 130K. Magnetisation curves fail to saturate even in fields of 20T, and always remain below that expected from Mössbauer measurements.

Alloys with Zr[4] and Hf[5] lie between these two extremes. Both systems exhibit ferromagnetic order, but T_c peaks at ~ 300K for x ~ 85 with an iron moment of $1.7\mu_B$. A ferromagnetic to spin glass transition well below T_c has been suggested on the basis of the appearance of irreversible magnetization[6], though others attribute this to the normal increase in coercivity expected in ferromagnets at low temperatures[7,8]. A quadratic temperature dependence of the magnetization has been cited as evidence for weak itinerant ferromagnetism in a-Fe-Zr[9], but this behaviour gives way to a $T^{3/2}$ spin wave like law for B > 0.4T[8]. The failure of the magnetisation at 4.2K to saturate, even in fields of 20T, for both a-Fe-Zr[9] and a-Fe-Hf[10] suggests an asperomagnetic spin structure, and support for random spin freezing comes from the absence of long-range ferromagnetic order below T_c as determined by neutron scattering[12].

Much of the uncertainty arises from interpreting a single measurement, frequently on one composition; in an attempt to resolve the ferromagnetic spin glass controversy, we have undertaken a systematic study of a number of amorphous iron-rich binary alloy over the ranges accessible by melt-spinning and using a variety of experimental techniques.

Alloys of the form a-Fe$_x$M$_{100-x}$ were prepared by melt spinning under helium. No traces of any crystalline phases were found on either surface of the ribbons. The composition limits were: $88 \leq x \leq 93$ for M=Zr, $90 \leq x \leq 92$ for M=Hf and $89 \leq x \leq 91$ for M=Sc.

Mössbauer spectra and magnetisation curves were obtained between 4.2K and 290K. Magnetic ordering temperatures were determined by ac-susceptibility, zero velocity thermal scans using the Mössbauer spectrometer, and the kink-point method from dc magnetisation. The same samples were used for all measurements, so that direct comparisons could be made without any possibility of error from variations in sample quality or composition along the ribbons.

Hydrogenation was also used as a further probe of the magnetic properties. Charging was achieved electrolytically[13], and contents were determined by thermo-piezic analysis (TPA)[14].

Ordering temperatures for a-Fe-Zr fall linearly from 265K at x=88 to 163K at x=93. A similar decline in T_c is also observed for a-Fe-Hf. By contrast, ordering temperatures for a-Fe-Sc are much lower and exhibit essentially no concentration dependence. These results are summarised in fig. 1.

All of the samples have a significant high field slope in their 4.2K magnetisation curves (fig. 2). The slope increases with x for alloys with

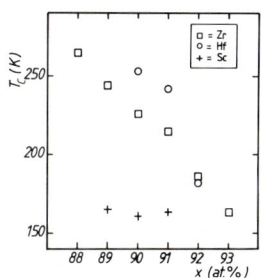

Figure 1. Variation of magnetic ordering temperature with composition for a-Fe$_x$M$_{100-x}$, M = Zr, Hf and Sc.

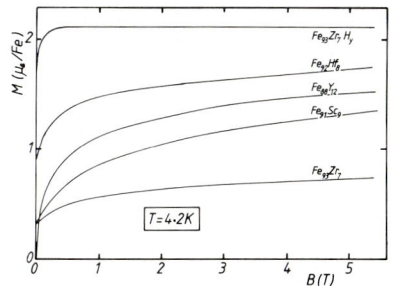

Figure 2. Magnetisation curves at 4.2K for a-Fe$_x$M$_{100-x}$ M = Zr, Hf, Sc and Y[3]. Also shown is a typical curve for a hydrogenated sample.

Zr and Hf. Alloys with Sc show little composition dependence in the range studied and are very similar to those with Y. The readily saturated component of the magnetisation M_z, defined by extrapolating the high field part of the curves (B>1T) to zero field, falls rapidly with increasing x for both a-Fe-Zr and a-Fe-Hf and is in all cases smaller than that expected from the moments obtained from Mössbauer measurements at 4.2K, (M_o); even in fields of 20T the magnetisation falls short of the expected values. Unlike M_z, M_o exhibits no composition dependence, the average values for each alloying element being : Zr and Hf $1.7\mu_B$, Y $2.0\mu_B$ [3] and Sc $1.6\mu_B$ [15].

The coercivity of alloys containing Hf and Zr decreases exponentially with temperature, and is in all cases negligible in comparison with the measuring field at the point where irreversibility develops in the thermomagnetisation curves (T_{xy}). In fact, the (much lower) temperature at which the coercivity is equal to the applied field is readily discerned in such measurements[6].

Low temperature specific heat measurements show that the linear component is unusually large, being $22mJ/mol/K^2$ in $Fe_{88}Y_{12}$, and $32mJ/mol/K^2$ in $Fe_{92}Hf_8$ (a-Fe-Zr exhibits non-ergodic behaviour below ~ 20K so no values could be obtained[16]). The electronic contribution is expected to be of order $5mJ/mol/K^2$ (the value found in $Co_{90}Zr_{10}$ [17], a simple ferromagnet), the remainder is attributed to magnetic excitations which result in a linear reduction in M_o with temperature.

Charging with hydrogen has a profound effect on the magnetic properties of all four series of samples[13]; loading levels of only 10 at. % of hydrogen are sufficient to transform them into ferromagnets with $T_c \geq 400K$, iron moments of $2.1\mu_B$ and very low coercivities even at 4.2K. Furthermore, the high field slope of the magnetisation curves is eliminated (fig. 2) and M_o and M_z are in agreement.

Materials that exhibit a large high field slope in their magnetization curves, and fail to attain the saturation magnetisation expected for the atomic moments deduced from their average hyperfine field cannot be collinear ferromagnets. Non-collinearity is confirmed directly by the persistence of lines 2 and 5 of the Mössbauer spectra in applied fields[10,11]. It follows that all of the alloys are asperomagnets. A measure of the non-collinearity may be obtained by assuming that the moments are distributed uniformly about some preferred direction over a cone of half-angle ψ. From M_z/M_o we find that $\psi \sim 35°$ in a-$Fe_{91}Hf_9$, 76° in a-$Fe_{90}Sc_{10}$ and 110° in a-$Fe_{93}Zr_7$. The large low field susceptibility corresponds to aligning the principal axes of these "cones", and the high

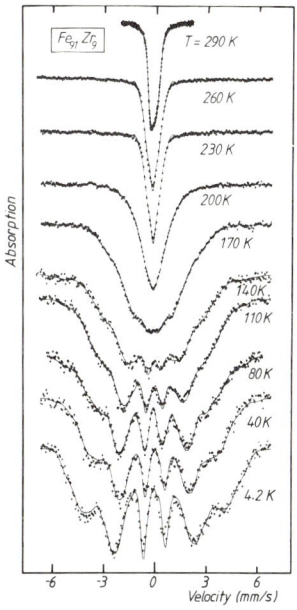

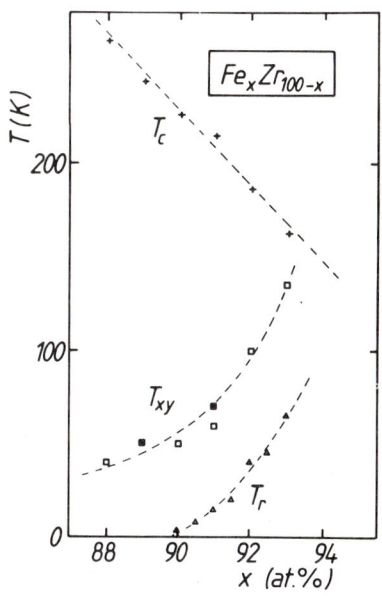

Figure 3. Mössbauer spectra for a-Fe$_{91}$Zr$_9$ showing the development of magnetic order. T_c for this alloy is 215K.

Figure 4. Proposed phase diagram for a-Fe-Zr[10]. Showing : T_c, onset of magnetic order; T_{xy}, transverse spin freezing; T_r development of strong irreversibility.

field slope, above ~1T to the slow compression of ψ by the applied field. The non-collinear spin structure results from mixed ferromagnetic and antiferromagnetic exchange, arising from the distribution of iron-iron distances, which, being centered around 0.25nm, lies between the values found in ferromagnetic and antiferromagnetic fcc iron[18].

The behaviour at T_c is also inconsistent with ferromagnetic ordering. Although T_c itself is well defined, reported critical exponents for the transition are far from those expected for a Heisenberg ferromagnet[19,20,21] (except in one case[22]). Furthermore, values for T_c obtained with an applied field (ac or dc) are larger than those in zero field. While this difference is only 5K in a-Fe$_{88}$Zr$_{12}$, it is nearly 40K in a-Fe$_{93}$Zr$_7$, demonstrating that a magnetic field has a strongly increasing effect on the development of magnetic order as the non-collinearity becomes more pronounced. Magnetically coupled moments have been observed above T_c in a-Fe-Zr [19]. The presence of magnetic broadening in the Mössbauer spectrum of a-Fe$_{91}$Zr$_9$ is clearly visible at 200K (fig. 3) almost 50K <u>above</u> T_c, and significant amounts of iron carry no ordered moment even at 170K almost 50K below T_c. It is clear that magnetic order in a-Fe-Zr is not established through a cooperative transition involving all of the moments, but rather through the growth of clusters of strongly coupled moments. T_c probably corresponds to the formation of an infinite percolating cluster.

Since the coercivity is negligible at the temperature at which the irreversible magnetisation appears (T_{xy}), it cannot be the cause of the irreversibility. From Mössbauer spectra in applied fields we can see that only the z-component of the moment is non zero for $T_c>T>T_{xy}$; the xy-component does not appear until $T<T_{xy}$ [10,11]. This is confirmed by the observation that M_z and M_o are in agreement above T_{xy}, the difference between them below T_{xy} being due to the frozen xy components of the iron moments.

The proposed phase diagram for a-Fe-Zr (fig. 4) is very similar to that predicted by Gabay and Toulouse[24] for the mean field theory of a mixed exchange system or ferromagnetic spin glass, with the appearance of coercivity, T_r being identified with the third predicted transition.

The key differences are : (i) the observed ordering at T_c does not occur via ferromagnetic phase transition (ii) the transition at T_{xy} involves a freezing of transverse degrees of freedom, not a canting of the moments (iii) it is not clear that a transition occurs at T_r (but it is unclear what the theory predicts here).

The magnetic properties of a-Fe_xM_{100-x} with M={Y,Sc} , {Zr,Hf} and {B} may be explained in terms of the above model. The magnetic order forms a steady progression being asperomagnetic, asperomagnetic turning ferromagnetic as x is reduced, and ferromagnetic respectively, and simply reflects the changing balance of competition between positive and negative exchange. It is interesting to note that the appearance of a high field slope and coercivity has been reported for a-$Fe_{90}B_{10}$[25] suggesting that all iron based amorphous alloys may be expected to develop non-collinear order as $x \to 100$. Differences in moment arise mainly from different degrees of hybridisation between the Fe-3d electrons and those of the alloying element.

The observed iron moment increase on hydrogenation[10,11,13] is due to a reduction in the hybridisation as a result of the hydrogen bonding to the M atoms. The moment in the hydrided alloys is uniform at 2.1μ_B/Fe, consistent with the reduction of chemical effects. The increase in T_c and the suppression of spin-glass behaviour follows directly from the spin glass model : increasing the Fe-Fe distance by a few percent is sufficient to eliminate all antiferromagnetic coupling and leads to simple ferromagnetic behaviour.

Since there is no evidence for a reduction in the iron moment as $x \to 100$, pure amorphous iron can be expected to exhibit a moment of 1.7-2.0μ_B, (depending on whether it tends to the Y series or Zr series limit). Extrapolation of T_c to zero at $x = 100^{26}$ cannot be reasonable in view of the large iron moment. Spin glass theory suggests that T_c is constant beyond the point where it meets T_{xy}, so an ordering temperature of ~150K is postulated for the x=100 limit. Amorphous iron will be spin glass like, it will order in a random non-collinear structure with no net magnetisation - a speromagnet.

This work was supported by grants from Trinity College, Dublin, and Trinity Trust in Ireland, and the National Science and Engineering Research Council in Canada.

REFERENCES

1. J.M.D. Coey and D.H. Ryan, IEEE Trans. Mag. MAG-20 (1984) 1280.
2. C.L. Chien and K.M. Unruh, Nucl. Inst. and Met. 199 (1982) 193.
3. J.M.D. Coey, D. Givord, A. Liénard and J.P. Rebouillat, J. Phys. F11 (1981) 2707.
4. K.M. Unruh and C.L. Chien, J. Magn. Magn. Mater. 31-34 (1983) 1587.
5. S.W. Liou, G. Xiao, J.N. Taylor and C.L. Chien, J. Appl. Phys. 57 (1985) 3536.
6. H. Hiroyoshi and K. Fukamichi, Phys. Lett. 85A (1981) 252.
7. D.A. Read, T. Mayo and G.C. Hallam, J. Magn. Magn. Mater. 44 (1984) 279.
8. W. Beck and H. Kronmüller, Phys. Stat. Sol. 132(b) (1985) 449.
9. S.N. Kaul, Phys. Rev. B 27 (1983) 6923.
10. D.H. Ryan, J.M.D. Coey, E. Batalla, Z. Altounian and J.O. Strom-Olsen, Phys. Rev. B35 (1987) (in press).
11. D.H. Ryan, J.M.D. Coey and J.O. Strom-Olsen, J. Magn. Magn. Mater. 67 (1987) (in press).
12. J.J. Rhyne and G.E. Fish, J. Appl. Phys. 57 (1985) 3407.
13. J.M.D. Coey, D.H. Ryan and Yu Boliang, J. Appl. Phys. 55 (1984) 1800.
14. D.H. Ryan and J.M.D. Coey, J. Phys. E19 (1986) 693.
15. R.K. Day, J.B. Dunlop, C.P. Foley, M. Ghafari and H. Park, Solid State Commun. 56 (1985) 843.
16. D.H. Ryan, J.M.D. Coey and R. Buder, Phys. Rev. Lett. 58 (1987) 385.
17. Y. Obi, L.C. Wang, R. Motsay, D.G. Onn and M. Nose, J. Appl. Phys. 53 (1982) 2304.
18. W. Kümmerle and U. Gradmann, Solid State Commun. 24 (1977) 33.
19. H. Yamamoto, H. Onodera, K. Hosoyama, T. Masumoto and H. Yamauchi, J. Magn. Mater. 31-34 (1983) 1579.
20. H. Yamauchi, H. Onodera and H. Yamamoto, J. Phys. Soc. Japan 53 (1984) 747.
21. K. Winschuh and M. Rosenberg, J. Appl. Phys. 61 (1987) 4401.
22. S.N. Kaul, A. Hofmann and H. Kronmüller, J. Phys. F 16 (1986) 365.
23. S.M. Fries, C.L. Chien, J. Crummenauer, H.G. Wagner and U. Gonser, Hyp. Int. 27 (1986) 405.
24. M. Gabay and G. Toulouse, Phys. Rev. Lett. 47 (1981) 201.
25. H. Hiroyoshi, K. Fukamichi, M. Kikuchi, A. Hoshi and T. Masumoto, Phys. Lett. 65A (1978) 163.
26. D.A. Read, T. Moyo and G.C. Hallam, J. Magn. Magn. Mater. 54-57 (1986) 309.

CORRELATION PROPERTIES OF INHOMOGENEITIES OF AMORPHOUS MAGNETS

V.A. IGNATCHENKO and R.S. ISKHAKOV
Kirensky Institute of Physics, Krasnoyarsk 660036, USSR

Review of further development of theoretical and experimental investigations of correlation properties of large-scale inhomogeneities in amorphous mediums is given.

1. INTRODUCTION

Theoretical-experimental methods of investigating large-scale inhomogeneities in real amorphous magnets based on the correlation theory of random functions are being developed beginning from our publications[1]. It is known that the main characteristic of inhomogeneities is a correlation function $K(r)$ which is characterized by two parameters: mean square fluctuation g and correlation radius r_c. Practically, there were no methods of measuring long correlation radii for amorphous magnets before our works. At present they are developed for inhomogeneities of exchange, magnetization and anisotropy. For this purpose:
(a) A phenomenological theory containing arbitrary correlation functions of inhomogeneities of exchange, magnetization and anisotropy is constructed;(b) within the framework of this theory the effects absent in an ideal crystal and strongly dependent on long correlations are found;(c) unknown correlation functions are modelled by simple expressions with arbitrary (unknown) correlation radii r_c and fluctuation values g, and analytical dependences of these effects on r_c and g are calculated;(d) two magnetostructural methods of measuring r_c and g are developed which consist in the corresponding carrying out of experiments on spin wave resonance (SWR) and on the law of approach of magnetization to saturation (LAS) and the corresponding processing of the data of these experiments;(e) values g and r_c, their concentration and temperature dependences are studied for a number of alloys.

The review of theoretical and experimental results obtained by 1984 has been published[2]. Here we dwell on further development of correlation methods of investigating large-scale inhomogeneities.

2. MANIFESTATION OF CORRELATION CHARACTERISTICS IN DIFFERENT PHYSICAL EFFECTS

In recent years methods of SWR and LAS were widely used for investigation of correlation characteristics g and r_c of amorphous alloys[3,4,5]. At the same time the theory of other methods for measuring g and r_c was being developed. It was shown that the corresponding modifications must manifest themselves in the dispersion laws of plasma, electromagnetic[6] and elastic[7] waves; in the field dependence of magnetic resonance frequency in ferrimagnets[8]; in the temperature dependence of magnetoresistance[9].

At present a number of approaches to theoretical description of SWR in multilayer systems[10,11] are being developed. Since the real multilayer systems are always partially stochastized (random fluctuations of layer thickness etc.) we have suggested to apply to these systems the correlation theory of random functions introducing periodicity into correlation function[12]. This enabled to measure a multilayer period by SWR method experimentally.

3. POSSIBILITY OF EXPERIMENTAL DETERMINATION OF A CORRELATION FUNCTION TYPE

3.1. Classification of correlation functions

In our works[13] it has been shown that in a number of cases the group of monotone decreasing functions which are usually applied for modelling inhomogeneities does not absolutely suit for this purpose. This group must be considerably widen.

Consider the situation corresponding to an inhomogeneously deformed magnetic crystal. Inhomogeneity in the location of atoms is described by the tensor of local structural deformations u_{ij} which exist at T=0 as a metastable state (in certain limits such a model is valid for an amorphous magnet too). Random functions $A(x)$ describing space fluctuations of all spin parameters - exchange, magnetization, anisotropy - will be determined by these structural deformations: $A(x) \sim u_{ij}(x)$. Let statistics of the vector components of atomic displacement u_i be described by the monotone correlation function $K_o(r)$ and the corresponding monotone spectral density (Fourier-transform) $S_o(k)$. Then statistics of u_{ij} and of all $A(x)$ will be determined by u_i derivatives. This corresponds to multiplication of $S_o(k)$ by k^2, and the new spectral density $S(k)$ will vanish to zero at k=o; $K(r)$ will have a negative half wave (i.e. they will not be monotone decreasing functions already).

It is clear that such situation is valid not only for deformations, but for all physical values which are derivatives of other physical values. Therefore, we have introduced a general classification of correlation functions of inhomogeneities having divided them into two types: 1st type if $S(0) \neq 0$, 2nd - if $S(0)=0$.

For small k we have $S(k) \sim k^{2n}$, where index 2n characterizes multiplicity of zero of the spectral density. Value n=o corresponds to the first type functions, values $n \geqslant 1$ - to different functions of the second type.

We have shown[13] that a number of physical effects considerably depends on the value of index n of correlation functions of inhomogeneities.

3.2. Magnetization curve

The magnetization curve in the region of sufficiently large fields for n=1 has been obtained earlier[14]; using the same method we obtain[13] for n=1

$$M_z / M_o = 1 - DH_a^2 / (H_c^{1/2} + H^{1/2})^4, \qquad (1)$$

where H_a is the anisotropy field, H_c is the correlation field.

Imre and Ma[15] were the first to pay attention to the fact that the dispersion of transverse components of magnetization diverges at H=0. They formulated the general conclusion about instability of a ferromagnetic state with respect to the action of as small as desired local anisotropy. From (1) it is seen that the result of Imre and Ma is not so general as it is accustomed to consider: it depends on index n of the correlation function and is valid only at n=o. For all functions with $n \geqslant 1$ the ground ferromagnetic state is stable.

However, differential magnetic susceptibility at H=0 diverges both for n=o and n=1. For the susceptibility to remain finite n must be more or equal to 2. There is a general regularity: for inhomogeneities whose spectral density $S(k)$ has zero of the 2n order at the point k=o, n-1 of the first derivatives of M_z in H are finite at H=0.

Thus, the magnetization curve caused by anisotropic inhomogeneities has a number of properties considerably determined by the type of the correlation function (i.e. by index n); this can make it possible to determine the correlation function type experimentally.

3.3. Spin wave damping

In papers[13,16] the general laws of modification of spin wave damping for the correlation functions of the second type have been obtained:

a) at fluctuations of exchange and magnetization

$$R(k) \sim \begin{cases} k^{2n+5}, & k \ll k_c \\ k^3, & k \gg k_c \end{cases} \quad (2)$$

b) at fluctuations of anisotropy

$$R(k) \sim \begin{cases} k^{2n+1}, & k \ll k_c \\ k^{-1}, & k \gg k_c \end{cases} \quad (3)$$

Damping due to scattering of spin waves on inhomogeneities was calculated in many papers, however, only the correlation functions with n=o being used.

From our general expressions it is seen that the laws of damping considerably depend on index n of the correlation function of inhomogeneities. Thus, the determination of the correlation function type is, in principle, possible from the experimental investigation of spin wave damping on inhomogeneities. Here, however, one should overcome difficulties connected with the extraction of this damping from the summary one due to different physical mechanisms. It is seen from the next section that it is not so easy to do.

4. EXPERIMENTAL INVESTIGATIONS OF SWR LINE WIDTHS

In our papers[1] it has been shown theoretically (for n=o) that in the region $k \sim k_c$ alongside with the dispersion law modification a characteristic modification of spin wave damping must be observed; as it is seen from (2) and (3) at n≠o it must manifest itself sharper. This modification was firstly investigated in the paper[17], where a change of the dependence of SWR line widths $\Delta H(k)$ was found in the vicinity of some k. However, the dependence of ΔH on k appeared to be in a sharp contradiction with the formulas (2), (3).

Being stimulated by the work[17] our experimentators have also conducted systematical investigations of SWR line widths. Fig. 1 shows the dependence of $\Delta H(k)$ for three alloys in logarithmic coordinates. It is seen that functions $\lg \Delta H$ experience a sharp break at some value $k=k_c$. Simultaneously, on the same alloys we observed the dispersion law modification with a characteristic break at the same point. Thus, the observation of the $\Delta H(k)$ modification is a reliable method of measuring correlation radii of inhomogeneities $r_c \sim k_c^{-1}$. However, as well as in the paper[17], there is a discrepancy between theoretical (2),(3) and experimental dependences $\Delta H(k)$. It means that it is not the mechanism of spin wave scattering on inhomogeneities that mainly contributes to the SWR line width.

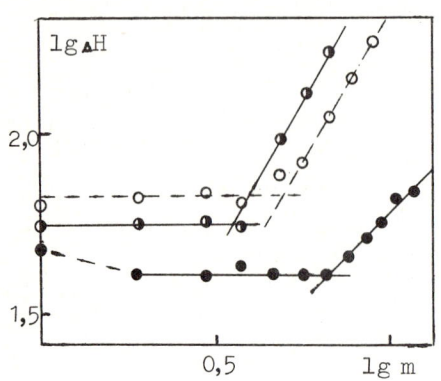

FIGURE 1
Dependence of ΔH on SWR mode number $m \sim k$ in logarithmic coordinates
o - double-phase(hcp & fcc) Co
◐ - fcc solid solution $Co_{93}P_7$
● - amorphous alloy $Co_{87}Zr_{13}$

It is clear that the experimental investigation of $\Delta H(k)$ will give much interesting information. In Fig. 1 two alloys have a crystalline structure, and one alloy has an amorphous structure. It is seen that if in the region $k \sim k_c$ the dependence of a line width is represented in the form: $\Delta H(k) \sim k^p$, then for crystalline alloys $p \approx 2$, and for an amorphous one $p \approx 1$. This effect, as well as the whole problem of SWR line width, still waits its theoretical explanation.

In conclusion of this section adduce a curious fact. We found in the literature that the break $\Delta H(k)$ similar to one shown in Fig. 1 was observed long ago[19,20]; when the dependence of frequency on k^2 was plotted according to the data of these authors we saw the dispersion law modification too. However, there was no theory[1] that time and these results were forgotten; now it may be shown that exchange fluctuations with $r_c \approx 200$ Å were observed in papers [19,20].

REFERENCES

1) V.A.Ignatchenko, R.S.Iskhakov, Zh. Eksp. Teor. Fiz. 72 (1977) 1005; 74 (1978) 1386; 75 (1978) 1438.

2) V.A.Ignatchenko, R.S.Iskhakov, in Physics of Magnetic Materials. Jadwisin' 84 (Poland) Part 1 - Invited Papers. Singapore-Philadelphia: World Scientific, 1985, p.527.

3) R.S.Iskhakov, M.M.Brushtunov, A.S.Chekanov, Preprint N429 F, Krasnoyarsk: Inst. of Phys. 1987.

4) R.S.Iskhakov et al. Doklady Akad. Nauk SSSR 284 (1985) 854.

5) R.S.Iskhakov et al. Fiz. Met. Metalloved. 61 (1986) 265.

6) V.A.Ignatchenko, Yu.I.Man'kov, F.V.Rakhmanov, Zh. Eksp. Teor. Fiz. 87 (1984) 228.

7) L.I.Deich, V.A.Ignatchenko, Fiz. Tverd. Tela 27 (1985) 1883; 29 (1987) 825.

8) I.V.Bogomaz, V.A.Ignatchenko, Preprint N324 F, Krasnoyarsk: Inst. of Phys. 1985.

9) Yu.I.Man'kov, F.V.Rakhmanov, Fiz. Met. Metalloved. 62 (1986) 1082.

10) R.P. van Stapele, F.J.A.M.Greidanus, J.W.Smits, J. Appl. Phys. 57 (1985) 1282.

11) K.Vayhinger, H.Kronmüller, J. Magn. Magn. Mat. 62 (1986) 159.

12) V.A.Ignatchenko, R.S.Iskhakov, A.S.Chekanov, L.A.Chekanova, Progr. and Abstracts. 3d ICPMM, Szczyrk-Bila (Poland) Sep. 9-14, 1986, p.170.

13) V.A.Ignatchenko, R.S.Iskhakov, Preprint N351 F, Krasnoyarsk: Inst. of Phys. 1985; N371 F, 1986.

14) V.A.Ignatchenko, R.S.Iskhakov, G.V.Popov, Zh. Eksp. Teor. Fiz. 82 (1982) 1518.

15) Y.Imry, S-K.Ma, Phys. Rev. Lett. 35 (1975) 1399.

16) I.V.Bogomaz, V.A.Ignatchenko, Preprint N363 F, Krasnoyarsk: Inst. of Phys. 1986.

17) L.J.Maksymowicz, D.Sendorek-Temple, R.Zuberek, J. Magn. Magn. Mat. 58 (1986) 303.

18) R.S.Iskhakov, A.S.Chekanov, L.A.Chekanova, Preprint N432 F, Krasnoyarsk: Inst. of Phys. 1987.

19) P.E.Wigen, Phys. Rev. 133 (1964) A1557.

20) T.G.Phillips, H.M.Rosenberg, Phys. Lett. 8 (1964) 298.

THERMOMAGNETIC ANALYSIS OF THE CRYSTALLISATION OF $Fe_{77}B_{15}Sc_8$

C.P. FOLEY, R.K. DAY, J.B. DUNLOP and R.B. ROBERTS

CSIRO Division of Applied Physics, National Measurement Laboratory, Lindfield, Australia 2070

A new amorphous alloy system, Fe-B-Sc, has been prepared by single roller melt quenching. On heating to crystallisation, these new alloys pass through a glass transition and various distinct crystallisation steps. As-quenched and annealed alloys were examined by X-ray diffraction, resistivity measurements, differential scanning calorimetry, thermal expansion and thermomagnetic analysis. The crystallisation process will be discussed in terms of the precipitation of metastable Fe-B-Sc compounds and of Fe_2Sc and alpha-Fe.

1. INTRODUCTION

We have recently discovered that by melt spinning, glassy alloys can be formed in the binary metal system, Fe-Sc, over a narrow composition range (1). We extended this study to consider a new amorphous alloy system, Fe-B-Sc, which is a combination of the binary alloys Fe-B and Fe-Sc. Routine thermal analysis of these materials indicates that complicated phase transformations occur during crystallisation. We can find no phase diagram for Fe-B-Sc alloys nor are there any magnetic or chemical data available on Fe-B-Sc ternary compounds.

As a first step, we have selected $Fe_{77}B_{15}Sc_8$ to illustrate the way in which a variety of techniques need to be used to gain some understanding of complex crystallisation processes. Samples were prepared by single roller quenching (1). Crystallisation processes and products were studied by the following techniques: differential scanning calorimetry (DSC), electrical resistivity, X-ray diffraction (XRD), scanning electron microscopy, thermal expansion and thermomagnetic analysis (TMA). In particular, TMA has been used for the first time to identify intermediate products during crystallisation by studying the variation of magnetisation and the magnetic ordering temperatures at each stage during crystallisation.

2. RESULTS AND DISCUSSION

The DSC trace for $Fe_{77}B_{15}Sc_8$ is shown in figure 1. There is a broad low energy exotherm at 500°C with a minimum at 560°C. This combination is usually attributed to a glass transition (2). The glass transition is immediately followed by a sharp exotherm of energy 93.5 mJg^{-1} at 567°C and less energetic

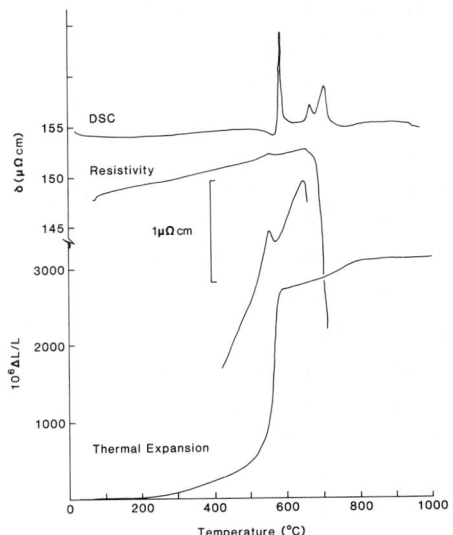

FIGURE 1
(a) DSC, (b) Electrical Resistivity, and (c) Thermal Expansion of $Fe_{77}B_{15}Sc_8$.

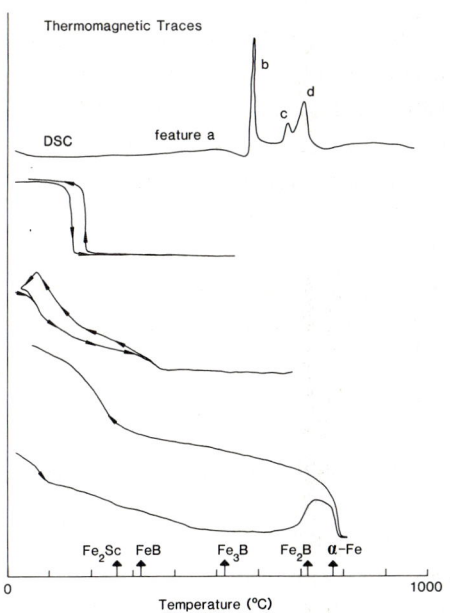

FIGURE 2
Thermomagnetic traces of $Fe_{77}B_{15}Sc_8$.

exotherms which occur at 673°C and 711°C. By sequentially heating to each of these features, we found them to be independent of each other.

2.3 Glass transition

In general, metallic glasses in the as-quenched state are not in internal equilibrium and relax structurally whenever atoms attain appreciable mobility causing many properties to change. Thermal expansion under slight tension (fig. 1) shows that the glass has a large increase in length which would be expected from reduced "viscosity". The rapid length extension ends sharply with the onset of crystallisation at 560°C. During the glass transformation, there is no apparent change in either the XRD traces or resistivity. After heating to 560°C, the Curie temperature was found to increase by 33°C. The behaviour described here is typical of glass transitions (2).

2.2 Onset of crystallisation

The DSC trace of $Fe_{77}B_{15}Sc_8$ shows an energetic exotherm at 567°C of a magnitude normally associated with crystallisation. As shown in figure 1 the resistivity of $Fe_{77}B_{15}Sc_8$ over the temperature range 300 to 590°C has a small peak at 560°C of about 0.2 µohmcm, not present on subsequent reheating, which corresponds to the onset of crystallisation. Moreover, the slopes of the resistivity before and after the maximum are similar. Identical behaviour has been observed in other glasses (3). Although not

not well understood, it has been attributed to the formation of microcrystallites with complex structures. In this environment the crystallites are so small that grain boundary scattering of electrons predominates. In the limit of very small grain sizes, the electrical resistivity of crystalline material will approach that of an amorphous alloy.

XRD shows a very fine, low intensity structure superimposed on a typical amorphous material trace. This is interpreted as either partial crystallisation or appreciable line broadening due to small crystallites. SEM micrographs show evidence of sub-micrometre structures.

TMA measurements in figure 2 show a small increase in magnetisation which possibly occurs due to initial nucleation of alpha-Fe or Fe_2B microcrystallites. The cool down cycle indicates the presence of materials with magnetic ordering transitions at $370^{\circ}C$ and $90^{\circ}C$. Over the temperature range of 20 to $400^{\circ}C$, the magnetisation shows a temperature hysteresis. O'Handley et al. (4) have reported similar behaviour. We attribute it to complex magnetic behaviour possibly associated with a magnetic ordering having a Neel temperature of $95^{\circ}C$. The magnetic ordering at $370^{\circ}C$ does not correspond with the Curie temperature of any known binary material and is possibly a metastable ternary of Fe-B-Sc not previously identified.

2.3 Further crystallisation (673 and $711^{\circ}C$)

DSC, resistivity, thermal expansion and XRD indicate further crystallisation or re-crystallisation steps at $673^{\circ}C$ and $711^{\circ}C$. TMA measurements associated with the second crystallisation exotherm (fig. 2) indicate the presence of materials with magnetic ordering temperatures of $100^{\circ}C$ and $440^{\circ}C$. The lower temperature would be associated with the micro-structured precipitate described above. The magnetic ordering occurring at $440^{\circ}C$ is possibly the Curie temperature of Fe_2B containing scandium in solution.

After the third crystallisation peak, the only products observable by the TMA and XRD measurements are alph-Fe and Fe_2Sc. We believe the boron is incorporated in the Fe_2Sc phase.

REFERENCES
1) R.K. Day, J.B. Dunlop, C.P. Foley, M. Ghafari and H. Pask, Solid State Comm. 56 (1985) 843.
2) H.S. Chen, Rep. Prog. Phys. 43 (1980) 353.
3) H.S. Chen and D. Turnbull, J. Chem. Phys. 48 (1968) 2560.
4) R.C. O'Handley, B.W. Corb, J. Megusar and N.J. Grant, J. Non-Cryst. Sol. 61 (1984) 773.

SURFACE EFFECTS AND MAGNETIC PROPERTIES OF AMORPHOUS Co_xY_{1-x} THIN FILMS

J.M.ALAMEDA, M.C.CONTRERAS, A.R.LAGUNAS†

Departamento de Fisica.Universidad de Oviedo.33007 Oviedo.Spain
†Instituto C.de los Materiales.CSIC.Madrid.Spain

Amorphous Co_xY_{1-x} films ($0.5<x<0.8$, thickness:100 nm) were grown on glass substrates maintained at room temperature. We use a special DC assisted triode sputtering system[1] (Ar pressure:4×10^{-4} torr) where simultaneous codeposition from two independently polarized Y and Co cathodes is available. Magnetic(hysteresis loops, transverse biased initial susceptibility(TBIS):$\chi_t(\beta)$) and magneto-optical spectrometry measurements were carried out on different experimental set-ups already described[2,3]. The magneto-optical transverse Kerr effect is usualy characterized by the phenomenological parameter: $\delta_k \equiv \Delta R/R$ [4].

In Figure 1, δ_k v.s. x for amorphous Co_xY_{1-x} films is shown. A monotonous decrease of δ_k with x up to $x=0.65$ is apparent. For higher Y concentrations δ_k remains constant in magnitude and a change in the polarity of the transverse Kerr effect is found (in the following this will be denoted as $\delta_k<0$). Figure 2 shows δ_k v.s. λ behaviour for films where $x=0.6$ and $x=0.75$. As it is seen from this figure, δ_k hardly varies in the spectral range considered even in Y-rich samples.

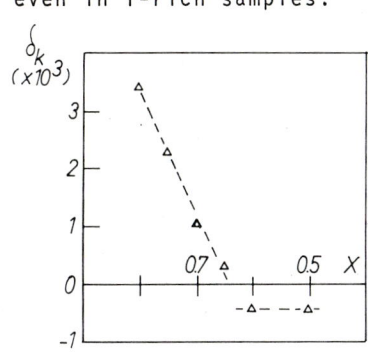

FIGURE 1
δ_k v.s. x for Co_xY_{1-x} films.

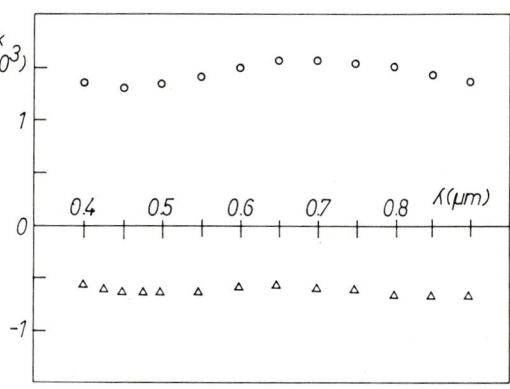

FIGURE 2
δ_k v.s. λ for Co_xY_{1-x} films. Open circles:$x=0.75$.
Open triangles:0.6

The change in Kerr effect polarity is related to a definite change in both the magnitude of the coercive force H_c and hysteresis loop shape. Figures 3 and 4(insets) illustrate these facts. The behaviour of TBIS is clearly differenciated for films where $\delta_k>0$ and $\delta_k<0$ (i.e.: x>0.65 and x<0.65 respectively). This is clear from Figures 4.a , 4.b and 4.c.

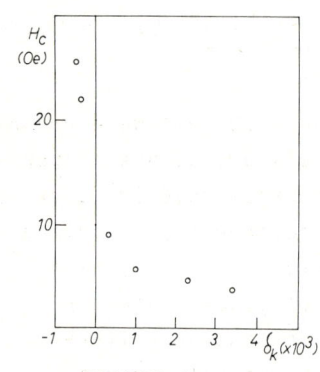

FIGURE 3
H_c v.s. δ_k for Co_xY_{1-x} films
(See also fig.1)

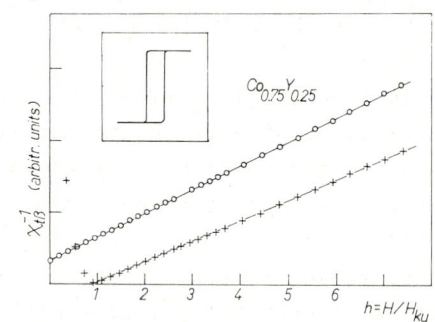

FIGURE 4.a
$1/\chi_t(\beta)$ v.s. H for Co_xY_{1-x} (x=0.75)

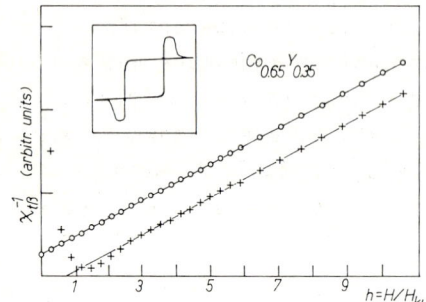

FIGURE 4.b
$1/\chi_t(\beta)$ v.s. H for Co_xY_{1-x} (x=0.65)

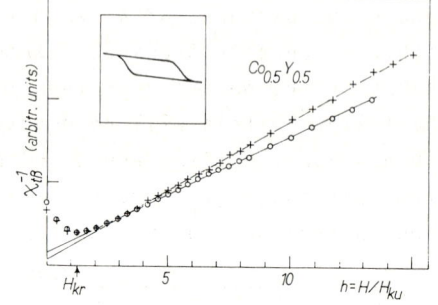

FIGURE &4.c
$1/\chi_t(\beta)$ v.s. H for $Co_{0.5}Y_{0.5}$

The magnetic and magneto-optical properties above described may be explained if we supose that on the top of the film surface, Y is segregated and forms there a protective layer of oxide. Below this top layer there is a subsurface region which is depleted with respect to Y, i.e.: enriched by Co. It has been suggested that the thickness of the subsurface region may be about 3-4 nm thick[5]. Then, from the magnetic point of view, the sample is constituted by Co-enriched isolated subsurface thin layers that may be exchange coupled to the underlaying bulk amorphous Co_xY_{1-x}.

In particular,for films where x>0.65 (See Figure 4a),both sub-surface and bulk are strong ferromagnets and magnetic coupling between them must be strong. In this case a contribution to the coercive force may arise from pinning of the wall by the magnetic inhomogeneities localized near the air film interface i.e.:random anisotropies of the Co-enriched sublayers. Furthermore,this kind of inhomogeneity may induce,through magnetic coupling,the magnetization ripple in the bulk of the sample and the related TBIS behaviour[6].(Figure 4a).

Otherwise in films where x<0.65 only Co-enriched subsurface clusters are magnetically ordered at room temperature. There is not contribution of the underlaying bulk to the overall magnetic behaviour and the clusters are magnetically isolated[7]. This gives high values for the coercive force ($H_c \approx 0.4 H_{kr}$) (See Fig.3) and the TBIS showed in Figure 4c.

Finaly,the magnetic behaviour of the film where x=0.65 (See Figure 4b) corresponds to the intermediate case where both the subsurface clusters and the underlying bulk are magnetically ordered but no strong magnetic coupling exists. Both hysteresis loop and TBIS behavior evidences the occurrence of two magnetic phases[7].

From the magneto-optical point of view, we conclude that a negative contribution to δ_k,which is related to the magneto-optical reponse of the compositive system: oxide layer-underlying (Co-enriched) region,is added to δ_k of the ideal non-oxidized amorphous Co_xY_{1-x} films,and gives rise to actual values given in Figure 1. This applies to the whole composition range studied.

REFERENCES

1) C.N.Afonso,A.R.Lagunas,F.Briones,S.Girón,J.Mag.Mag.Mat.15-18, (1980)833.
2) J.M.Alameda, F.López,Phys. Stat.Sol.(a),69(1982)757.
3) F.Briones Ph.D.Thesis. Univ. Complutense (Madrid) 1972
4) A.V.Sokolov "Optical properties of metals" Ed.Blackie &Son (London) (1967)
5) A.P.Malozemoff,J.P.Jamet,R.J.Gambino IEEE Trans.Mag..MAG-13 (1977) 1609
6) J.M.Alameda,M.C.Contreras,H.Rubio,F.Briones,D.Givord,A.Liénard J.Mag.Mag.Mat.67 (1987)115
7) J.M.Alameda,M.C.Contreras,A.R.Lagunas J.Mag.Mag.Mat.(submitted)

TRANSVERSE SUSCEPTIBILITY IN AMORPHOUS NdFeB THIN FILMS

J.M. ALAMEDA, F. BRIONES*, M.C. CONTRERAS, F. FUERTES,
D. GIVORD† and A. LIENARD†
Departamento de Física, Universidad de Oviedo, 33007 Oviedo, Spain
* Instituto de Física de materiales, C.S.I.C., Madrid, Spain
† Lab. Louis Néel, C.N.R.S., Grenoble, France

Amorphous $Nd_{15}Fe_{77}B_8$ and $Nd_{13}Fe_{81}B_6$ thin films(thickness range: 100 to 300 nm) were grown on glass substrates by triode sputtering. Two experimental set-ups were used. In the first one codeposition from three independent polarized Nd, Fe and $Fe_{75}B_{25}$ cathodes was performed. Substrates were held at room temperature and a in-plane magnetic field($\approx$100 Oe) was present during film deposition. In the second case, single sputtering targets, of nominal composition $Nd_{15}Fe_{77}B_8$, obtained by melting in an electromagnetic induction furnace was used. A turnable substrate holder(TSH) held at room temperature(RT) or 77 K was used in this case. The Argon residual pressure was close to 4×10^{-4} torr in both cases. Thermal anneals of the samples were performed in UHV at temperatures above T_c. The amorphousness of the films was checked by X-Ray diffraction. Transverse biased initial susceptibility (TBIS):$\chi_t(\beta)$ measurements were performed using a Kerr effect arrangement already described[1]. TBIS behaviour was studied at both interfaces; film-air (f/a) and glass-film (g/f).

Both macroscopic magnetic anisotropy and magnetization dispersion(ripple) of the films may be extracted from TBIS. H_k is obtained from the linear extrapolation to the value $1/\chi_t=0$ of $1/\chi_t(\beta)$ v.s.H curves in the high field range(H is the bias field and β is the angle between H and the easy axis of induced anisotropy). In real films, these extrapolations being applied to $1/\chi_t(0)$ and $1/\chi_t(\frac{\pi}{2})$ cut the abcissa at asymmetrical points separated appart by $2H_k$. We denote this asymmetry by ΔH. The normalized value $\Delta H/H_k$ is phenomenologically related with the magnetization dispersion ripple [2,3].

Figure 1 shows $1/\chi_t(\beta)$ v.s. H curves obtained for typical as-prepared and annealed films at both interfaces f/a and g/f.

As-prepared and annealed NdFeB amorphous films present non-

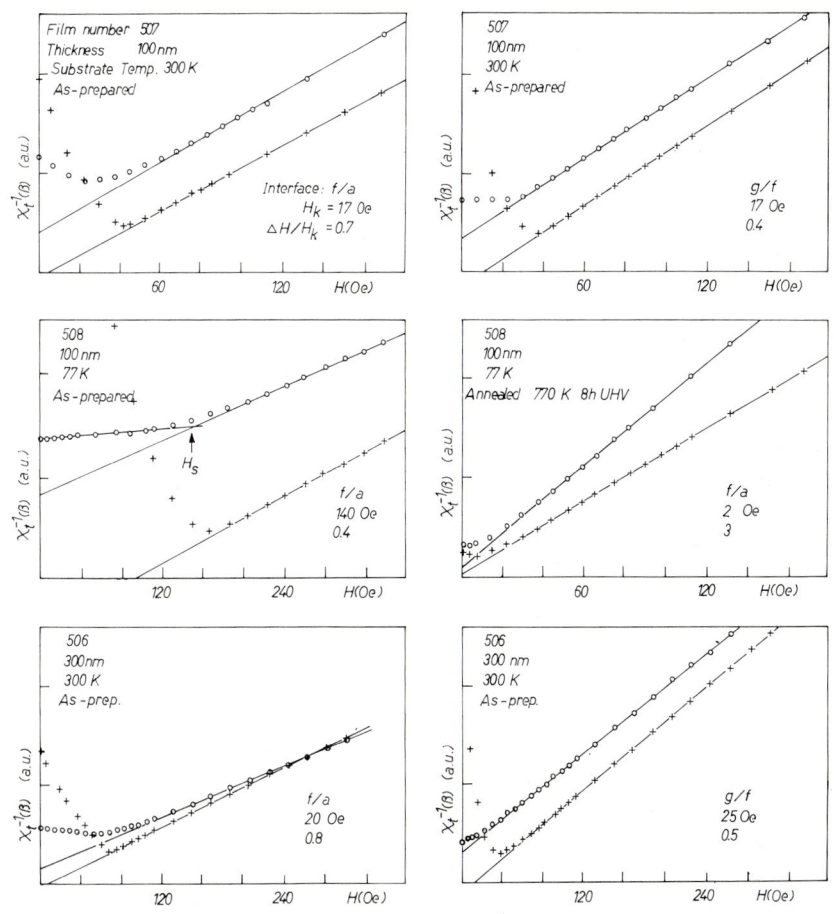

FIGURE 1
$1/\chi_t(\beta)$ v.s. H for $Nd_{15}Fe_{77}B_8$ amorphous thin films (see text)

vanishing values of $\Delta H/H_k$, hence of magnetization dispersion.

Influence of growth parameters on the magnetic properties:
The magnitudes of H_k and $\Delta H/H_k$ for MCS-films deposited at RT ($H_k \approx 9$ Oe, $\Delta H/H_k \approx 0.2$) are lower than for SCS-films($H_k \approx 18-20$ Oe, $\Delta H/H_k \approx 0.7-0.8$). In the first case H_k is mainly controlled by the applied field present during deposition. In the second case, the main influence on H_k is the effective oblique incidence of the vapor beam. Otherwise, H_k and $\Delta H/H_k$ for SCS-films deposited on cooled substrates (77 K) were 140 Oe and 0.4 respectively. So it is concluded that a better definition of the induced anisotropy

is obtained for cooled substrates. Moreover, hysteresis loop shapes and TBIS behaviour seem to indicate that weak stripe domains, related with perpendicular magnetic anisotropy, may be present in as-prepared films. The magnitude of H_s (above which the magnetization lies in the film plane) may be deduced from $1/\chi_t(\beta)$ v.s. H curves (see figure). $H_s \simeq 30$ Oe and 150 Oe for films deposited at 300 K and 77 K respectively. Then, both in-plane and perpendicular anisotropies increase when substrate temperature decreases.

Influence of thermal annealing on magnetic properties:
H_k decreases in annealed amorphous films to a value close to 2 Oe. Moreover, hysteresis loop shapes and TBIS behaviour correspon to films with in-plane magnetization. All the above is valid even for as-prepared films with H_k close to 10^2 Oe. Accordingly the magnetization dispersion $\Delta H/H_k$ increases with annealing. We conclude that the main contribution to H_k arises from structural inhomogeneities, induced by the growing process, that disappear in a great extent with annealing. However, remaining inhomogeneities at a microscopic scale give rise to magnetization ripple even in annealed samples.

Influence of the interface on magnetic properties:
For as-prepared films (thickness$\simeq$1000nm): $H_k(g/f) \simeq H_k(f/a)$ and $\Delta H/H_k(g/f) < \Delta H/H_k(f/a)$.
For annealed films (thickness$\simeq$100nm): $H_k(g/f) \simeq H_k(f/a)$ and $\Delta H/H_k(g/f) < \Delta H/H_k(f/a)$ for 620 K (24 h) anneals and $H_k(g/f) < H_k(f/a)$, $\Delta H/H_k(g/f) > \Delta H/H_k(f/a)$ for 770 K (8 h) anneals.
For thicker as-prepared films (300nm): TBIS behaviour in the low field range shows marked differences for g/f and f/a interfaces (See Figure 1). In particular perpendicular anisotropy (K_N) seems to be only present in the f/a interface. This is not the case for lower thickness.

REFERENCES
1) J.M. Alameda, F. López, Phys. Stat. Sol. (a)69(1982)757
2) H. Hoffman, Phys. Stat. Sol. (a)33(1969)175
3) J.M. Alameda, M.C. Contreras, H. Rubio, Phys. Stat. Sol.(a)85 (1984)511
4) J.M. Alameda, J.M. González, F. López, J.L. Vicent, J.Mag.Mag.Mat.38 (1983)105

ON THE CRYSTALLISATION OF $Pd_{40} Ni_{40} P_{20}$

A. GARCÍA ESCORIAL, M.C. CRISTINA, P. ADEVA and A.L. GREER[+]
CENIM, Madrid, Spain
+ Department of Materials Science and Metallurgy, University of Cambridge, UK

Crystallisation studies on $Pd_{40} Ni_{40} P_{20}$ and $Co_{76} Fe_4 B_{20}$ melt-spun ribbons have been carried out to illustrate their different behaviours, by means mainly of scanning electron microscopy and X-ray diffraction.

Understanding the micromechanisms of crystallisation of metallic glasses to impede or control crystallisation is a prerequesite for most applications, as the stability against crystallisation determines their effective working limits. On the other hand, controlled crystallisation of metallic glasses can be used for designing very special partially or fully crystallised microstructures than can not be obtained from the liquid or crystalline state.

In the present work, crystallisation has been studied on two extreme amorphous alloys: $Pd_{40} Ni_{40} P_{20}$ and $Co_{76} Fe_4 B_{20}$ to compare the variety of possible crystallisation behaviours.

$Pd_{40} Ni_{40} P_{20}$ is an exceptional amorphous alloy[1] with an unusually high reduced glass transition temperature T_{rg} = 0.67 , showing a remarkable resistance to crystallyse, therefore it has been widely studied[2,3] but its crystallisation mechanisms are not still completely understood. Meanwhile $Co_{76} Fe_4 B_{20}$ has a comparatively low T_{rg} = 0.5.

Previous work on $Pd_{40} Ni_{40} P_{20}$ melt-spun ribbons in air[3] shows that crystallisation is heterogeneous predominantly from the surface and is more prevalent on the wheel side. This behaviour appears to be motivated by mild oxidation during production, as Auger electron spectroscopy has pointed out. Crystallisation causes the development of noticeable relief on the sample surfaces due to the volume change accompanying crystallisation, as Figure 1 shows. Liquid behaviour during its crystallisation is also shown in Figure 2. Nuclei have ternary eutectic morphology and the difficulty of three phases growing cooperatively contributes to the exceptional glass-formability of this alloy.

Crystallisation has been studied in the temperature range 540K-920K and the same unidentified crystalline phases are present all over it. Subsequent anneals above crystallisation change the morphology from eutectic cells to microcrystals accorging Koster[4]. Figure 1 shows a small nucleus presumably growing into a previous one and this could keep some relationship with this change.

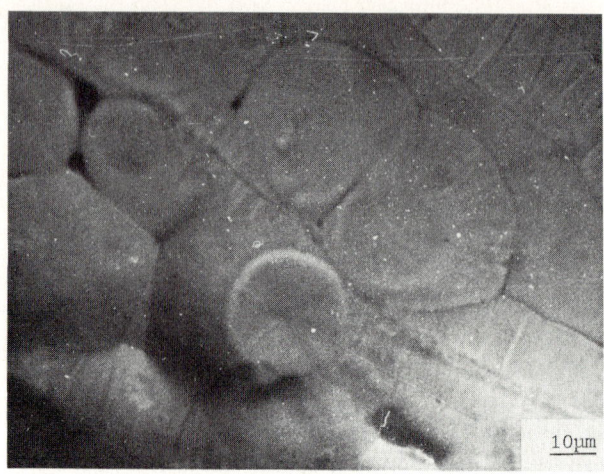

FIGURE 1
Optical micrograph on the air side of $Pd_{40}Ni_{40}P_{20}$ annealed in argon 3 h at 653K.

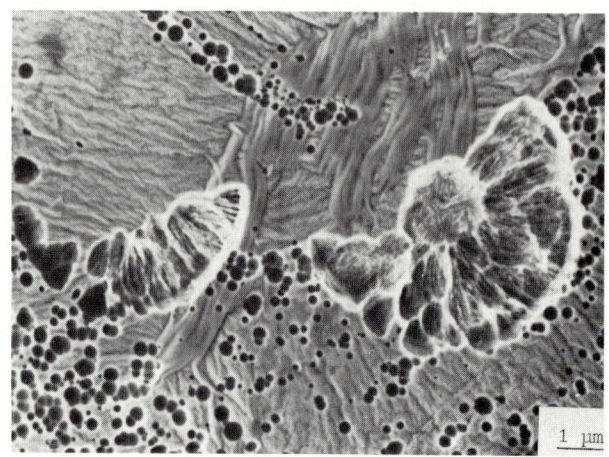

FIGURE 2
SEM micrograph on the air side of $Pd_{40}Ni_{40}P_{20}$ annealed in air 4 h at 613K.

$Co_{76}Fe_4B_{20}$ crystallisation takes place from quenched-in nuclei, avoiding the wheel side, which is faster cooled. Crystallisation is eutectic with a mixture of CoFe, fcc, and Co_3B, orthorhombic, being the fcc the first nucleation step. There is also a contribution of surface heterogeneous nucleation on the air side, where active sites are those poorer in boron, probably due to evaporation. This effect has been observed is accompanied by surface relief in a similar way than in $Pd_{40} Ni_{40} P_{20}$, as it is seen in Figure 3.

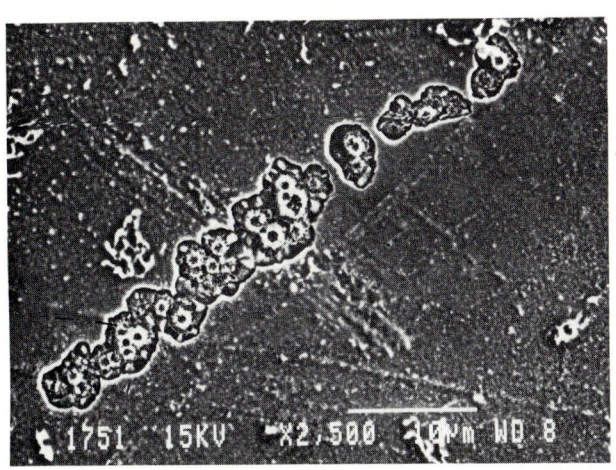

FIGURE 3
SEM micrograph on the air side of $Co_{76} Fe_4 B_{20}$ annealed in air 5sec. at 913 K.

REFERENCES

1) H.W.Kui, A.L.Greer and D.Turnbull, Appl.Phys.Let. 45 (1984) 615.

2) A.J.Drehman and A.L.Greer, Act.Met. 32 (1984) 323.

3) A.García Escorial and A.L.Greer, J.Mat.Sci. " to be published".

4) U.Koster and U.Herold, Crystallization of Metallic Glasses, in: Glassy Metals I, eds.H.J.Guntherodt and H.Beck (Springer, Berlin, 1981) p.225-259.

CRYSTALLIZATION OF THIN AMORPHOUS $Fe_{1-x}Si_x$ FILMS

Tadeusz LUCINSKI

Institute of Molecular Physics, Polish Academy of Sciences,
ul. Smoluchowskiego 17/19, 60-179 Poznan, Poland.

Crystallization of amorphous $Fe_{1-x}Si_x$ films was investigated by analysing temperature dependences of resistivity and thermoelectric power. The results are presented together with the crystal phases that appear during crystallization. The minimum in the crystallization temperature was obtained for x=0.40.

$Fe_{1-x}Si_x$ films were obtained by d.c. cosputtering method, using a two facing targets system described by Strzyzewski[1]. The glow-discharge was stabilized by a magnetic field and the value of the magnetic induction component parallel to the substrate table was 20 mT. The working pressure of argon was $6.6*10^{-2}$Pa.

X-ray diffraction patterns were taken for the films of Si concentration 0.14<x<0.70 annealed for one hour at every 50 deg starting from room temperature to 740 K in a vacuum of 10^{-1} Pa. X-ray diffraction studies revealed that the films were amorphous in the whole concentration range considered and no sharp maximum was found in the x-ray diffraction pattern.

Crystal phases appeared after the samples had been annealed at the characteristic temperatures T_x and T_y determined from temperature dependences of resistivity and defined in ref.2. The characteristic temperatures versus silicon concentration in the films are presented in Fig.1 and the appropriate crystal phases are specified in Table .

TABLE

x	T_x	T_y
<0.20	Fe	Fe_3Si
0.24	Fe_3Si	Fe_5Si_3
0.40	Fe_5Si_3	Fe_3Si
0.52	amorphous	$Fe_3Si, Fe_5Si_3, FeSi$
0.56	amorphous	$Fe_3Si, Fe_5Si_3, FeSi$
0.67	amorphous	$FeSi_2$

As follows from a comparison between Table and Fig.1 the appropriate crystalline phases appear at T_x for silicone concentrations up to x<0.50 and above this concentration the films remain amorphous at T_x. Above x=0.50 the

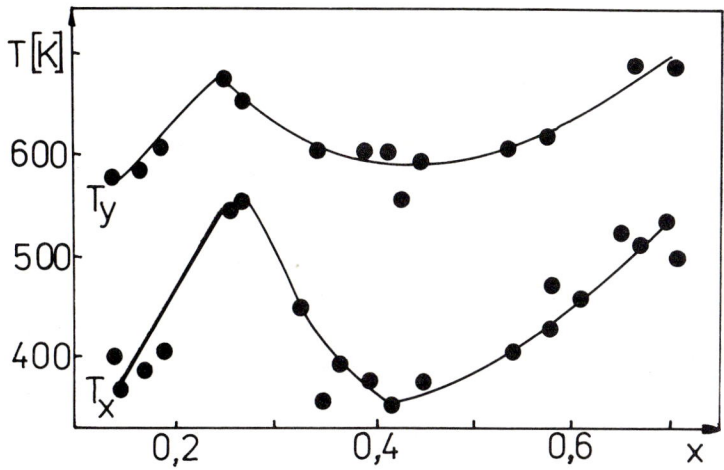

FIGURE 1
Characteristic temperatures determined from temperature dependences of electrical resistivity.

crystalline phases appear no sooner than at T_y. Additional information on the crystallization of the samples were obtained from the temperature measurements of thermoelectric power [3]. From the obtained dependences we could determine the characteristic temperatures T_1 and T_2 (Fig.2) at which a change in slope of the thermoelectric power versus temperature occurred Above x=0.50 only one of these characteristic temperatures, T_2, could be determined. As follows from the above presented results the crystallization temperature for x<0.50 is T_x and for x>0.50 is T_y. The latter may be the temperature at which one amorphous structure is reconstructed into another amorphous one.

The observed minimum in the temperature of crystallization of $Fe_{1-x}Si_x$ films observed near x=0.40 may be explained as follows.

From the temperature measurements of magnetization we know that for x=0.40 the Curie temperature of the amorphous state is 350 K [4] whereas the Curie temperature of a crystalline phase Fe_5Si_3 that appears slightly above 350 K, is 385 K [5]. Since $T_c \sim J_{Fe-Fe}$ where J_{Fe-Fe} is the exchange interaction, we may expect that in the amorphous state, for x$\sim$0.40, the short-range ordering is close to the Fe_5Si_3 type ordering. In this case a little portion of thermal energy supplied to the system may result in formation of a short-range ordering

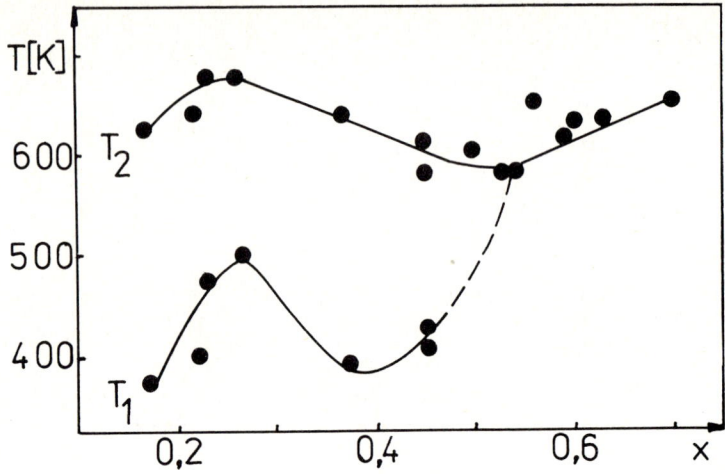

FIGURE 2
Characteristic temperatures determined from temperature dependences of thermoelectric power.

corresponding to that of Fe_8Si_3 phase via a short-range diffusion. Then, we obtain a low crystallization temperature. When the short-range orderings are much different from the orderings of the appropriate crystalline phases the crystallization temperatures are higher.

REFERENCES

1) P. Strzyzewski, Japan. J. Appl. Phys. 21 (1982) 192.

2) T. Lucinski and J. Baszynski, phys. stat. sol. a84 (1984) 607.

3) T. Lucinski and J. Baszynski, phys. stat. sol. a87 (1985) K191.

4) T. Lucinski and J. Baszynski, to be published.

5) C.E. Johnson, J.B. Forsyth, G.H. Lander and P.J. Brown, J. Appl. Phys. 39 (1968) 465.

MAGNETIC PROPERTIES OF RING-SHAPED ELECTRODEPOSITED Co-P AMORPHOUS ALLOYS

Guillermo RIVERO

Laboratorio de Magnetismo, Facultad de Física, U. Complutense, Madrid, Spain

Ring - shaped Co P ferromagnetic amorphous alloys have been obtained by electrolysis. Two different electrode arrangements (single or multiple Co anode) with bias or square - pulsed electrolytic current have been used. Samples exhibit squared azimuthal hysteresis loops corresponding to magnetization processes ruled by wall displacements. Optimum soft magnetic properties are reached using multiple anode arrangements and square - pulsed electrolitic current.

1. EXPERIMENTAL

Samples were electrodeposited on a Cu cylindrical substrate rotating round its axis with 1 to 100 rev./min[1]. The anode consisted of Co plates (30 x 8 mm.) with two different arrangements:

a) eigth plates surrounding the cathode on a circle of 100 mm. in diameter.

b) only one plate at a distance of 50 mm. from the cathode.

Independently of the anode arrangement, two different types of electrolytic current have been used:

a) a constant current density of 250 mA/cm^2.

b) a square-pulsed current density[2] as shows in figure 2.

Samples have dimensions of 20 mm. in diameter and 3 mm. in length, with $\sim$50 μm. thickness. The X-Ray diffraction analysis shows a complete absence of peaks corresponding to crystalline structures.

The compositional analysis, attained by X-ray fluorescence, gave an average composition of 15% at. P for D.C. electrodeposited samples. According to previous work[3], samples obtained with square-pulsed current consisted of ferromagnetic layers ($\sim$16% at. P content and few hundred Å thickness) corresponding to positive pulses T_+, and non - ferromagnetic boundaries ($\sim$34% at. P content and $\sim$ 40 Å thickness) corresponding to negative pulses T_-.

The coercive field H_c, remanent magnetization M_r and maximum total permeability, have been measured from $B\phi - H\phi$ hysteresis loops at very low frequency (0.1 H_z), by means of a toric coil around the ring. The azimuthal magnetizing field was produced by a triangular alternating current flowing through a conductor coaxial with the rings.

Distribution of the magnetization in the samples surface has been studied by using the Bitter Technique.

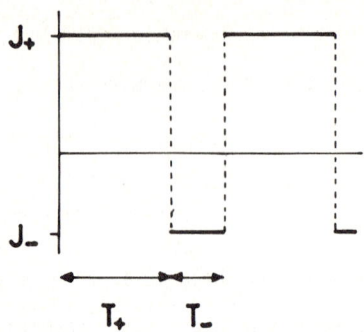

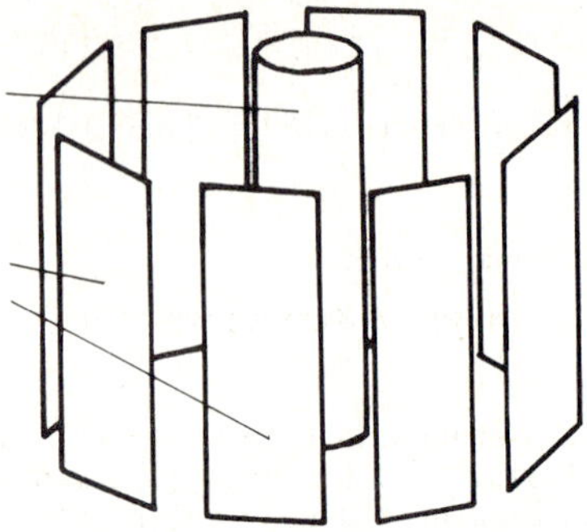

FIGURE 1
Square - pulsed electrolytic current density. $T_- = 60$ ms, $J_- = 100$ ms, T_+ and J_+ are variables

FIGURE 2
Electrode arrangement used for electrodeposition. Dimensions are arbitray.

2. RESULTS AND DISCUSSION

The samples obtained with D.C. electrolytic current and multiple anode arrangement present a high radial anisotropy, with a distribution of magnetization in strong stripe domains, as can be seen in figure 3. Consequently, their permeability is very low, and the coercive field is in the order of 10 Oe.

By using single anode arrangement or square - pulsed current density, the above mentioned anisotropy is destroyed[3-5]. In this case, the samples have a magnetization distribution on the surface, with circular 180º Bloch walls, as can be seen in figure 4. They present a high permeability ($\sim 10^5$) and a coercive field values between 20 and 100 mOe.

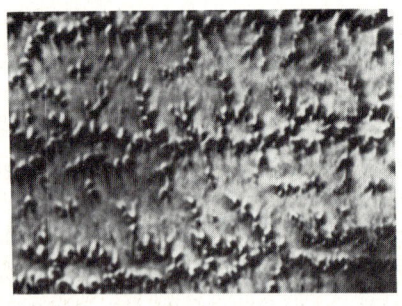

———>> azimuthal direction
FIGURE 3
Bitter patern of a sample electrodeposited with constant current density

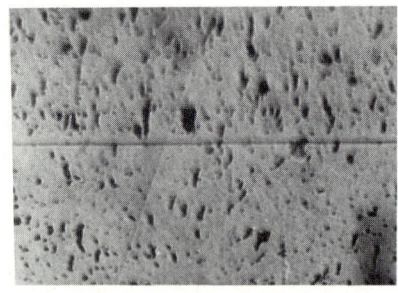

———>> azimuthal direction
FIGURE 4
180º Bloch wall in a ring electrodeposited with square - pulsed current.

B_ϕ - H_ϕ Hysteresis loops of these samples are square - shaped, as is expected from magnetization process ruled by 180º Bloch wall displacements (fig. 5).

Optimum soft magnetic properties are reached by using eigth Co plates arrangement and square - pulsed electrolytic current, with parameters: T_+ = 0.5 s. and J_+ = 250 mA/cm^2. In this case the rings are always magnetized, even in absence of magnetizing field, with remanent magnetization higher than 0.96 M_s and a coercive field value of 20 mOe. The magnetization process takes place by only one step, suggesting that the nucleation field of the wall is higher than the propagation field (Fig. 6). The maximum value of total permeability for these samples is 4.5 x 10^5.. These results are independents of rotation speed of the substrat.

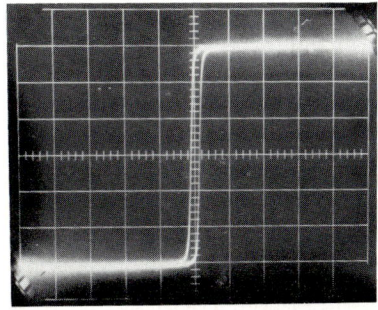

FIGURE 5
B_ϕ - H_ϕ hysteresis loop of a ring without radial anisotropy.
H_{max} = 2 Oe.; ν = 50 Hz.

FIGURE 6
B_ϕ - H_ϕ hysteresis loop of a ring obtained with square pulsed current.
ν = 0.1 Hz.

No improvement of shoft magnetic properties has been achieved by using a single anode arrangement combined with square pulsed current density. This suggests that the destruction of radial anisotropy produced by both methods separately, has a similar origin: the fluctuation of the composition with time during the electrolysis. This effect seems to be stronger by using square pulsed electrolytic current.

REFERENCES
1) G. Dietz and Klett, J. Mag. Mag. Mat. 8 (1978) 57.
2) C. Aroca, J.M. Riveiro and G. Rivero, J. Phys. E 15 (1982) 504.
3) J.M. Riveiro and G. Rivero, IEEE Trans. Mag. 6 (1981) 3082.
4) J.M. Riveiro, M.C. Sánchez and G. Rivero, IEEE Trans. Mag. 3 (1981) 1281.
5) G. Dietz, H. Bestzen and J. Hungenberg, J. Mag. Mag. Mat. 9 (1978) 208.

MAGNETIZATION NOISE IN AMORPHOUS FERROMAGNETIC ALLOYS

M. Celasco(°), A. Masoero and A. Stepanescu

Istituto Elettrotecnico Nazionale Galileo Ferraris
Corso M. D'Azeglio 42, I-10125 TORINO, ITALY
Gruppo Nazionale di Struttura della Materia and Centro Interuniversitario di Struttura della Materia, Torino, Italy.
(°) Physics Dept., Università of Genova, Italy

In the present paper the presence of a strong correlation between magnetic aftereffect and Barkhausen noise is put in evidence. Several experimental results concerning the behaviour of the noise power spectrum as a function of temperature and of the magnetizing frequency are given. It is observed that the influence of the magnetic aftereffect on the Barkhausen noise in amorphous materials is quite different from the one observed in crystalline materials. A discussion of this point is also given.

1. INTRODUCTION

In a preceeding paper, several experimental results are reported about Barkhausen (B.) noise power spectra in amorphous alloys. These results put in evidence for the first time a correlation between the magnetic aftereffect and the B.noise. Noise in ferromagnetic crystalline materials is generated by the pinning of the Bloch wall during their motion due to defects, impurities, localized stresses, i.e. the same sort of inhomogeneities which are responsible for the coercive field and the hysteresis losses. Thus the noise can in principle give information on the magnetization dynamics even if its relation with the macroscopic magnetic characteristics of the material is not a direct one. In amorphous materials there are also other sources of noise, actually: a), the existence of intrinsic fluctuation of the local atomic arrangement and thus of the local wall mobility and thickness; b), the presence of a strong magnetic aftereffect which is generated by the interaction of diffusing atoms with the domain walls.

The aim of this paper is to give an interpretation of the experimental B. noise in amorphous materials taking into account the contribution of the magnetic aftereffect.

2. EXPERIMENTAL RESULTS

Owing to the absence of the crystalline anisotropy, amorphous ferromagnetic ribbons are characterized by the possibility of obtaining a very simple structure of antiparallel domains through the application of a suitable mechanical tension. In these conditions the material presents high mobility of the Bloch walls, as well as very low dissipative effect (owing to the high resistivity and small thickness of the specimens), and is very suitable for the study of the B.noise power spectrum up to a rather high analysis frequency ($\sim$ 100 kHz) in a broad range of magnetization frequencies.

The experimental results refer to strips of Metglas 2605 SC alloy 0.050 mm thick and 5mm wide, submitted to a mechanical tension of about 100 MPa: in

Magnetization noise in amorphous ferromagnetic alloys

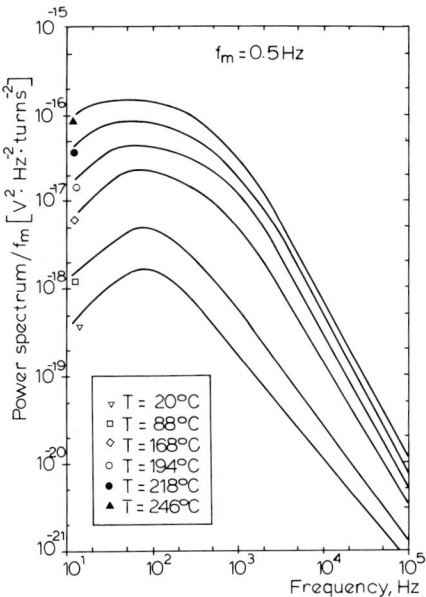

FIGURE 1
Power spectra of the B.noise for different values of the temperature T at the magnetizing frequency f_m =0.5 Hz. Wall oscillation amplitude=68 um peak to peak. The spectra are divided by f_m

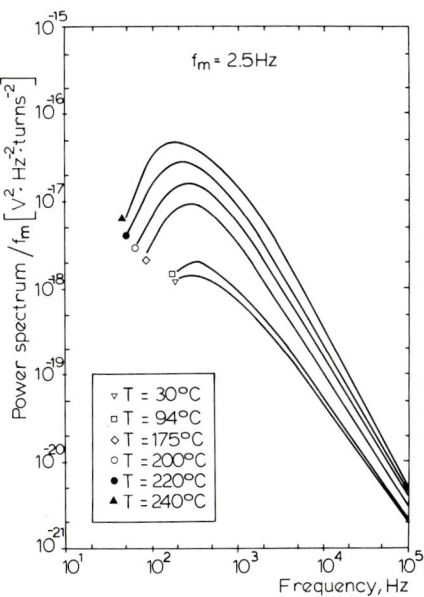

FIGURE 2
Power spectra of B.noise for...(as in fig. 1), at the magnetizing frequency f_m =2,5 Hz

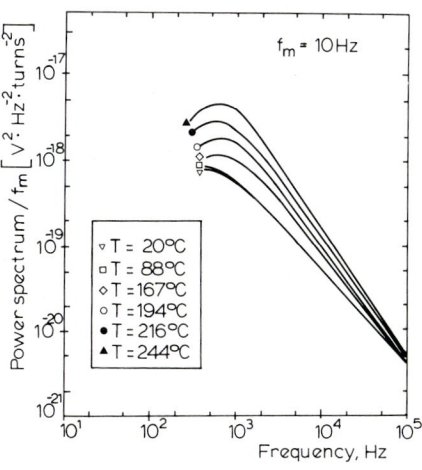

FIGURE 3
Power spectra of B.noise for...(as in fig. 1), at the magnetizing frequency f_m =10 Hz

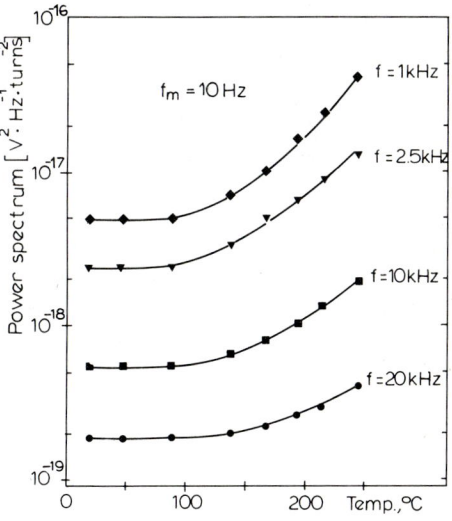

FIGURE 4
B.noise power spectral density for different values of the analysis frequency in a Metglas 2605 SC amorphous ribbon as a function of temperature. Magnetizing frequency f_m =10 Hz

these conditions a simple structure of 6 antiparallel domains separated by Bloch walls oscillating under the action of an external field, was generated.

By an optical method making use of the longitudinal magnetic Kerr effect, the oscillation amplitude of the Bloch wall was also monitored. The experimental results reported below were detected with a Bloch wall oscillation amplitude kept constant at a value of about 68 um peak to peak. The magnetizing frequency was changed within broad limits and the waveform was kept triangular in order to have a nearly constant value of the wall speed between field inversions. The noise was detected in a small coil of 20 turns linked to the specimen, amplified with a low noise Ithaco Amplifier and sent to a FFT Nicolet Mod. 660 B Signal Analyzer for the detection of the power spectrum.

In fig. 1-3 are reported the results concerning the power spectra of the B. noise taken at different temperature and for different values of the magnetizing frequency f_m. In fig.4 the spectral density for different values of the analysis frequency is given as a function of the temperature.

The main feature of the experimental results reported in fig. 1-4 are:
i) an increase of the power spectral density with temperature, particularly in the low frequency range. This fact is a sort of an anomalous B. effect for the first time observed in amorphous materials;
ii) a sharp drop of the B.noise density in the low frequency range;
iii) a shift of the position of the maximum of the power spectrum towards low frequency by increasing the temperature;
iv) a shift of the maximum of the spectra towards high frequencies as the magnetizing frequency is increased (see fig. 1-3).

3. DISCUSSION AND CONCLUSIONS

In the present paper an experimental study of the Barkhausen noise power spectrum in a range between 10 Hz-100 kHz is presented. The noise is generated under different physical conditions, by changing the magnetizing frequency f_m, the wall displacement amplitude, and the magnetic aftereffect. The spectra are all renormalized to 1 Hz by dividing the power spectrum by f_m.

As it was stressed in the introduction, in amorphous ferromagnetic materials the viscosity field effect, always play a specific role in noise generation. Actually with reference to figure 5, in amorphous materials the initial rate of increase of the viscosity field H_t defined as the field necessary to move a Bloch wall from the position where it is stabilized from atomic diffusion processes, and which strongly depends on temperature is always faster than that external field: the walls do not move as long as the external field H_e does not cross, the viscosity field H_t at the time t_c. As a consequence the large B. discontinuities do not occur until the crossing time t_c is reached and thus the large B.discontinuities are further correlated by effect of the viscosity field. Because the aftereffect increases with temperature, according to fig. 5 the crossing time t_c also increases with the temperature, while on the contrary t_c decreases when the magnetization frequencies is increased.

It is thus expected that the correlation effect between large B. discontinuities due to the viscosity field may explain the points i)-iv) previously stressed in sect. 2.

It is worth to note that these correlation effects due to viscosity field are essentially deterministic, in contrast with the stocastic correlation due to local demagnetizing fields responsible of the clustering of large B.

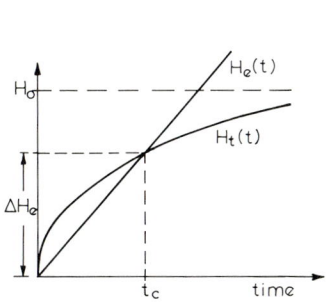
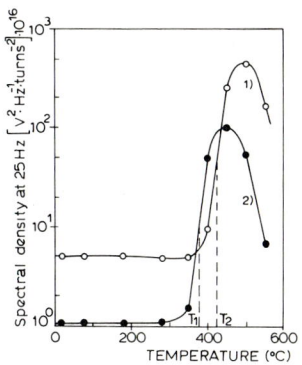

FIGURE 5
Viscosity field $H_t(t)$ and external field $H_e(t)$ acting on a Bloch wall after a Barkhausen jump vs.time

FIGURE 6
Temperature shift of the peak of the spectral power density of the Barkhausen noise at 25 Hz analysis frequency in Fe-Al 12% weight alloy when the magnetizing frequency f_m is changed. Curve 1), magnetizing frequency 0.025 Hz; curve 2) magnetizing frequency 0.003 Hz. (From reference 4))

discontinuities and previously studied[2,3].
A theory which account for the behavior of the power spectrum of the noise in the presence of a magnetic aftereffect has been developed in[4]. It should be pointed out that this theory explains the experimental results in crystalline materials only. The strong difference which exist between the magnetic aftereffect in crystalline and amorphous materials is related to the fact that diffusional effects in crystalline materials are characterized by a single or a few time constants, while in amorphous materials the process is characterized by a wide spectrum of time constants. Consequently the contribution of the aftereffect to the B. noise in crystalline materials takes place in a rather narrow temperature interval (see fig.6), while in amorphous materials the effect is spread over a wide range of temperature (see fig.4). The theory developed in[4] must be modified to take into account this difference.
  The development of this theory which is now in progress, will be published elsewere.

REFERENCES
1) M.Celasco, A.Masoero, P.Mazzetti and A.Stepanescu: "Experimental study of the magnetization noise in amorphous ferromagnetic alloys" 9th International Conference on Noise in Physical Systems. Montreal 25-29 May 1987.
2) M.Celasco and A.Stepanescu: J. Appl. Phys. 48, 3635 (1977)
3) M.Celasco, F.Fiorillo and P.Mazzetti: Il Nuovo Cimento, 23B, 376 (1976)
4) A.Ferro, P.Mazzetti and G.Montalenti: Il Nuovo Cimento, 56B, 111 (1968)

APPLICATION OF AMORPHOUS ALLOYS IN GROUND FAULT INTERRUPTERS*

J. GUZ, G. MATRAS, A. NAFALSKI, A. WAC-WŁODARCZYK

Lublin Technical University, 20-618 Lublin, Poland

The paper presents the operating principle of a ground fault interrupter. More and more such devices are used nowadays for protection from electric shock in the home. The suitability of amorphous material for the sensing transformer core is discussed.

1. INTRODUCTION

Electrical shock protection in households by means of interrupting circuit breakers becomes more and more commonly used in many countries. Ground fault interrupters (GFI) are to detect the ground fault current I_s and, when its value exceeds what is conventionally assumed to be safe, automatically disconnect the protected circuit from the supply mains. The operation principle of the GFI is shown in Fig.1.

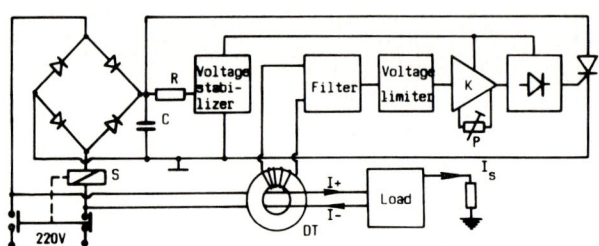

FIGURE 1
GFI block diagram

The wires supplying the load are threaded through the toroidal core of the differential transformer DT. When the circuit supplying the load operates correctly, current I_+ flowing into the

---

*This work was supported by Central Research Project CPBR 2.4.

load is equal to the current I_- flowing from the load. The magnetic field produced by these currents counterbalance each other and the secondary voltage of DT is zero. If current I_B flows to earth through faulty insulation or through a human who touches a live conductor, the currents in the pair threading the transformer are unequal. Therefore, a magnetic field proportional to current I_B is generated in the transformer core. For the current I_B greater than the operation threshold value, the secondary voltage induced is, after amplification, sufficient to operate the electromagnet which disconnects the supply circuit. The operation threshold I_B of breakers in most European countries is standarized and equals 30 mA. Recently, in the USA this value has been reduced to 5 mA, which permits protecting more than 99% of the population from electric shocks big enough to make it impossible to "let go" of the live conductor. Such a low value of ground fault current requires exact knowledge of differential-current transformers performance and therefore of core material properties. In practice, ground fault currents can be as much as several dozen ampers (short circuit) and can have the form of current impulses of any shape. They change the position of the operating point on the magnetization characteristic of the DT core and this affects the sensitivity of the DT. Therefore, the usability of amorphous material for the core in a breaker is above all determined by the small-signal dynamic permeability which has to satisfy the following requirements:
- largest possible value
- low dependence on operation point of magnetization curve
- as low as possible dependence on ambient temperature
- low influence of ageing processes.

2. MEASUREMENTS

Measurements of the small-signal magnetic permeability were carried out for cores of amorphous material AMM ($Co_{88}Si_5Fe_3B_2Mo_2$) type. The specimens of toroidal form were made from amorphous ribbons without heat treatment and in the second case after heat treatment which consisted of annealing in air at a temperature of 410°C for 10 minutes. The results, averaged for several specimens, are presented in Fig.2 and Fig.3.

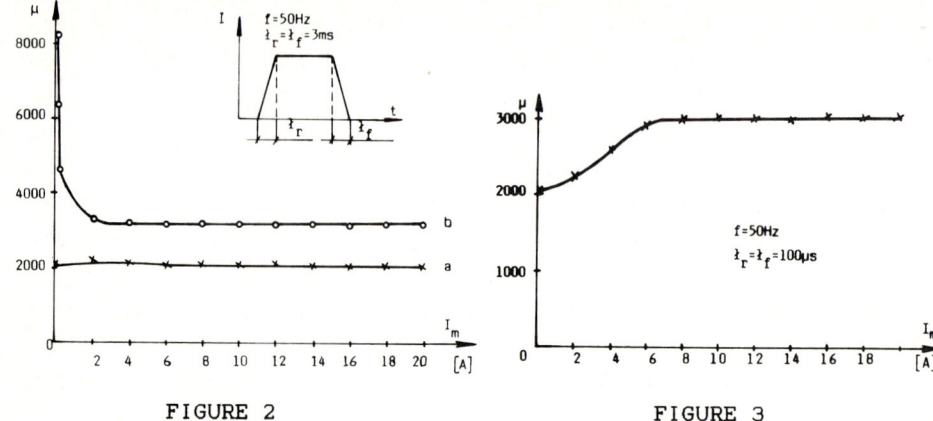

FIGURE 2
Magnetic permeability as a function of impulse current with long pulse rise time and pulse decay time for the core before annealing (a) and after annealing (b)

FIGURE 3
Magnetic permeability as a function of impulse current with short pulse rise time and pulse decay time

3. CONCLUSIONS

On the basis of experimental investigations of the properties of transfomer core material the following conclusions can be derived:

- small-signal dynamic permeability of amorphous material AMM has value in the range 2000-3100 for frequency f=50 Hz
- for slowly changing current impulses (l_r, l_f >3ms) the magnetic permeability is practically unchanged but is changed by about 50% for fast-changing current impulses
- the magnetic permeability of the specimens investigated decreases as the frequency of the supply voltage increases
- the influence of temperature changes on circuit breakers operating in the standard range for closed rooms, i.e. 10°C to 30°C, is negligible
- the above properties are satisfactory and advantageous. They ensure that the changes of the operation threshold due to temperature and the effects of current impulses are within admissible limits.

REFERENCES

1) G.B. Finke, IEEE Trans. on Magn. MAG-10 (1974) 144.

EXPERIMENTAL STUDY OF THE DOMAIN WALL

E. LÓPEZ, L. de PEDRO (*), P. SÁNCHEZ (*), C. AROCA,
C. MUÑOZ (*) and M.C. SÁNCHEZ
Dept. Materia Condensada, Fac. Ciencias Físicas, Univ. Complutense, Madrid.
(*) Dept. Física, E.T.S.I. Telecomunicación, Univ. Politécnica, Madrid.

It has been developed a computer controlled experimental system to obtain the differential susceptibility at any applied field, and by successive integrations to get the magnetization and the domain wall energy versus the domain wall position in the sample. This method let us to study accurately the domain wall stabilization processes and the interaction between the domain wall and any kind of defects in the sample.

1. BASIC CONSIDERATIONS AND RESULTS

It has been developed a based computer method that let us to study the hindrances to the domain wall (d.w.) motion, particularly it has been applied to improve the study of d.w. stabilization. This method requires ferromagnetic samples with magnetization processes due to a single d.w.. Assuming a prismatic sample (length l, width d, thickness e), the total energy, W, of the sample when the d.w. is at the x position, is the magnetostatic energy

$$W_n = \frac{1}{2} \frac{4 \mu_0 M_s^2 N}{d} \cdot x^2$$

plus the d.w. superficial energy

$$W_s = (A \cdot K)^{1/2} \cdot l \cdot e$$

(A, exchange constant, and K anisotropy constant)
plus the d.w. energy variations produced by factors such as surface roughness, inclusions, internal stresses, structural defects...

If a d.w. at the x position is shaken by an small a.c. field such as the amplitude of the d.w. vibration is smaller than the length wave of the energy perturbation, the induced e.m.f. in a secondary coil around the sample, E_0, will be proportional to d^2W/dx^2. By measuring E_0 at different positions of the d.w. in the sample (which correspond to different d.c. applied field H_1), we got the E_0 vs. H_1. By integrating two times it has been obtained W vs. x . Varying H_1 very slowly, the d.w. position can be obtained from $x = M(x) \cdot d/2M_s$ being $M(x) = \int \chi(H_1) \, dH_1$, where $\chi(H_1)$ is proportional to E_0, and the energy is obtained from $W = \int H_1 \, dM$.

The above described study can easily be done by a computerized system, be-

---

This work was supported by the Comisión Asesora de Investigación Científica y Técnica of the Department of Education of Spain (nº 2863/83).

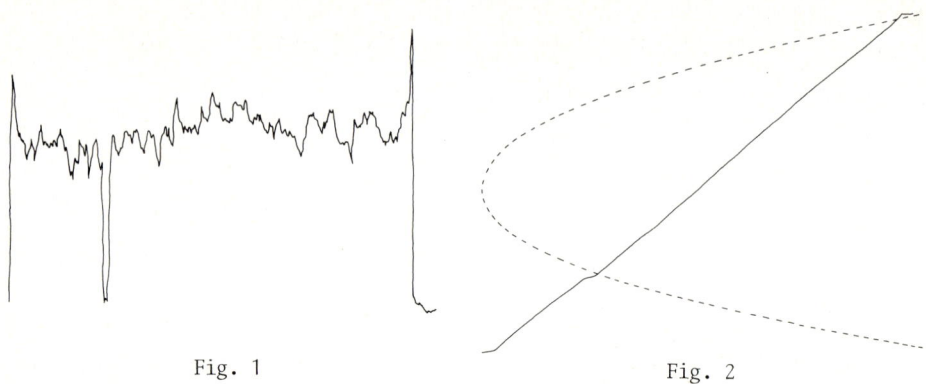

Fig. 1   Fig. 2

cause it is necessary to handle a great number of data and to perform two numerical integrations. Fig.1 shows E_o vs. H_1 with 4095 data of 10 bits of resolution. We can see the variation of d.w. energy and specially, the big one, induced by a d.w. stabilization (1). Fig. 2 shows in continuous line x vs. H_1 or M vs. H_1, really this curve is close to a classical hysteresis loop perturbed by the big defect above mentioned and in dashed line W vs. x . Energy fluctuations can not be observed because the magnetostatic energy variation is much bigger than any perturbation of the d.w. energy. The advantage of this method is that we can substract the effect of the magnetostatic energy to the E_o vs H_1 curve before to perform the integrations and in this way to obtain only the d.w. energy fluctuations.

Other possibility is to study induced defects by substracting E_o vs. H_1 curves before and after to induce the defect. Fig.3 shows E_o vs. H_1 (a) before, and (b) after, to induce a defect by a d.w. stabilization, (c) shows the result of substract both curves (the fig.3 only shows the zone where the defect was induced). All these results were obtained in samples of metglas 2826.

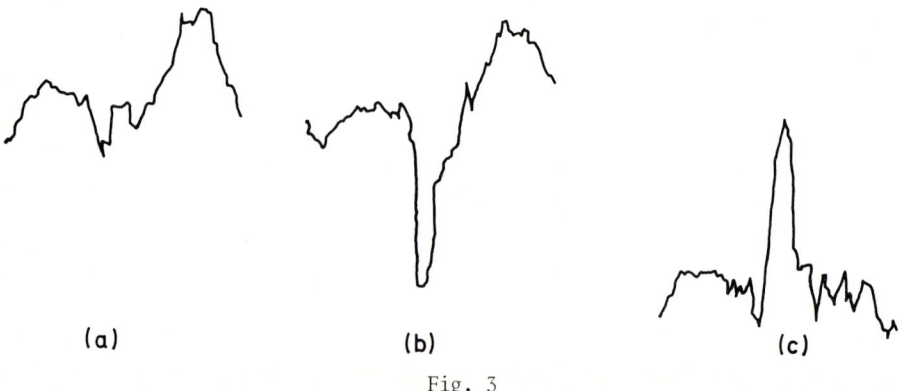

(a)   (b)   (c)

Fig. 3

2. EXPERIMENTAL SYSTEM

The basis of this experiment is described in (2). Actually we have computerized the system by using an IBM PC xt. The induced e.m.f. is measured by a lock-in amplifier, then the speed of the system is not crucial (the time constant normally used in the experiment is about 1s., so RS232 communications are fast enough. Fig.4 shows the scheme of the experimental system. The programable gain amplifier (64 step of gain) is used to control the amplitude of the a.c. magnetic field that produces the d.w. vibration. The D/A conversor (4095 steps) is used to control the d.w. position, that is measured by the A/D conversor (10 bits or resolution).

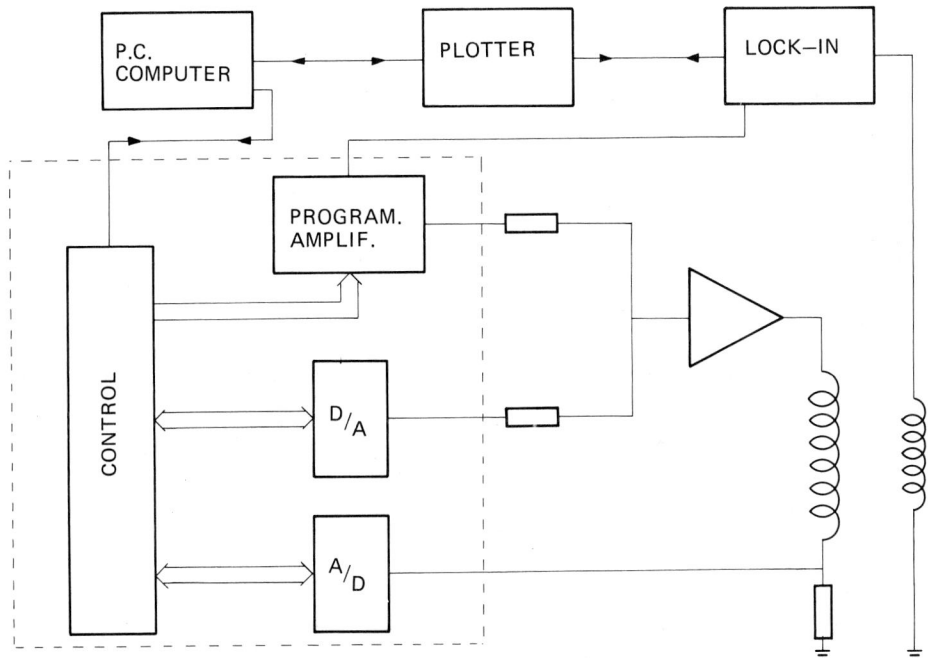

Fig. 4

This system is versatile enough to measure not only E_o vs. d.w. position at different a.c. fields but also any kind of other experiments such as to measure the viscous damping parameter of d.w. motion or to find the minimum energy position of the d.w. vs. H_1.

REFERENCES

(1) C. Aroca, P.Sánchez, E.López, Phys. Rev. B 34 (1986), 490
(2) C. Aroca, E. López, P. Sánchez, Phys. Rev. B 30 (1984), 4026.

"CLUSTER-in-SHELL" MODIFICATION OF THE MICRO-CRYSTALLINE MODEL
OF AMORPHOUS ALLOY STRUCTURE

A.V. SEREBRYAKOV

Solid State Physiscs Institute, Chernogolovka, Moscow
distr., 142432, USSR

Recently much attention has been paid to the microcrystalline models of the structure of the so-called amorphous alloys. As a rule most metallic glasses are binary (or multicomponent) alloys; the composition of these alloys usually corresponds to the two (or more than two)-phase field of the equilibrium phase diagram. So one should expect that the amorphous state is some frozen-in intermediate state on the way of an alloy descomposition and crystallization. There is quite a wide range of such states from fully topological and chemical randomness, where random dense packed structure models can work, to various extents of descomposition and ordering. The main problem here is to understand if there is or there is not any preferable position of an alloy within this range.

The "cluster-in-shell" modification of the micro-crystalline model is based on an assumption that one of the phases resulting from the descomposition of an alloy forms a thin two-dimensional layer in boundary regions between the ordered clusters of another phase in a binary system (or other phases in a multicomponent system). This can be the case if the formation of such a layer (or shells) around clusters leads to the free energy decrease of the intercluster boundaries. In the transition metal-metalloid glasses the role of such a phase can be played by a solid solution on a metallic basis, the ordered clusters with rigid bonds being "greased" by the solution in this case.

The amount of a solid solution (V) required for shells to form around clusters decreases with an increase of the cluster sizes (r) as 1/r, and in the case of spherical clusters containing the same number of atoms of about 50, separated by one or two atomic distances, the number of atoms in the boundary regions is 3 to 4 times greater than in clusters[1]. Concentration and, hence, the amount of a solid solution are also dependent on the cluster size but as

$\exp(\alpha/r)$ provided Gibbs-Thomson equation is valid in the range of sizes of interest. The estimates obtained from the value of the surface energy involved in α and assumed to be equal to 10^2 ergs/cm^2 show that for typical metal-metalloid glasses an appreciable deviation of a metalloid concentration in the solid solution from an equilibrium one for the case of a planar interface commences at r values equal to (1.5-3) nm. Thus in a certain range of small r values the amount of a solid solution in equilibrium with clusters (Scheme, curve 1) can be expected to exceed the amount of the so-

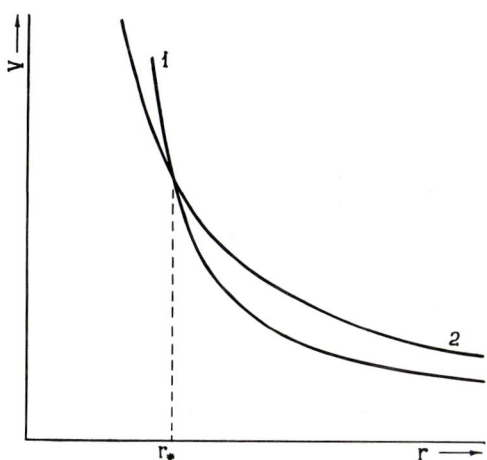

lution necessary for a two-dimensional boundary layer to form between clusters (curve 2). To grow beyond the above mentioned range clusters have to overcome the energy barrier, the latter value being the greater, the greater is the difference between the surface energies with or without an intermediate layer between the clusters. It should be noted that the r_* value in the situation considered does not exceed several atomic distances. This and the results obtained by Fujita el al.[1] suggest that in an X-ray diffraction experiment this model system would behave as an amorphous one.

The above model allows many results of amorphous alloy investigations to be accounted for, in particular such as a relatively high thermal stability of alloys and reversible and irreversible changes in their properties at thermal cycling, as being due to a reversible change of r with temperature and irreversible increase of a mean cluster size.

Of interest are also the implications of the model with respect to magnetic properties of amorphous alloys. Thus anelastic anisotropy in the samples under stress annealing$^{2-6}$ can be explained, in addition to generally recognized mechanisms, as due to internal stresses developing after unloading because of the difference in local elastic moduli in clusters and intercluster layers. Development of the preferential cluster orientation ("texture") under creep may lead to a plastic anisotropy component if clusters are not isotropic. The non-recoverable field-induced anisotropy may result from the "texture" formation when clusters nucleate and grow in a magnetic field.

REFERENCES

1) T. Hamada, F.E. Fujita, Jap. J. Appl. Phys. 21 (1982) 981

2) O.V. Nielsen, J. Magn. Magn. Mat. 36 (1983) 81

3) J.M. Barandiaran, A. Hernando, O.V. Nielsen, IEEE Trans. Magn. MAG-22 (1986) 1864

4) H.Warlimont, H.R. Hilzinger, Proc. 4th. Internat. Conf. Rapidly Quenched Metals (Sendai 1982) p.1167

5) M. Vazquez, W. Fernengel, H. Kronmuller, Phys. Stat. Sol. (a) 87 (1985) 609

6) V. Madurga et al, J. Appl Phys. 61 (8) (1987) 3228

LOW AND HIGH FIELD MAGNETIC STUDIES OF HEAT TREATED AMORPHOUS $(Fe_{0.5}Ni_{0.5})_{83}P_{17}$

R. Puźniak[E], J. S. Muñoz[&] and K. V. Rao

Dept. of Solid St. Phys., Royal Institute of Technology, Stockholm, Sweden

The crystallization of an amorphous Fe-Ni-P alloy during isothermal annealing has been monitored by measuring the evolution of the saturated magnetic hysteresis loop every eight minutes for time scales extending over a few orders of magnitude ($10^2 \div 10^5$ s). From our data we found that the spontaneous magnetization σ_s, the reduced initial susceptibility $(\partial\sigma_{init}/\partial H)/\sigma_s$, the reduced retentivity σ_r/σ_s as well as the saturation field H_s can be rescaled along the time axis and to collapse into universal curves for all four parameters as a function of "scaled time". This suggests a more universal behavior in crystallization of an amorphous material. Coercivity, however, is found to be practically independent of the annealing temperature.

1. INTRODUCTION

The stability of amorphous magnetic materials is of central importance to their technical applications[1]. At the same time, the metastability of amorphous phase and the manner in which it return to equilibrium are of fundamental interest. Recent work[2,3] on the development of order in systems quenched far from equilibrium has explored the possibility that universality and scaling, concepts central to the understanding of second-order transitions, may have relevance for first-order transitions as well. Especially the work of Roig et al.[4] demonstrated that the transformation kinetics obeys a time scaling law. In that work the crystallization of an amorphous alloy during isothermal annealing has been studied by measuring the magnetization of the sample, which is directly proportional to the fractional crystallization X(t). It was shown that by rescaling the time axis by the time $t_{1/2}(T_A)$ at which crystallization is 50 % complete for each annealing temperature, the results of the fractional crystallization X(t) fall on one universal curve of Johnson-Mehl-Avrami form[5].

In present work the crystallization of an amorphous Fe-Ni-P alloy during isothermal annealing has been monitored by measuring of the evolution of the saturated magnetic hysteresis loop. We find that the scaling concept is more general and meaningfully explains the evolution of the various magnetic

[E]Visiting scientist from Institute of Physics, Polish Academy of Sciences, Warsaw, Poland
[&]Visiting scientist from Autonoma University of Ballaterra-Barcelona, Spain

parameters that characterize the hysteresis loops, during long time annealing of metastable magnetic systems.

2. EXPERIMENT

Our method[6] relies on the fact that recrystallization occurs at measurable rate only above the Curie temperature of the glass, but well below that of the crystalline product. Measurements were carried out using a vibrating sample magnetometer (PAR model 155) with a resolution of about 10^{-4} emu. With a few rectangular pieces of ribbon (typically 1.5 · 2.5 mm) our experimental set-up permits a heating rate of about 0.5 K/s followed by an isothermal anneal. The temperature during the annealing was stabilized to better than ± 0.2 K. In our experiment we measured the evolution of the saturated magnetic hysteresis loop every eight minutes with a maximal applied magnetic field 10 kOe for time scales extending over 3 orders of magnitude ($10^2 \div 10^5$ s).

In this study samples of $(Fe_{0.5}Ni_{0.5})_{83}P_{17}$, typically 20 μm thick and 1.5 mm wide were prepared by a single-roller melt-spinning method. X-ray diffraction measurements confirmed they were amorphous.

Measured samples were rapidly heated to annealing temperatures T_A above the Curie temperature T_C = 483 K of the "glassy" state[4]. On crystallization, which proceeds at increasing rates as T_A increases, fcc Fe-Ni crystallites[7] in a phosphide glassy matrix are formed. Spontaneous magnetization of the sample provides a measure of the degree of crystallization, and the "sum effect" of nucleation, etc.

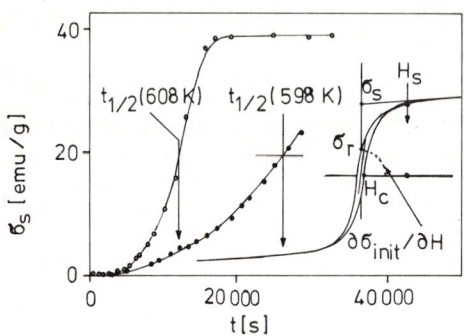

FIGURE 1
A typical hysteresis loop with marked spontaneous magnetization, initial susceptibility ($\partial\sigma_{init}/\partial H$), retentivity σ_r, saturated field H_s and coercivity H_c, and time dependence of the spontaneous magnetization σ_s for two different annealing temperatures T_1 = 598 K and T_2 = 608 K with marked $t_{1/2}(T_A)$

3. RESULTS AND DISCUSSION

From the hysteresis loops the spontaneous magnetization σ_s, the initial susceptibility ($\partial\sigma_{init}/\partial H$), the retentivity σ_r, the saturation field H_s and the coercivity H_c were determined. Figure 1 presents the time dependence of the spontaneous magnetization σ_s for two different temperatures. In the same figure is shown a typical hysteresis loop in which the parameters determined in our experiment are indicated. As was shown in the work of Roig et al.[4] by rescaling the time axis by the time $t_{1/2}(T_A)$ at which crystallization is 50 % complete, the presented data collapse into one universal curve. Rescaling the time axis by the same time $t_{1/2}$ determined from temperature dependence of spontaneous magnetization we found that reduced initial susceptibility $(\partial\sigma_{init}/\partial H)/\sigma_s$, reduced retentivity σ_r/σ_s and saturation field H_s fall on new universal curves (see figures 2, 3, 4). For presentation we used data of initial susceptibility and retentivity reduced by spontaneous magnetization because both of these quantities are proportional to the degree of crystallization. Results presented in figures 2, 3, 4 suggest a

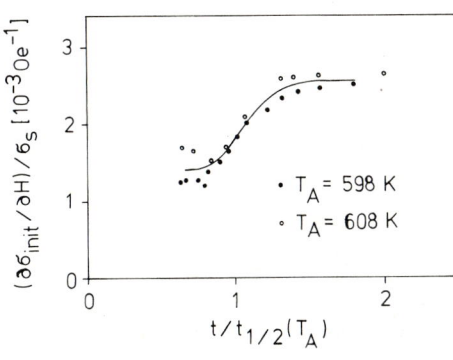

FIGURE 2
Reduced initial susceptibility $(\partial\sigma_{init}/\partial H)/\sigma_s$ versus scaled time

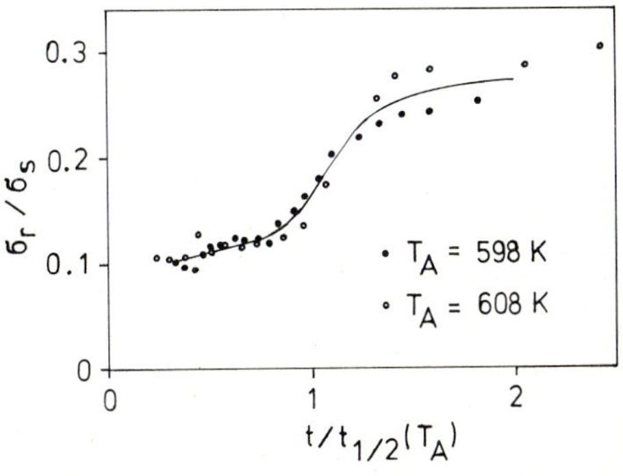

FIGURE 3
Reduced retentivity σ_r/σ_s versus scaled time

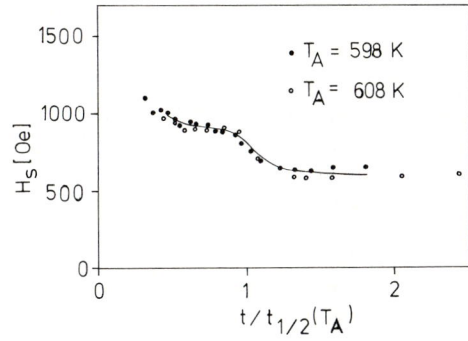

FIGURE 4
Saturated field H_s versus scaled time at two annealing temperatures $T_1 = 598$ K and $T_2 = 608$ K

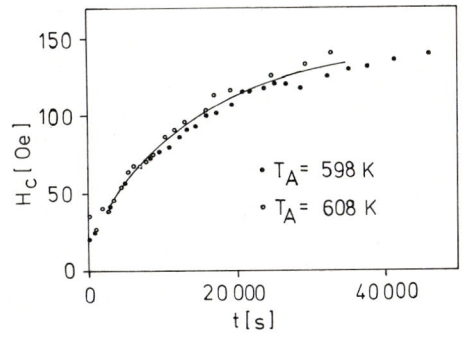

FIGURE 5
Time dependence of coercivity H_c at two annealing temperatures $T_1 = 598$ K and $T_2 = 608$ K

more universal behavior in crystallization of an amorphous material. Coercivity is found to be practically independent of the annealing temperature.

4. CONCLUSIONS

The spontaneous magnetization σ_s, the reduced initial susceptibility $(\partial \sigma_{init}/\partial H)/\sigma_s$, the reduced retentivity σ_r/σ_s as well as the saturation field H_s of annealed amorphous alloys rescale along the time axis and in each case this parameters collapse into the universal curves as a function of "scaled time". From our analysis of experimental data we may conclude that the crystallization process of an amorphous materials is governed by a more universal law.

REFERENCES

1) T. Egami, Rep. Prog. Phys. 47 (1984) 1601-725.

2) G. F. Mazenko, O. T. Valls and F. C. Zhang, Phys. Rev. B31 (1985) 4453-64; F. C. Zhang, O. T. Valls and G. F. Mazenko, Phys. Rev. B31 (1985) 1579-89.

3) N. Hamaya, Y. Yamada, J. D. Axe, D. P. Belanger and S. M. Shapiro, Phys. Rev. B33 (1986) 7770-6.

4) A. Roig, J. S. Muñoz, M. B. Salamon and K. V. Rao, J. Appl. Phys. 61 (1987) in print.

5) P. Cotterill and P. R. Mould, Recrystalization and Grain Growth in Metals (John Wiley & Sons, New York, 1976) pp. 43-7.

6) W. Minor , R. Malmhall , A. Roig , A. Inoue and K. V. Rao , in : Rapidly Solidified Alloys and Their Magnetic Properties, eds. B. C. Giessen, D. E. Polk and A. I. Taub (Materials Research Society, Pittsburgh, 1986) pp. 109-12.

7) T. Watanabe and M. Scott, J. Mat. Sci. 15 (1980) 1131-9.

CRYSTALLIZATION BEHAVIOUR OF THE $Ni_{81}Cr_{15}B_4C$ (wt. %) ALLOY

A. Criado, M. Millán, A. Conde and R. Márquez

Departamento de Física del Estado Sólido. Instituto de Ciencias de Materiales. Universidad de Sevilla - C.S.I.C. Apartado 1065. Sevilla 41080. Spain.

A characterization of the crystallization of the $Ni_{68}Cr_{14}B_{18}C$ (at. %) quenched alloy is carried out by calorimetric (DSC) and X-ray diffraction techniques. The two crystallization stages are resolved in isothermal DSC records as overlapped exotherms. The onset of crystallization is 689±1 K (10 K/min) and, after completion, Ni and Nickel boride phases are revealed by X-ray diffraction. The Johnson-Mehl-Avrami approach is used to derive isothermal kinetic parameters.

1. INTRODUCTION

Metallic glasses lose their beneficial properties during heating and so, for many applications it is important to know when and how fast these amorphous alloys crystallize. However, the crystallization temperature as obtained from quick DTA or DSC experiments depends very much on the used heating rate. So, an understanding of the crystallization mechanisms and kinetics of thermal evolution is important to explain the thermal stability of these materials.

In this paper we study both the isothermal and non-isothermal kinetics of a NiCrB alloy and the crystalline phases formed during the transformation are identified.

2. EXPERIMENTAL

The alloy metglas MBF80 of nominal composition Ni(Balance), Cr (14.5-16.0), B (3.17-4.2), C (0.06 max.) was supplied by Allied Chemicals Co. (USA). DSC experiments were carried out in a Perkin-Elmer DSC-2C on specimens of mass 2-5 mg. Heating rates between 2.5 and 80 K/min were used in continuous heating experiments. To prevent unwanted annealing the specimens were heated up to the maximum available rate (320 K/min) from room temperature. In isothermal scans the specimens were also heated up to the annealing temperature at maximun available rate. The fraction transformed at any time t was determined as described elsewhere[1].

X-ray diffraction was performed at room temperature on specimens previously heated in the calorimeter chamber to allow direct correlation of the phase composition with calorimetric results. X-ray spectra were recorded with Cu K_α radiation on a Philips diffractometer, at a scanning rate of 2° (2θ)/min.

3. RESULTS

3.1. Non-isothermal crystallization

Typical calorimeter traces under non-isothermal and isothermal conditions are shown in Fig.1. Crystallization occurs in two stages as revealed by the two overlapped exotherms displayed in the isothermal records but these two stages are not resolved in the non-isothermal runs. The onset of the crystallization is at 689±1 K, for a heating rate of 10 K/min. The enthalpy of the first crystallization stage is considerably lower than that of the second one, the overall enthalpy of transformation being 4.6±0.1 kJ/mol. The activation energy for the second crystallization event was derived according to Kissinger[2], yielding a value of 4.5±0.5 kJ/mol.

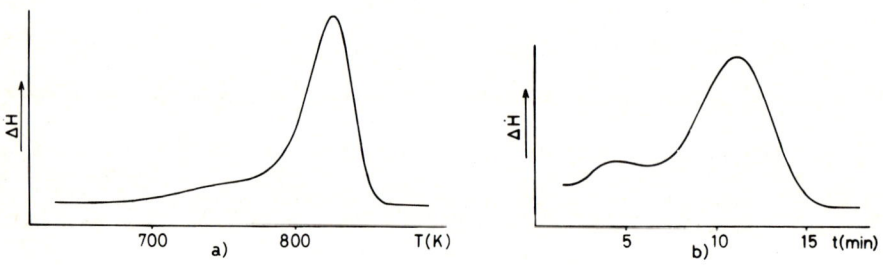

Fig.1. DSC records: a) 10 K/min, b) annealing at 685 K.

X-ray results suggest that the initial crystallization stage is connected with a precipitation of Nickel (fcc), but the small fraction of precipitated crystals should give characteristic lines of very low intensity. In the second stage the amorphous phase is exhausted and Nickel and Nickel boride $(Ni,Cr)_3B$ phases coexist.

3.2. Isothermal kinetics

The kinetics of either stages of transformation was analyzed in terms of the Johnson-Mehl-Avrami equation[3]. Fig.2.shows the ln-ln plots of $-\ln(1-x)$ against t for both crystallization exotherms and the experimental points can be roughly fitted in all cases to a straight line. Values obtained for the Avrami exponent n, for the different annealing temperatures, are n_1=2.5±0.1 for the first crystallization stage, and n_2=4.0±0.1 for the second one. These values, in spite of the lack of microstructural observations, should suggest a three-dimensional diffusion controlled growth with a constant nucleation rate, as expected for a primary crystallization, for the first stage of the transformation. For the second stage the n value should indicate an interface controlled growth with a constant nucleation rate. These values are also found an an analogous NiCrP alloy[4].

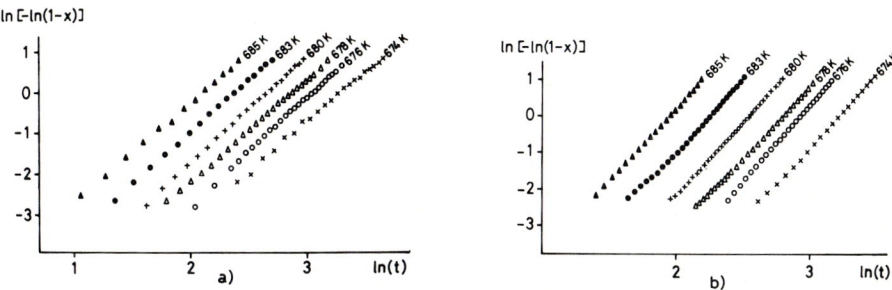

Fig.2. JMA plots: First (a), and second (b) exotherms.

The activation energy for the overall crystallization process can be obtained from the Arrhenius temperature dependence of the rate constant. The straight line fits are well achieved and the slopes yield the values E_1=4.0±0.1 kJ/mol for the first stage and E_2=4.6±0.1 kJ/mol for the second one, the last in good agreement with that obtained by Kissinger's method from non-isothermal data.

REFERENCES

1) H. Miranda, C.F. Conde, A. Conde and R. Márquez, Mat. Lett. 4 (1986) 442.

2) H.E. Kissinger, Anal. Chem. 29 (1957) 1702.

3) M.G. Scott, Amorphous Metallic Alloys, ed. F.E. Luborsky (Butterworth, London, 1983) p. 162.

4) C.F. Conde, H. Miranda, A. Conde and R. Márquez. To be published.

MAGNETOELASTIC PROPERTIES OF MULTILAYERED Co P AMORPHOUS ALLOYS

G. RIVERO, M. LINIERS, E. ASCASIBAR* and J.M. GONZALEZ**
Laboratorio de Magnetismo, Facultad de Física, U. Complutense
* C.I.E.M.A.T. Division de Fusion
** I.C.M.M. (C.S.I.C.), Madrid, Spain

Magnetoelastic behaviour of Co-P electrolytic alloys has been studied. ΔE effect and magnetoelastic coupling coefficient have been measured in tubular samples of different compositions. Remarkable improvement of this behaviour has been obtained by azimuthal magnetic annealing.

1. EXPERIMENTAL

Tubular samples of Co-P amorphous alloy have been obtained by electrolysis. Besides the rotating Cu cathode and eight-fold Co anode a square-pulsed electrolytic current has been used[1] . In the resulting multilayered configuration the layer thickness and composition can be controlled through the amplitude and width of the positive and negative pulses[2].

Samples are 45mm long and 22μm thick with 10.5 mm in diameter. The average compositions range from 11.5% to 18% at. P, as obtained from chemical analysis. The amorphous character of the samples was confirmed by X-ray diffraction.

M_z-H_z and B_ϕ-H_ϕ DC hysteresis loops have been obtained by using a conventional set-up. ΔE values resulted from the measurement of the resonance frequency excited by an azimuthal alternating field as a function of the longitudinal bias field[3].

Magnetic annealings have been performed in controlled pure Ar atmosphere at 183±1°C for 30 minutes. The azimuthal saturating applied field was produced by a DC current flowing through a conductor coaxial with the tubes.

2. RESULTS AND DISCUSSION

B_ϕ-H_ϕ and M_z-H_z DC hysteresis loops indicate the presence of a magnetic anisotropy with azimuthal easy axis. Figure 1 corresponds to 14% at. P content sample. The anisotropy field, H_k and saturation magnetization, $\mu_o M_s$, increase with Co content from 11 to 18.5 Oe, and 0.65 to 1.1 T, respectively.

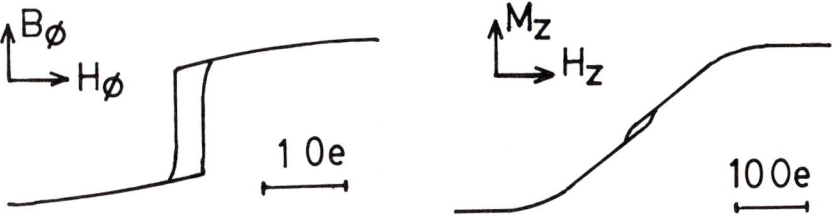

Fig.1. Azimuthal and longitudinal loops of 14% at. P sample.

As obtained samples present values of $\Delta E/E^h$ and magnetoelastic coupling coefficient k_{33}, almost independent of composition. After annealing both magnitudes increase remarkably and reach values up to 20% ($\Delta E/E^h$) and 41% (k_{33}) for 12% at. P (Fig.2).

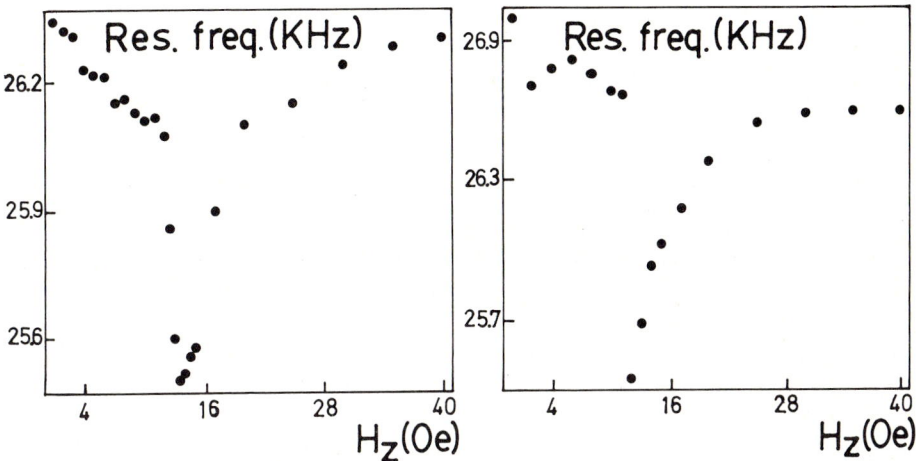

Fig.2. Resonance frequency vs. longitudinal bias field as obtained and after annealing (14% at. P).

The following table summarizes the obtained results as a function of composition:

% at.P	$\mu_o M_s$(T)	H_k(Oe)	$\Delta E/E^h$ as o.	k_{33} as o.	$\Delta E/E^h$ ann.	k_{33} ann.
18	0.65	11	0.05	0.22	0.09	0.29
14	0.86	15	0.06	0.25	0.13	0.34
12	0.95	17	0.06	0.25	0.20	0.41
11.5	1.10	18.5	0.07	0.26	0.15	0.36

Anisotropy field values cannot be solely interpreted in terms of demagnetizing field ($H_d \simeq 6$ Oe). Therefore, a composition-dependent directional pair ordering anisotropy must be present. The higest anisotropy is achieved at 11% at. P content[4]. This fact could explain the increase of the anisotropy field with Co content. Moreover, magnetic annealing does not affect H_k, suggesting that the maximum value has been reached during fabrication.

Nevertheless, magnetic annealing can contribute to stress relaxation and partial elimination of local anisotropy fluctuations, giving rise to a higher susceptibility and, consequently, to a higher ΔE effect. Both, ΔE and k_{33}, increase with Co content except for the 11.5% at. P sample, in good agreement with magnetostriction measurements carried out by Hüller et al[5].

ACKNOWLEDGEMENT

Authors wish to thank Dr. A. Garcia Escorial for the compositional analysis of the samples.

REFERENCES

1) G. Rivero, Magnetic properties of ring shaped electrodeposited Co-P amorphous alloys, this volume.
2) J. M. Riveiro and G. Rivero, IEEE Trans. Magn. 8 (1981) 3082.
3) A. Hernando, V. Madurga, J.M. Barandiaran and M. Liniers, J. Magn. Magn. Mat. 28 (1982) 109.
4) J. M. Riveiro, G. Rivero and M. C. Sanchez, J. Magn. Magn. Mat. 31-34 (1983) 1551.
5) K. Hüller, M. Sydow and G. Dietz, J. Magn. Magn. Mat. 53 (1985) 269.

INDUCED ANISOTROPY BY LASER ANNEALING

C. AROCA, M.C. SANCHEZ, M. GARCIA (*), E. LOPEZ, P. SANCHEZ (**)
Dept.Física de Materiales, Fac. Físicas, Univ. Complutense, Madrid.
(*) Dept. Óptica, Fac. Físicas, Univ. Complutense, Madrid
(**) Dept. Física, ETSI Telecomunicación, Univ. Politécnica, Madrid.

It has been studied the induced anosotropy in metglas 2705Xa from Allied Co. The samples were locally pulse laser annealed in spots of 10 μm of diameters creating regulars or irregulars dots matrix. The induced anisotropy as a function of the distance between dots, and pulse duration, has been studied. The results can be interpreted assuming the behaviour of the laser impacted zones like magnetic dipoles.

1. INTRODUCTION

It has been studied the magnetization processes and induced anisotropy in metglas 2705Xa amorphous samples, locally pulse laser annealed; after every laser pulse the sample was displaced to obtain all the sample covered of a matrix of equally annealed zones. Two kings of matrix were used: a regular one, distance between annealing zones 1 mm and 1 mm, and a irregular one, distance between annealed zones .5 mm and 2 mm. Several samples with both kind of matrix were obtained varying the pulse duration. The magnetization processes and induced anisotropy has been studied. The result can be justified with a simple model that assumes much more lower permeability for the zones impacted with laser pulses than in the non impacted zones.

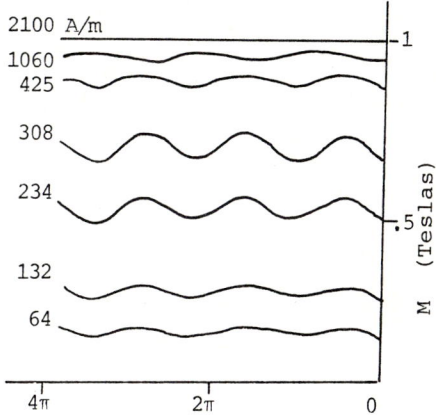

Fig. 1

2. EXPERIMENTAL PROCEDURE

Samples 50 μm thick, were in form of disk (20 mm diameter) in order to avoid the shape factor. The samples covered with an adequate mask, were cut by immersion in a nitric acid solution.

This work was partially supported by the Comisión Asesora de Investigación Científica y Técnica of the Department of Education of Spain (nº 2863-83).

The samples were locally annealed by a continous argon laser with an output laser power of 5 W (1). The laser light was focused by a suitable microscope objective up to get spot size of 10 µm. The sample was mounted on a micropositioning base with ultraprecise two-axis positioning which provide the displacement between successive irradiations.

Histeresis loops and curves of maxima magnetization versus applied field direction at different applied field like fig. 1 were obtained with an original sampling and hold system (2). From these curves it is easy to obtain commutation curves at any direction of the applied field. The commutation curve in the easy axis direction is given by the maximum points of the fig. 1 curves and the one in the hard axis by the points of minima. So the magnetization work and anisotropy directions can be easily obtained from these kinds of curves.

3. EXPERIMENTAL RESULTS

The anisotropy energy obtained from the magnetization work for samples annealed with different pulse duration and for both matrix configuration are:

pulse duration (s)	0	.0166	.033	.5	1
symmetrical (J/m^3)	46	44	36		44
asymmetrical (J/m^3)		42	55	132	252

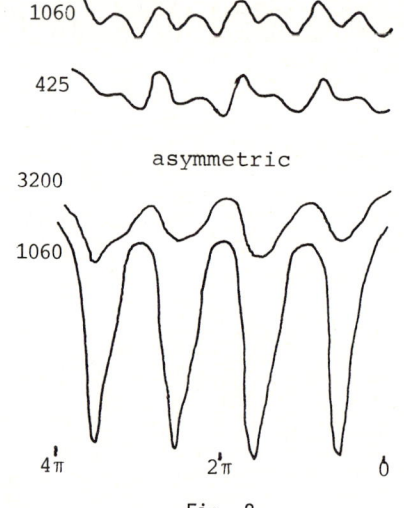

Fig. 2

As can be seen the anisotropy energy for samples with symmetrical destribution of the annealed zones is almost independent of the annealing time, and increase with the annealing time for the asymmetrical one. The samples with asymmetrical distribution of the annealed zones shows higher anisotropy energy than symmetrical ones with the same annealing time and same density of laser impacts. Fig. 2 shows the maxima magnetization for a fixed applied field versus the applied field direction in two samples after long time laser impact, being symmetrical and asymmetrical the distribution of annealed zones.

The samples show the hysteresis loops with two distinct regions. One with high coercive field and apparent low magnetization, that we assume is connected with the impacted zones, and other with very low coercive field which is corresponding to the rest of the sample. From fig. 2 it is easy to observe the uniaxial behavior of the asymmetrical sample with easy axis parallel to the .5 mm rows. However the symmetrical annealed sample shows an anisotropy that looks like the addition of initial uniaxial anisotropy plus a biaxial one with easy axis at 45 degrees with the rows of annealed zones. This effect is only observed in samples with long annealing time and increases with it.

4. DISCUSSION

The results obtained can be interpreted assuming that the radiation effect in the samples, for a pulse duration larger than .0166 s is to reduce the permeability of the impacted area, probably due to its crystallization. Then these crystalized areas do not contributed to the magnetization processes at low field, and will be sourrounded by in plane magnetization closure domains. When a magnetic field is applied to these samples, the magnetization processes at low fields take place by displacement of these domain walls. For higher magnetic fields magnetic poles will appear in the borders of the annealed areas and they will control the mangetization processes. The apparition of the observed anisotropy of the samples can by justified with a simple calculus with this assumption. If the distance between annealed zones is great enough, theirs closure domains will be separated, but if the distance is short enough it will appear common closure structures, being the magnetization processes mainly due to them, as happen in the asymmetrical sample, in which the absolute value of $\sin 2\theta$ shape of the fig. 2 curves, can be explained by a magnetization processes due to displacement of domain wall parallel to the .5 mm rows.

In conclusion the magnetic behavior of a magnetic material can be widely modified by these kind of thermal threatments

REFERENCES

1) M. García, M.C. Sánchez   Revue Phys. Appl. 21 (1986) 207
2) C. Aroca, R. Jimenez, P. Sánchez, E. López. To be published

THE RESONANCE MAGNETIC FIELD SHIFT INDUCED BY DC CURRENT IN AMORPHOUS ALLOYS.

Andrzej Wadas

Institute of physics Polish Academy of Sciences,
al.Lotnikow 32/46.Warsaw,Poland

A shift of the Curie point induced by current in $(Co_{0.6}Ni_{0.4})_{78}Si_8B_{14}$ has been observed. The electron- phonon interaction or the single-particle excitation is the origin of the resonance-magnetic-field shift induced by current.

1. INTRODUCTION.

Recently, the influence of dc current on magnetic properties such as hysteresis loop , permeability and resonance magnetic field has attracted the attention of several authors[1-3]. The detailed studies of the resonance magnetic field shift caused by dc current are of interest to the author. The aim of this paper is to report and interpret the results obtained after applying the high current density.

2. EXPERIMENT.

The $(Co_{0.6}Ni_{0.4})_{78}Si_8B_{14}$ amorphous alloy was chosen for this experiment. The amorphous ribbon was prepared using a single-roll quenching technique. The sample was 30μm thick, 1 mm wide and 3 mm long. The dc current was applied parallel to the long axis of the ribbon. The sample was placed in the microwave cavity and the ferromagnetic resonance at f=9.36 GHz was observed at room temperature (RT). The dc current was applied and the resonance field was measured when the magnetic field was perpendicular to the current direction and to the ribbon surface. This field shift is marked ΔH.

The situation when the resonance magnetic field is perpendicular to the ribbon surface can be described by the resonance equation presented below;

$$\omega/\gamma = H - 4\pi M - H_u \quad (2.1)$$

where H_u is the uniaxial perpendicular anisotropy.

So, when the dc current causes the change of the resonance field we can directly calculate the change of the effective anisotropy $4\pi M - H_u$. The resonance field shifts to lower values and the amplitude of the ferromagnetic signal decreases. The simplest solution of this phenomenon is to attribute the usual heating process as the origin of the observed effect. So, one should measure the temperature of the sample during the flowing of current. This was done with a copper-constantan thermocouple placed at the ribbon. We have compared the plots of the reduced magnetization M(T)/M(RT) versus T obtained in our experiment with those obtained by the magnetic balance method in Fig.1. The latter experiment was done at the external magnetic field H = 3600 Oe - the lowest resonance magnetic field.

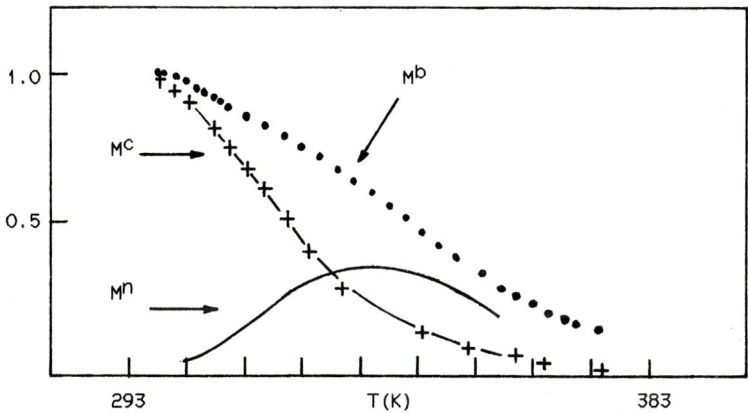

FIGURE 1.
The reduced magnetization M^b/M(RT) measured by the magnetic balance method at H= 3600 Oe. The values of the reduced magnetization M^c/M(RT) measured by the "current" experiment and calculated values of $M_n = M^b - M^c$.

3. DISCUSSION

One can imagine that flowing current induces the interactions between conduction electrons from one side and ions from other side. The direct scattering effect on electrons which are responsible for magnetization is possible also. So, these "magnetic" electrons become more itinerant and we can expect that they are thermally excited. Therefore, one of the ways to describe our results is by treating our case as an itinerant-electron ferromagnet forced by current. We can relate the temperature dependence of reduced magnetization with temperature dependence of D (spin-wave-stiffnes constant).
Then $D=D_0(1-D_1T^2)$, where T^2 term is due to the interaction between spin-waves and thermally excited itinerant electrons. Then the reduced magnetization can be described in the following way;

$$M(H,T)/M(H,RT) = a*T^{3/2}*(1-DT^2)*er\{3/2,T/T_g\} + b*T^{5/2}*er\{5/2,T/T_g\} \quad (3.1)$$

Fitting parameter T_g has a physical meaning and reflects the presence of an effective internal field arising from the applied field, the anisotropy field and demagnetization. Thus parameter T_g is known as a temperature gap $T_g = H_{int}/k_b$. Fitting the equation (3.1) to the experimental points we can obtain dependence of the standard deviation versus parameter T_g. Figure 2 presents these results for different cases; datapoints obtained from the balance method and from "current" experiment. The minimum observed by the balance method is shifted to zero in "current" case. It means that the best fit of equation (3.1) exists when the internal field tends to zero.

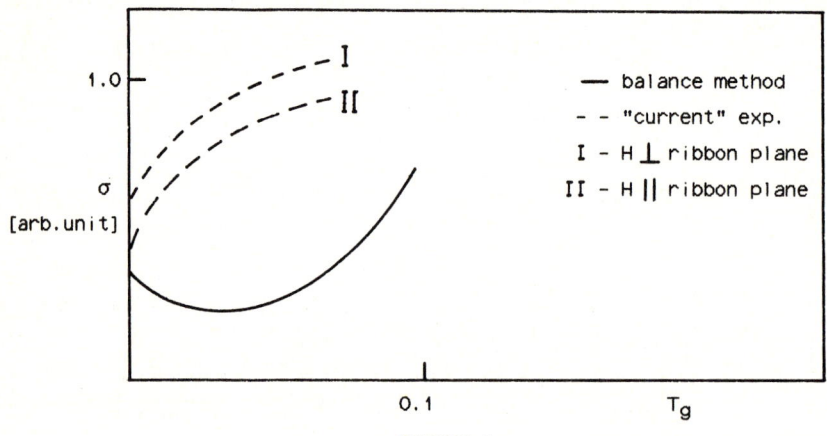

FIGURE 2.
The standard deviation versus parameter T_g.

One can also imagine a second explanation of the present experiment. When we pass the current through the sample we must expect that we disturb the screening behaviour of conduction electrons. We can say that we modify the electron-phonon interaction. As stated previously[3] this interaction influences the sample magnetization. This approach was performed earlier[2] and the important is result that the electron-phonon interaction decreases the magnetization as $0.3\mu_b$ per atom. At this time it is very difficult to state which of the proposed mechanisms is more valid and decides about the magnitude of resonance-magnetic-field shift.

REFERENCES

[1] C. Aroca et al. J. Magn. Magn. Mater. 23, (1981) 193

[2] A. Wadas, Phys. Rev. B. 34, (1986) 6493.

[3] D. J. Kim, Phys. Rev. B. 25, (1982) 6919.

A SEMI-INFINITE TRANSVERSE ISING MODEL WITH SURFACE
AMORPHIZATION: PHASE DIAGRAMS*

T.KANEYOSHI[§] and E.F.SARMENTO[+]

[§]Department of Physics, Nagoya University, Nagoya 464, Japan
[+]Departamento de Física, Universidade Federal de Alagoas,
57000 Maceió-AL, Brazil

Phase Diagrams of a semi-infinite Ising model with surface amorphization in the presence of a transverse field are studied within the framework of an effective field theory. Reentrant phenomena on the surface are obtained for appropriate values of the surface interaction and transverse field.

Our model consists of a semi-infinite system, which is described by the Ising Hamiltonian of spin variable $S_i = \pm 1/2$ with nearest-neighbor interaction in the surface $\bar{J}_s$, $\bar{J}_1$ between spins on the surface and its nearest-neighbor at the first layer, and the bulk ferromagnetic interaction J_b otherwise:

$$H = -1/2 \sum_{ij} J_{ij} S_i^z S_j^z - \Omega_b \sum_m S_m^x - \Omega_s \sum_n S_n^x \qquad (1)$$

where Ω_b and Ω_s are transverse fields in the bulk and on the surface, respectively. $\bar{J}_s$ and $\bar{J}_1$ are random variables governed by following probability distribution functions

$$P(\bar{J}_s) = \frac{1}{2}[\delta(\bar{J}_s - J_s - \Delta J_s) + \delta(\bar{J}_s - J_s + \Delta J_s)] \qquad (2)$$

$$P(\bar{J}_1) = \frac{1}{2}[\delta(\bar{J}_1 - J_1 - \Delta J_1) + \delta(\bar{J}_1 - J_1 + \Delta J_1)]$$

in which we assume $0 \leq \Delta J_b \leq J_b$.
Following Sá Barreto et al[1], the longitudinal $\langle \sigma_i^z \rangle = 2\langle S_i^z \rangle$ site magnetization is given by

(3)

$$\langle \sigma_i^z \rangle_r = \prod_\delta [\langle \cosh(DJ_{i,i+\delta}) \rangle_r + \langle \sigma_{i+\delta}^z \rangle_r \langle \sinh(DJ_{i,i+\delta}) \rangle_r] f_i(x)|_{x=0}$$

where $D = \partial/\partial x$, $\langle ... \rangle_r$ means the random bond average, $\prod_\delta$ is the product over the nearest-neighbor of the site i, and the function $f(x)$ is given by

---

* Work partially supported by CNPq and FINEP (Brazilian Agencies).

$$f(x) = (x/\Gamma_i(x)) \tanh[(\beta/4) \Gamma_i(x)]$$

where $\beta = 1/k_B T$ and $\Gamma_i(x) = [(2\Omega_i)^2 + x^2]^{1/2}$.

Now, let us apply equation (3) to our layered simple cubic system with a disordered surface. By performing the configurational averages, and taking into account that we are concerned with the calculation of the surface and bulk ordering near the transition temperature that allow us to consider only terms linear in m_i, we find

$$m_s = 4k_1 m_s + k_2 m_1 \tag{4}$$

$$m_1 = 4k_3 m_1 + k_4 m_s + k_3 m_2 \tag{5}$$

and

$$m_n = k(m_{n-1} + 4m_n + m_{n+1}) \quad \text{for } n \geq 2 \tag{6}$$

where $m_i = \langle\sigma_i\rangle_r$, and m_n, m_{n-1} and m_{n+1} are the z components of the magnetizations in the nth, (n-1)th and (n+1)th layers, respectively. The k's coefficients are expressed as

$$k_1 = \langle\sinh(D\bar{J}_s)\rangle_r \{\langle\cosh(D\bar{J}_s)\rangle_r\}^3 \langle\cosh(D\bar{J}_1)\rangle_r \, f_s(x)|_{x=0}$$

$$k_2 = \langle\sinh(D\bar{J}_1)\rangle_r \{\langle\cosh(D\bar{J}_s)\rangle_r\}^4 \, f_s(x)|_{x=0}$$

$$k_3 = \langle\cosh(D\bar{J}_1)\rangle_r \sinh(DJ_b) \cosh^4(DJ_b) \, f_b(x)|_{x=0} \tag{7}$$

$$k_4 = \langle\sinh(D\bar{J}_1)\rangle_r \cosh^5(DJ_b) \, f_b(x)|_{x=0}$$

$$k = \sinh(DJ_b) \cosh^5(DJ_b) \, f_b(x)|_{x=0}$$

In order to obtain the phase diagrams and transition temperatures for the bulk and surface ordering as functions of the transverse field and $\delta_s = \Delta J_s/J_s$, according to Binder and Hohenberg[2] we assume that $m_{n+1} = a\, m_n$ for $n \geq 2$. Equation (4) and (5) yield the following equation

$$(4k_1 - 1)[(4+a)k_3 - 1] = k_2 k_4 \tag{8}$$

where the parameter a is given by, upon using (6)

$$a = \frac{(1-4k) - \sqrt{(1-4k)^2 - 4k^2}}{2k} \tag{9}$$

The bulk transition temperatura T_c^b can be determined by putting $m_n = m_{n-1} = m_{n+1} = m$ into (6), and in figure 1 T_c^b is plotted as a function of Ω_b. As can be seen T_c^b reduces to zero at the critical value of Ω_b, $\Omega_c^b = 2.35\ J_b$. This is an improvement on the traditional MFA, which provides $\Omega_c^b(\text{MFA}) = 3.0\ J_b$.

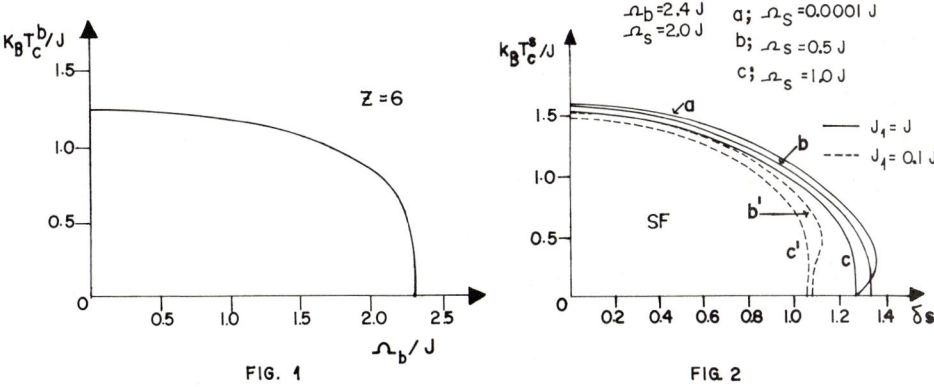

FIG. 1 FIG. 2

In figure 2, the changes of T_c^s with δ_s are plotted for the case with $\Omega_b = 2.4\ J_b$, $J_s = 2.0\ J_b$ and $\delta_1 = 0.0$, selecting the values of Ω_s and J_1. For the case of $J_1 = J_b$ (solid lines), the reentrant phenomenon is observed only for a very weak transverse field (curve (a)). For the case of $J_1 = 0.1\ J_b$ (dashed lines), the curve (b') with $\Omega_s = 0.5\ J_b$ expresses a weak reentrant phenomenon, although the corresponding curve (b) does not show it. When J_1 becomes weak, the surface behaves like a 2D amorphous ferromagnet, which can exhibit the reentrant phenomenon for the values of δ larger than $\delta = 1$.

We conclude that the reentrant phenomenon is possible on the surface of the semi-infinite transverse Ising model with a surface amorphization, when J_s becomes rather larger than the bulk J_b and J_1 is taken as a value smaller than the bulk J_b.

REFERENCES

1) F.C. Sá Barreto, I.P. Fittipaldi and B. Zeks, Ferroelectrics 39 (1981) 1103.

2) K. Binder and P.C. Hohenberg, Phys.Rev.B 9 (1974) 2194.

DISTRIBUTION AND EVOLUTION OF TEMPERATURE DURING FLASH-ANNEALING OF METALLIC GLASSES

J.M.BARANDIARAN and N.ZABALA

Dpto. de Electricidad y Electrónica, Facultad de Ciencias, U.P.V. Lejona, Apdo. 644, 48080 Bilbao (Spain).

We propose a theoretical method to calculate the temperature of metallic ribbons during "flash-annealing" performed by means of current pulses. The calculation is based on energetic considerations and gives the time evolution of the temperature and the final steady state value. The predictions are in good agreement with experimental temperature determinations, based on the value of the saturation magnetization during the heating of metallic glasses.

1. INTRODUCTION

In the last few years, annealing of amorphous ribbons by means of electric current has been reported as an appropriate procedure to achieve optimum magnetic properties of metallic glasses. It also allows magnetic anisotropies to be induced when stress is applied during the treatment[1,2,3,5].

The rapid heating rate ($500°C\ s^{-1}$) allows one to reach higher final temperatures avoiding the nucleation and growth of crystals. Furthermore, the rapid cooling rate at the end of the pulse contributes to freezing in the attained state.

In any case, important obstacle of this technique is the accurate determination of temperature during the heating, which in most cases is not better than $\pm 30°C$[1,2,6].

2. EXPERIMENTAL

The experiments were performed with amorphous ribbons of composition $(Co_{94}Fe_6)_{75}Si_{15}B_{10}$ and $Co_{80}Nb_8B_{12}$ (the latter one prepared with two different cross sections[5]) Of 8 cm long and about $10^{-4} cm^2$ in cross section. The Curie Temperatures were 350°C for the first sample and greater than 500°C for the others. The experimental set-up allowed the visualizing the hysteresis loop at 50Hz, so that the evolution of the spontaneous magnetization (M_s) could be obtained at 10 ms intervals. The values of M_s measured in that way were converted into temperature by comparing them with the temperature dependence of M_s obtained in a conventional furnace up tp 500°C. The temperature as well as the spontaneous magnetization stabilized in about 1.5 s, after an exponentially shaped approach (see Fig. 1 and 2).

3. STEADY STATE TEMPERATURE

Far from the end of the clamps, the temperature may be assumed uniformly distributed throughout the ribbon. Taking into account the equation of conduction of heat when the steady state is reached, we have estimated the error due to this approximation between 0.1% and 1%.

When the current pulse is started, the temperature of the sample increases until it reaches a final temperature T_f determined by the following heat balance: the supplied power, $R I^2$, equals the speed of dissipation of heat by conduction, convection and radiation, i.e.:

$$0.24\ RI^2 = H A (T_f - T_a)^{5/4} + 2 A \sigma \alpha(T_a) (1 + \beta(T_f - T_a)) (T_f^4 - T_a^4) \qquad (1)$$

where R is the resistance of the sample and A its surface per unit lenght, I is the intensity supplied, T_f the final temperature reached, T_a the ambiance temperature, the convection coefficient[4] has been taken as $H(T_f - T_a)^{1/4}$, σ is the Stefan-Boltzmann's constant and α the emisivity of the ribbon, that has been considered to increase linearly with a temperature coefficient β. Giving numerical values to the parameters $H, \alpha(T_a)$ and β and supposing the others are known we way plot T_f (I) with the aid of a computer and determine their values by adjusting the theoretical curves so obtained to experimental results. An example is shown in Fig. 1.

On the other hand, from eq. (1) one can deduce that samples with higher cross section need more intensity to reach the same final temperature, in agreement with measurements in $Co_{80}Nb_8B_{12}$ samples.

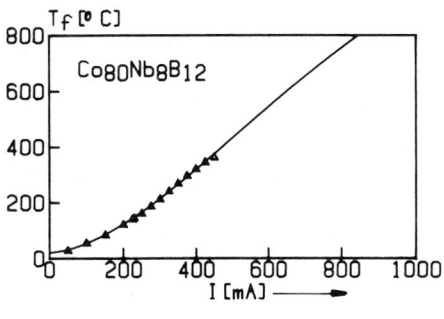

FIGURE 1

Experimental ($\triangle$) and theoretical (-) values of the final temperature as a function of the intensity. Parameters are: $H= 1.02\ 10^{-3}$cal s^{-1} cm^{-2} k^{-1} and $\alpha = 0.42 + 5.5\ 10^{-4}\ (T_f - T_a)$.

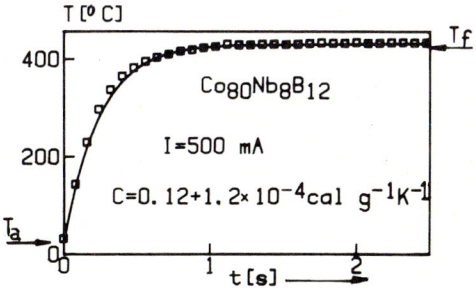

FIGURE 2

Experimental and theoretical evolution of the temperature during the pulse. Eq.(2) and the simple exponential behaviour cannot be distinguished in this scale.

4. TRANSITORY STATE TEMPERATURE

In order to determine the temperature evolution since the time at which the pulse is switched on until the steady state is reached, the energy supplied in a differential time dt is supposed to be shared between heat stored in the sample and heat dissipated by the mechanism mentioned previously:

$$0.24RI^2 dt = mcdt + \{HA(T-T_a)^{5/4} + 2A\sigma\alpha(T_a)(1+\beta(T-T_a))(T^4-T_a^4)\}dt \qquad (2)$$

where all the parameters are known from expression (1) except for the specific heat c, which may be determined by comparison with experimental results.

When so doing, it has been necessary to introduce a linear dependence of c with the temperature. In Fig. 2 temperature is plotted versus time as calculated from (2). Although the analytical expression T(t) derived from that expression turns out to be very complex, it may be approximated to an exponential behaviour with a single relaxation time given by $\tau_0 = (T_f - T_a)mc/0.24\ RI^2$ \qquad (3)

In conclusion the evolution of temperature during "flash-annealing" has been determined experimentally and by comparison with the theoretical curves and a precision of ±5°C can be claimed. This allows us to extrapolate to temperatures higher than the Curie temperature with good precision.

ACKNOWLEDGEMENTS

The authors wish to thank Dr.V. Madurga of the University Complutense for the amorphous ribbons used in this work.

REFERENCES

(1) T.Jagielinski, IEEE Trans.Magn. MAG-19 (1983) 1925.
(2) R.J. Gibbs et al., IEEE Trans. on Magn. MAG-20 (1984) 284.
(3) J.M. Barandiarán et al., IEEE Trans. Magn. MAG-22 (1986) 1964.
(4) M.W. Zemansky, "Calor y Termodinámica", (Aguilar, Madrid 1968)
(5) V. Madurga et al., in: Current Topics on Non Crystalline Solids, Eds. M.D. Baró and N.Clavaguera (World Scientific, Singapore, 1986) pp.373.
(6) A. Zauska et al. in: Rapidly Quenched Metals, Eds. S. Steeb and H. Warlimont (Elsevier, Amsterdam, 1985) pp. 235.

LASER SURFACE TREATMENT OF AMORPHOUS AND CRYSTALLINE $Fe_{40}Ni_{38}Mo_4B_{18}$

J.V. Armstrong, J.M.D. Coey, J.G. Lunney

Department of Pure and Applied Physics,
Trinity College, Dublin 2, Ireland

The vitrification of a crystalline surface layer on melt-spun $Fe_{40}Ni_{38}Mo_4B_{18}$ (Metglas 2826MB) ribbon by pulsed KrF (249 nm) laser irradiation is demonstrated. Conversion-Electron Mössbauer Spectroscopy (CEMS) and X-ray diffraction (XRD) are used to determine the laser-induced microstructural changes of both as-quenched and crystallised ribbon.

1. INTRODUCTION

Laser glazing of a metallic alloy can be achieved by using a short laser pulse to melt a thin layer on the surface: heat conduction into the bulk then gives a rapid quench rate and a thin amorphous surface layer is formed[1]. The melt depth is of order $\sqrt{(D\tau)}$ where D is the diffusivity and τ the laser pulse length; for a 20 ns pulse a melt depth of ~1 μm is expected. Analytical techniques used to study the laser treatment should only probe the surface layer [2,3]. Here CEMS and XRD were applied to study the effects of pulsed KrF (249 nm) laser irradiation of an $Fe_{40}Ni_{38}Mo_4B_{18}$ alloy.

2. EXPERIMENT

A 40 μm thick ribbon of the alloy was used. Crystalline samples were obtained by annealing in vacuum up to 800 °C for 2 hours. A KrF Excimer laser (τ = 22 ns) was used to irradiate an area ~1 cm^2 by rastering the sample in the focussed beam. Transmission and conversion-electron ^{57}Fe Mössbauer spectra were recorded at room temperature. The CEM spectra were obtained using a proportional counter sealed in a 1 bar 95%He-5%CH$_4$ gas mixture. CEM data were binomially smoothed and all spectra were fitted to a histogram distribution of hyperfine fields. XRD (Cu K$_\alpha$) was used to investigate the effects of laser surface treatment since 90% of the diffracted signal at 2θ ~ 45° is due to the uppermost 3 μm of the sample.

3. RESULTS AND DISCUSSION

   3.1. As-quenched amorphous ribbon

   Figure 1 shows the XRD from the shiny side and dull side (in contact with the melt-spinning wheel) of the ribbon. The shiny side is amorphous, while

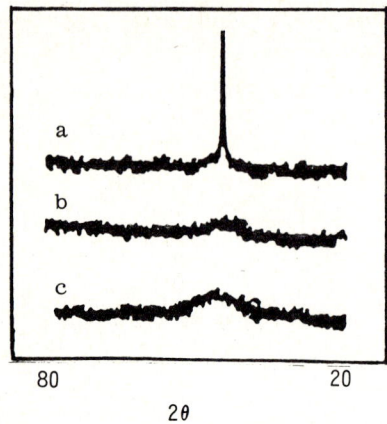

FIGURE 1

X-ray diffraction patterns from as-quenched ribbon:

a: dull side
b: shiny side
c: dull side after laser treatment of 3 J/cm²

the dull side shows a crystalline peak at $2\theta = 44.3°$ superimposed on a broad amorphous halo. The crystalline diffraction pattern corresponds to a textured $M_{23}B_6$ τ-phase[4], where M = (Fe,Ni,Mo), with the (111) planes parallel to the ribbon surface. It has been shown that the crystalline material is restricted to the top 1 μm of the dull side[5]. We find that laser irradiation at 3 J/cm² eliminates all trace of the crystalline phase. There is no appreciable difference between the CEM spectra of the dull and shiny sides of the as-quenched ribbon, nor between the spectra of the dull side before and after laser glazing. It follows that the spectrum of the crystalline $(Fe,Ni,Mo)_{23}B_6$ phase is quite similar to that of the amorphous alloy.

3.2 Annealed crystalline ribbon

When the amorphous ribbon is annealed at 500 °C the crystalline phases $M_{23}B_6$ and γ(Fe,Ni) are formed, with $M_{23}B_6$ predominant. Annealing at 800 °C leads to an increase in the fraction of γ(Fe,Ni). There is no indication of α-Fe in agreement with a previous study[6]. Mössbauer spectra for the 500 °C annealed sample are shown in Figure 2: there is a marked difference in the hyperfine field distribution in the surface and bulk. Laser irradiation at 3 J/cm² on the dull side removes all trace of crystallinity in XRD. The corresponding CEM spectrum shows a broadened hyperfine field distribution also suggesting vitrification of the surface. This spectrum is similar to that obtained from the surface of the untreated amorphous ribbon. The transmission Mössbauer spectrum shows no observable change suggesting that the bulk of the ribbon is unaffected by this laser treatment. This contrasts with a previous study where picosecond laser irradiation of a glassy metal[7] caused reorientation of the bulk magnetisation. Similar laser irradiation on the sample annealed at 800 °C showed a reduction in intensity of the crystalline XRD peaks corresponding to absorption of the X rays in a 0.9 ± 0.2 μm amorphous surface layer[2].

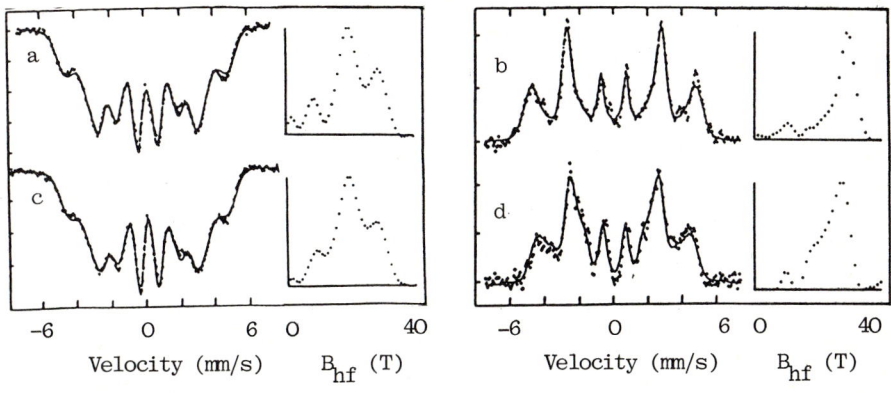

FIGURE 2

Mössbauer spectra, with their corresponding hyperfine field distributions for the 500 °C annealed sample: a & b before c & d after laser treatment at 3 J/cm^2. CEMS spectra were fitted assuming a perpendicular spin texture, whereas that in the bulk is approximately isotropic.

4. CONCLUSIONS

Laser surface vitrification of both as-quenched and annealed metallic glass ribbon has been demonstrated. The bulk is unaffected. This technique might be of value since surface crystallinity on melt-spun ribbons is known to degrade their mechanical, magnetic and corrosion resistance[8] properties.

ACKNOWLEDGEMENTS

The authors thank Dr. J.M. Cadogan for his help with the Mössbauer fitting. Support for this work was provided by the NBST and the CEC.

REFERENCES

1) C.J. Lin and F. Spaepen, Appl. Phys. Lett. 41 (1982) 721.
2) J.W. McCamy, M.J. Godbole, A.J. Pedraza and D.H. Lowndes, J. Mater. Res. 1 (1986) 629.
3) P.P. Patil et al., Phys. Rev. Lett. 58 (1987) 238.
4) J.E. Rout and J.A. Leake, in the Proceedings of the 5th International Conference on Rapidly Quenched Metals, eds. S. Steeb and H. Warlimont (North-Holland, Amsterdam, 1985) p. 291.
5) Hang Nam Ok and A.H. Morrish, J. Appl. Phys. 52 (1981) 1835.
6) E.E. Kamzeeva, N.A. Khatanova and G.S. Zhdanov, Sov. Phys. Crystallogr. 29 (1984) 338.
7) G.L. Whittle, A. Calka, A.P. Radliński and B. Luther-Davies, J. Magn. Magn. Mat. 50 (1985) 278.
8) R. Schulz, N.L. Lee, B.M. Clemens, J. Mater. Res. 2 (1987) 46.

MAGNETOVOLUME EFFECTS IN AMORPHOUS TRANSITION METAL BASED ALLOYS

J. SCHNEIDER, A. HANDSTEIN, J. ARNOLD[+] and J. KAMARAD[+]
ZFW, AdW der DDR, 8027 Dresden, Postfach, GDR
[+]Institute of Physics, CSAV, Prague 8, Na Slovance 2, CSSR

1. INTRODUCTION

As well known, magneto-volume properties are very helpful to describe the character of the local magnetic moments and their interaction in crystalline and amorphous alloys. In this respect it is interesting to look for Invar characteristics. As shown by Wohlfarth[1], weak itinerant ferromagnetism accompanies and causes Invar behaviour. However, Invar characteristic may appear also for strong magnetic systems, as in the case of ordered Fe_3Pt and amorphous Fe-based alloys. In this paper the experimental results for dT_c/dp and for the magnetoelastic properties (Young's modulus E, volume magnetostriction ω) of amorphous Fe- and NiFe-based alloys are briefly regarded. The relevance of existing theoretical models for describing the magnetic behaviour of these systems is discussed.

2. PRESSURE DEPENDENCE OF THE CURIE TEMPERATURE

Within a Landau-Ginzburg model of ferromagnetism the following equations are found[2]:

(1) $\quad dT_c/dp \sim a\, T_c + b/T_c$

for the magnetically homogeneous case and

(2) $\quad dT_c/dp \sim - d_1\, T_c + d_2\, T_c^2$

for the inhomogeneous case, where the constants a, b, d_1, $d_2 > 0$.

Within a model of localized and itinerant electrons one obtains for $J_{AA} \simeq 0$ and $(\chi(T_c) - \chi(0)) \ll 1$:

(3) $\quad dT_c/dp = T_c\,(2\, d\ln J_{AB}/dp + d\ln\chi(0)/dp)$,

where J_{AB} is the exchange energy between the localized A and the itinerant B electrons and $\chi(T)$ gives the susceptibility of the itinerant electrons. For $T_c \longrightarrow 0$ (dilute and giant moment like systems) eq. (3) gives $dT_c/dp \sim \pm\, T_c$, depending on the signs of $d\ln J_{AB}/dp$ and $d\ln\chi(0)/dp$ [2].

The experimental results for some of the amorphous FeNi-based and Fe-based alloys are summarized in Fig. 1. None of the theoretical models describe the behaviour of the Ni-rich FeNi-based alloys. An empirical law of the form

(4) $\quad dT_c/dp = d_1 T_c - d_2 T_c^2$

(d_1, $d_2 > 0$) was found to fit the data very well[3]. Taking the data for amorphous Co-rich (FeCo)SiB[4] (see Fig. 1) and amorphous FeZr, FeHf[5] alloys with low Fe-content this formula is found to be quite well obeyed. For amorphous (FeCr)SiB, (FeMn)SiB[7], Fe-rich FeZr, FeHf[5] alloys and perhaps also for amorphous FeSiB alloys[8] with low Fe-content, Fe-rich (FeCo)SiB alloys[4] as well as for crystalline Fe-rich Ti(FeCo), Co-rich Si(feCo) alloys[6] and NiRh alloys[9] (magnetically heterogeneous crystalline alloys) the observed dependence of dT_c/dp on T_c may be described by eq. (2) (see also Fig. 1). The data for amorphous FeNi- and FeCo-based alloys as well as for amorphous FeZr and FeHf alloys indicate that the character of the curve $d\ln T_c/dp$ vs. T_c changes near the maximum of T_c vs. the Fe-concentration.

3. MAGNETOELASTIC EFFECTS

Beside large values of dT_c/dp also large values for the forced $d\omega/dH$ and spontaneous magnetostriction ω_s as well as an anomaly of the Young's modulus (Δ E-effect) on passing through T_c are characteristic for Invar alloys. The largest values of $d\omega/dH$ are found for the Fe-rich FeNi[4]- and the Fe[10]-based alloys. Measurements of ω for the magnetically inhomogeneous Ni-rich FeNi-based alloys in comparison with those for the Fe-rich alloys would be of interest to seek for the influence of a spatial inhomogeneity in the magnetization M on ω. Within the Landau-Ginzburg model of ferromagnetism for magnetically inhomogeneous alloys one finds at zero pressure for the intrinsic magnetic contribution to ω [2, 11]

(5) $\quad \omega = K (c + c') <M>^2$,

$c \sim dA/d\omega$, $c' \sim <\Delta c^2>$, where $<\Delta c^2> = (c - <c>)^2$ and ($<\Delta c^2> \sim <\Delta M^2>$) describe the fluctuations of the concentration of the magnetic atoms, K gives the compressibility and $A = A(T, H, \omega)$ is one of the parameters in the model.

A large ΔE-effect was found for amorphous FeB[10], FeP[10], Fe BSi[12], and FePGa[13] alloys. But only for amorphous FePGa a large

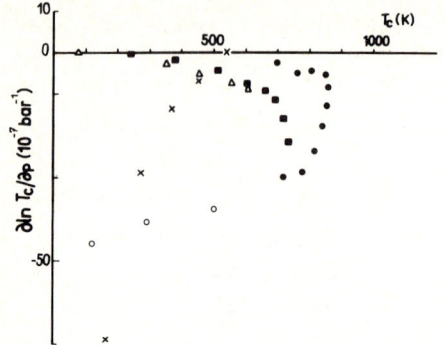

FIGURE 1
$d\ln T_c/dp$ vs. T_c for amorphous FeNi- and Fe-based alloys:

△ $(Fe_xNi_{1-x})_{80}P_{10}B_{10}$ [3]

■ $(Fe_xNi_{1-x})_{77}Si_{10}B_{13}$ [4]

● $(Fe_xCo_{1-x})_{77}Si_{10}B_{13}$ [4]

○ $(Fe_xCr_{1-x})_{77}Si_{10}B_{13}$ [7]

× crystalline NiRh alloys[9].

intrinsic contribution $\Delta E_m = \Delta E_\omega + \Delta E_M$ was observed ($\Delta E_m = E_p - E_s$, where E_s and E_p are the Young's modulus at magnetic saturation and in the paramagnetic state, respectively).

REFERENCES

1) E.P. Wohlfarth, J. Magn. Magn. Mater. 10 (1979) 120.
2) D. Wagner and E.P. Wohlfarth, J. Phys. F 11 (1981) 2417.
3) J. Kamarad, Z. Arnold et al., J. Magn. Magn. Mater. 15-18 (1980) 1409.
4) H. Tange, M. Goto et al., in: Proc. Rapidly Quenched Metals IV (The Japan Inst. of Metals, Sendai, 1982) p. 811.
5) K. Fukamichi et al., IEEE Trans. Magn. MAG-22 (1986) 424.
6) J. Beile, D. Bloch et al., J. de Phys. 38 (1977) 339; N. Buis, J.J.M. Franse et al., Inst. Phys. Conf. Ser. No. 39 (1978) 393.
7) H. Tange, M. Goto et al., Physica 119 B+C (1983) 188.
8) J. Kamarad and Z. Arnold, Physica 139/140 B (1986) 382.
9) H. Kadomatsu and H. Fujiware, Solid State Commun. 29 (1979) 255.
10) K. Fukamichi, in: Amorphous Metallic alloys, ed. F.E. Luborsky (Butterwoths, London, 1983) p. 317.
11) J. Schneider, A. Handstein, et al., to be published.
12) G. Hausch and E. Török, in: Proc. Rapidly Quenched Metals V (Elsevier Science Pub., Amsterdam, 1985) p. 1135.
13) L.T. Baczewski et al., Solid State Commun. 60 (1986) 505.

MAGNETIC AND STRUCTURAL CHARACTERIZATION OF $TbDyFe_2$ SPUTTERED FILMS

E T M LACEY, A D BIRCHENOUGH, D G LORD and P J GRUNDY

Department of Pure and Applied Physics, University of Salford, Salford, M5 4WT, UK.

Preliminary studies are reported on the fabrication and characterization of $TbDyFe_2$ sputtered films. We describe the microstructure and magnetic properties in both as-deposited and heat treated states through observation by optical and electron microscopy.

1. INTRODUCTION

Rare earth iron alloys are known to exhibit a wide variety of magnetic properties when in both bulk and thin film form.[1] In particular, the compound $Tb_xDy_{1-x}Fe_2$, with x = 0.27 (Terfenol-D), is of interest in bulk form because as well as possessing low magnetocrystalline anisotropy energy, it exhibits a room temperature value of magnetostriction of the order of 2×10^{-3}. Previous work on sputtered films of $TbFe_2$[2] and $DyFe_2$[3] have reported that they both possess amorphous structure at room temperature, with $TbFe_2$ exhibiting an easy axis of magnetization normal to the film plane while $DyFe_2$ has an easy axis parallel to the film plane. Preliminary studies on the growth and characterization of $Tb_xDy_{1-x}Fe_2$ metal thin films, with x varying nominally between 0.2 and 0.4, are reported in this paper. We describe the microstructure and magnetic properties in both as-deposited and heat treated states through observation by electron and optical microscopy.

2. EXPERIMENTAL

A 130 dc Trimag sputtering source enabled alloy films of TbDyFe to be deposited at a rate of 1000 Å/min (sputtering power 500 Watts) onto various substrates (glass, carbon and NaCl) in an argon atmosphere of 8 mTorr. Typical base pressure attainable was 7.5×10^{-7} Torr. A composite-type target was used for ease of compositional variation and was comprised of an Fe backing plate and symmetrically disposed pieces of Dy and Tb of surface dimension 10 mm by 10 mm.

3. RESULTS AND DISCUSSION

A number of deposits of $Tb_xDy_{1-x}Fe_2$ alloy have been made around the composition of x = 0.27. X-ray microanalysis, using an energy-dispersive system, has been used to determine film composition (x) to within 10%.

We compare here three types of thin film, with thicknesses varying between 1 μm and 6 μm, with compositions given by $x = 0.2$, 0.3 and 0.4. All deposits were found to be amorphous in structure at room temperature.

Films with $x = 0.3$ and $x = 0.4$ were found to possess a dominant magnetization component parallel to the film normal (Figures 1(a) and 1(b)); this direction is the easy axis. The planar saturation fields were estimated to be approximately 20 kOe which are of the same magnitude found for TbFe deposits of similar composition [2].

Films with $x = 0.2$ were found to exhibit large values of planar remanence (Figure 1(c)). The planar saturation field for films of this composition was approximately 7 kOe. The torque curves of Figure 2 indicate the presence of a planar easy axis.

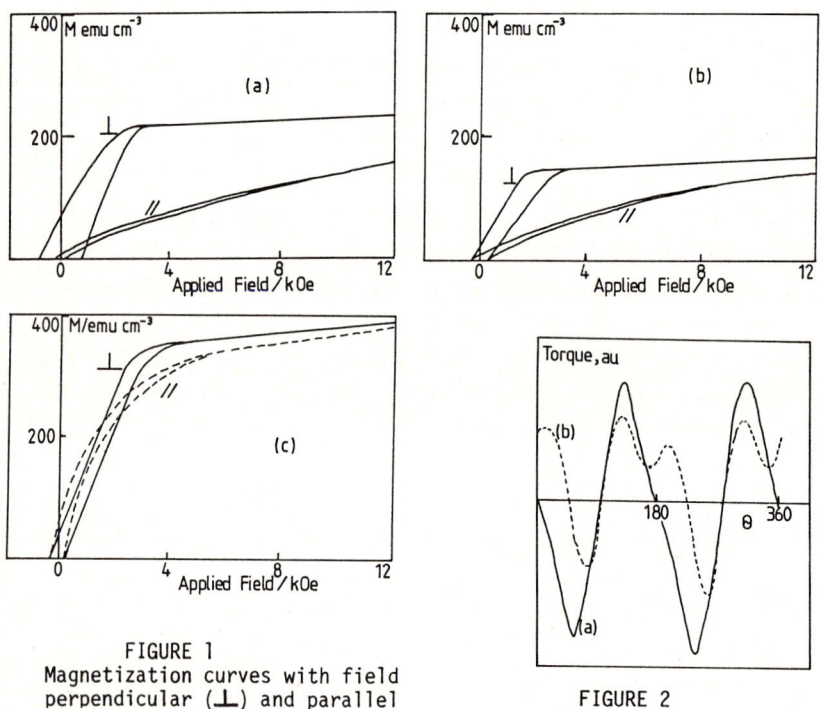

FIGURE 1
Magnetization curves with field perpendicular ($\perp$) and parallel ($//$) to film plane for
a) $x = 0.4$, b) $x = 0.3$ and c) $x = 0.2$

FIGURE 2
Torque curves for a) $x = 0.3$ and b) $x = 0.2$ where θ is angle between film surface normal and applied field of 9 kOe.

In-situ isochronal annealing experiments in the optical microscope indicate a Curie, or compensation, temperature (T_c) for films with $x = 0.3$ of 140°C. For films with $x = 0.4$, a value of 190°C was estimated.

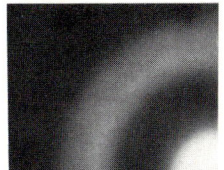 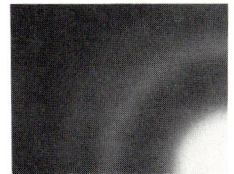 

FIGURE 3
Electron diffraction patterns of films at a) 20°C, b) 150°C and c) 400°C.

Electron diffraction studies during isochronal experiments in the TEM on films of composition with x = 0.3 have shown good film stability up to 150°C. At this elevated temperature, the broad diffuse amorphous halo (Figure 3(a)) resolves itself into a two diffuse ring system (Figure 3(b)). Two additional outer diffraction rings begin to appear at 400°C (Figure 3(c)) and a shift in position of the original rings suggests a significant microstructural change has occurred. The presence of crystalline phases, at this temperature, corresponding to Dy/Tb Fe_2 and Dy/Tb Fe_3 results in a loss of perpendicular anisotropy and the development of a large planar anisotropy. At a temperature of 550°C, crystallization of the film is almost complete, with an additional Dy/Tb Fe_5 phase emerging.

4. CONCLUSIONS

Isothermal and isochronal experiments have demonstrated good stability of these amorphous deposits at elevated temperatures. The crystallised films exhibit a planar rather than a perpendicular anisotropy indicating an ability to modify the magnetic properties, with respect to structure, to satisfy contrasting requirements. The variation of magnetic character with composition in this system may prove to be of interest with respect to the problems associated with magnetic ordering in the amorphous state and a detailed investigation is currently being pursued.

ACKNOWLEDGEMENT
   The Authors would like to thank the UK Science and Engineering Research Council for their support of this work.

REFERENCES
1) A.E. Clark, Magnetostrictive rare earth-Fe compounds, in: Ferromagnetic Materials, Vol. 1, ed. E.P. Wohlfarth (North-Holland, Amsterdam, 1980) p.531.

2) P.J. Grundy, E.T.M. Lacey and C.D. Wright, J. Magn. Magn. Mat. 54-57 (1986) 227.

3) Y. Mimura and N. Imamura, Japan J. Appl. Phys. 15 (1976) 937.

POSITRON LIFETIME STUDY AND MAGNETIC PROPERTIES OF $Fe_{81.5}B_{14.5}Si_4$ GLASS

S. LINDEROTH, C. HIDALGO, J.M. GONZALEZ*, M. LINIERS+ and J.L. VICENT+
Laboratory of Applied Physics,Technical University of Denmark,Lyngby,Denmark
* Instituto de Ciencia de Materiales de Madrid, C.S.I.C. Madrid, Spain
+ Departamento de Fisica de la Materia Condensada. Facultad de C. Fisicas. Universidad Complutense, Madrid, Spain

1. INTRODUCTION

Defects in amorphous alloys, understood as local variations of free volume[1], can interact with the magnetization via magnetoelasticity, making magnetic measurements very sensitive to thermally induced relaxation and crystallization.

On the other hand, these local density fluctuations can act as traps for positrons[2,3]. Positrons localized in the larger holes have a larger lifetime than those in the smaller holes.

In this work we try to correlate both magnetic and positron lifetime measurements on samples with different structural states.

2. EXPERIMENTAL

We have studied an amorphous alloy of composition $Fe_{81.5}B_{14.5}Si_4$ produced by quenching from the melt.

Positron lifetime measurements were performed with a standard fast-slow time spectrometer[3], in samples as-quenched and annealed during 30 minutes at 400, 500 and 600ºC. The measuring temperature range was 10-300 K.

As a function of the isothermal treatments, saturation coercive force (Hc) and magnetization work (W) were measured. These magnitudes were obtained from DC magnetic hysteresis loops performed with a conventional induction set-up.

3. RESULTS AND DISCUSSION

In figure 1 is shown the variation of coercive force, relative to the as-quenched state value Hc(0), as a function of the annealing time at 400ºC. Figure 2 shows the magnetization work results.

In all studied samples only one positron lifetime component could be resolved. The positron mean lifetime as a function of measuring temperature is shown in figure 3. The lifetime is, at 300 K, around 145 ps, which is larger than in well annealed Fe (113 ps)[4] and shorter than for positrons trapped at mono-vacancies (170 ps)[5]. These results suggest that positrons annihilate at structural defects with a mean volume smaller than one atomic (Fe) size.

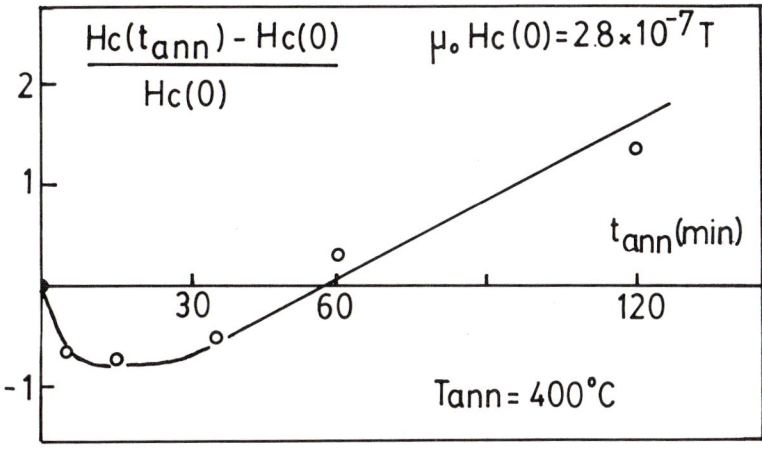

FIG. 1

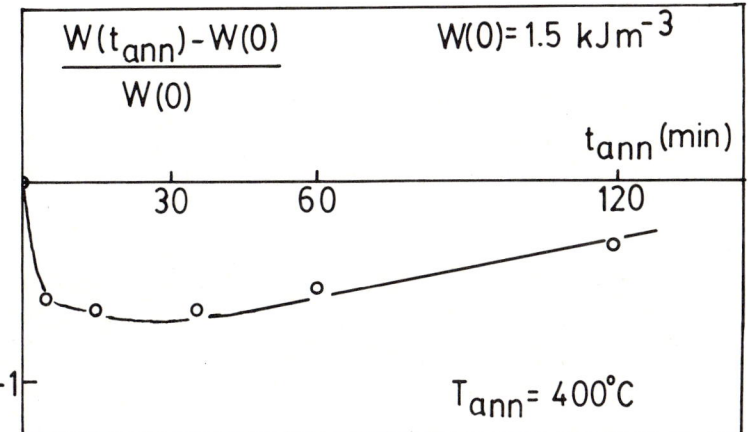

FIG. 2

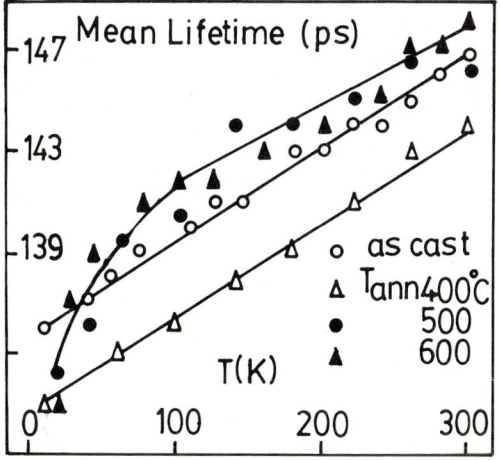

FIG 3.

FIGURE 1. Normalized evolution of the coercive force with the thermal treatments.

FIGURE 2. Normalized evolution of the magnetization work with the thermal treatments.

FIGURE 3. Temperature dependence of the positron mean lifetime.

The positron mean lifetime decreases strongly when lowering the temperature, which can be interpreted as a competing trapping between two or more types of defects with different positron binding energies.

As it is clear from figure 1, we can distinguish two different regimes in the Hc dependence with the annealing time. Up to approximately 30 minutes, coercive force diminishes with annealing time. This result corresponds to a relaxation process which leads to an improvement of soft magnetic behavior via the elimination of wall displacement (Hc) or magnetization rotation (W) hindrances. This magnetic softening agrees with the lower values of the positron lifetime (relative to the as-quenched state) observed when measuring the sample annnealed at 400ºC.

At 50 minutes annealing time, the relative variation of the Hc shifts sign, which is related with the magnetic hardening processes characteristic of the onset of crystallization. We can see in figure 3 that magnetization work, although sensitive to the relaxation processes, does not vary drastically at the onset of crystallization.

The rather similar positron lifetime values for amorphous and crystallized samples would suggest a similar mean size of positron traps in both states. On the other hand, the different temperature dependence of the positron mean lifetime would imply a different trap size distribution.

REFERENCES

1) M.Fahnle,H.Kronmuller.J.Magn.Magn.Mater. 8 (1978) 52

2) P.Hautojarvi and Yli-Kauppila. Nucl.Instr. and Meth. 199 (1982) 75

3) "Positron Solid State Physics" Eds. W.Brandt and A.Dupasquier (North-Holland Amsterdam, 1983) .

4) S.Linderoth and A.V.Shishhin .Phil.Mag. A55 (1987) 291

5) P.Hautojärvi, T.Judin, A.Vehanen, J.Yli-Kauppila, J.Johansson, J.Verdi and P. Moser.Sol.Stat.Commun. 29 (1979) 855.

IRREVERSIBLE CHANGES OF RESISTIVITY INDUCED BY STRESS, APPLIED AT ROOM TEMPERATURE, IN SOME AS QUENCHED AMORPHOUS RIBBONS

A. Hernando, V. Madurga, M. Vázquez, E. Ascasibar and A. García Escorial*

Laboratorio de Magnetismo. Facultad de C. Físicas. Universidad Complutense. 28040 Madrid. Spain

*C.E.N.I.M. Ciudad Universitaria. 28040 Madrid. Spain

Recently Haimovich et al[1] and Suzuki et al[2] have observed bond orientational anisotropy induced by stress annealing, in $Co_{67}Fe_4Mo_{1.5}Si_{16.5}B_{11}$ and $Fe_{49}Ni_{40}Mo_3Si_{12}B_{12}$ compositions, by using energy dispersive X-ray diffraction technique. Presumably this topological anisotropy of bonds is closely related to the origin of two striking phenomena observed in Co-rich alloys, the magnetic anisotropy induced by stress annealing[3] and the dependence of λ on the tensile stress[4-5]. However we need further analysis to elucidate how the anisotropy of bonds does or does not produce magnetic anisotropy depending on the chemical composition.

In order to collect new experimental data dealing with the structural rearrangements, induced by applying tensile stress, σ, the dependence of the electrical resistivity, ρ, on σ has been determined, at room temperature, for several compositions, in the as quenched state. A standard four terminal technique was used.

In the framework of the diffraction model the change produced by the stress σ on the electrical resistance, R, can be written as[6]:

$$\frac{\Delta R}{R} = \frac{\Delta S_{TT}(2K_F)}{S_{TT}(2K_F)} + \frac{5}{3}\frac{\sigma}{Y} \qquad /1/$$

where $S_{TT}(2K_F)$ is the structure factor of the metallic atoms for $K = 2K_F$, K_F being the radii of the Fermi sphere of the conducting electrons and Y the Young modulus. Since at room temperature the changes in CSRO among the metallic atoms can be disregarded we have considered only two types of atoms, metallic and metalloid ones. The second term of the right hand side in /1/ reflects the geometrical contribution to the resistance change originated by the elastic strain. This term can easily be evaluated from the stress-strain tests which were performed by using an Instron machine. Therefore the difference between the experimental value of $\frac{\Delta R}{R}$ and $\frac{5}{3}\frac{\sigma}{Y}$ gives us information about the relative change of the metallic structure factor. The initial metallic pair distribution function $P_{TT}^0(r)$ should evolve as the stress is increased towards $P_{TT}(r, \theta)$ which can be written as

$$P_{TT}(r, \theta) = P_{TT}(r) + \alpha(r) |3 \cos^2 \theta - 1| \qquad /2/$$

$\alpha(r)$ is a function of σ which accounts for anisotropic bonds reorrientations and θ is the angle made by $\vec{r}$ and the applied stress.

Equation /2/ immediately leads to:

$$\Delta S_{TT}(2K_F) \propto \int 4\pi r^2 (\Delta P_{TT}^0(r) - 1) \frac{\sin(2K_F r)}{2K_F r} dr + \int 4\pi r^2 \alpha(r) J_2(2K_F r) dr \qquad /3/$$

where $J_2(2K_F r)$ is the spherical Bessel function[2].

The isotropic first term of the right hand side accounts for the relaxation of ρ observed by annealing at low temperatures and which has been found to be correlated to the free volume annihilation[7]. The thermodynamic character of the processes contributing to $\Delta P_{TT}^0(r)$ can be elucidated from the sign of $\Delta \rho$ through the thermal coefficient $\alpha = \frac{1}{\rho} \frac{d\rho}{dT}$. Structural relaxation would produce $\Delta \rho < 0$ in samples for which $\alpha > 0$, and viceversa. The second term reflects the influence on resistivity of the anisotropic bonds orientation of the metallic atoms.

The compositions which have been studied are $(Co_{0.94}Fe_{0.06})_{75}Si_{15}B_{10}$, $(Co_{0.95}Fe_{0.05})_{75}Si_{10}B_{15}$, $(Fe_{0.5}Ni_{0.5})_{75}Si_{15}B_{10}$, $(Co_{0.95}Fe_{0.05})_{80}B_{20}$; $Co_{80}Ni_8B_{12}$ and $Pd_{40}Ni_{40}P_{20}$ which is paramagnetic. The coefficient α is positive for all of them (typically $\alpha \sim 2 \cdot 10^{-4}$ K^{-1}). The cross section of the samples range between 1.1 and 1.8 $10^{-8} m^2$ and the maximum load applied during measurements was 1.2 Kg. The stress-strain test carried out up to 2 Kg load show a linear and reversible behaviour for all the samples.

The more important results can be summarized as follows i) the resistivity shows an irreversible decrease with loading and unloading. The kinetics of the resistivity drop is identical for either loading or unloading processes. The total decay depends on the loading time and strength and reaches values higher than 1% for all the compositions. Figure 1 shows R at $\sigma = 0$ after applying σ during 5 minutes. ii) Once the ρ value is more stabilized the dependence of ρ on σ exhibits large hysteresis effects as illustrates

Figure 2. The hysteresis loops as well as the minor loops can be well fitted to the following equations:

for increasing σ: $R = a\sigma + b \sinh \gamma\sigma$ /4-a/

for decreasing σ from σ_m; $R = a\sigma + b \sinh \gamma\sigma_m - b \sinh \gamma(\sigma_m - \sigma)$ /4-b/

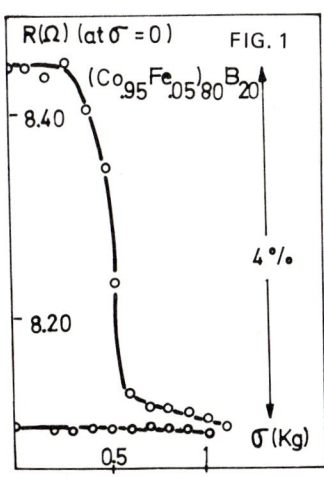
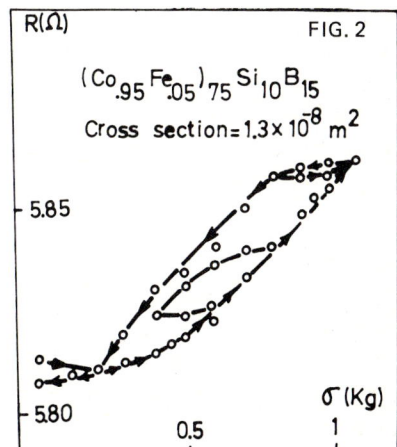

It will be shown in a detailed work in preparation how the aspect i) is related to structural relaxation or annihilation of defects so affecting $\Delta p_{TT}^0(r)$ in eq./3/ while aspect ii) is related to anelastic anisotropy in the bond orientation giving rise to a typical dependence of $\alpha(r)$ in eq./3/ with sin $(\frac{\gamma\Omega\sigma}{KT})$ In fact eq./4/ point out that the stress polarizing anelastic jumps coincides with the applied stress σ, while depolarizing effects are governed by the "back stresses" (eq. 4-b).

REFERENCES

1) J. Haimovich, T. Jagielinski and T. Egami, J. App. Phys. 57 (1985) 3581.

2) Y. Suzuki, J. Haimovich and T. Egami, Phys. Rev. B, 35, 5 (1987) 2162.

3) O.V. Nielsen, IEEE Trans. on Magn. MAG-21, 5 (1985) 2008.

4) G. Herzer, Proc. on Soft Magnetic Mat., 7, Ed. Wolfson Centre, Cardiff, England Blackpool (1985).

5) J.M. Barandiarán, A. Hernando, V. Madurga, O.V. Nielsen, M. Vásquez I and M. Vázquez II, Phys. Rev. B, 35, 10 (1987) 5066.

6) S. Takayama and R. Maddin, Scripta Met. 9 (1975) 343.

7) T. Komatsu, K. Matusitu and R. Yokota, J. Non-cryst. Solids, 85 (1986) 358.

STRESS-FIELD INDUCED MAGNETIC ANISOTROPY IN Co-Fe-Ni METALLIC GLASSES

M. Vázquez, J. González*, V. Madurga, J.M. Barandiarán**, A. Hernando and O.V. Nielsen***

Lab. Magnetismo, Facultad C. Físicas. Un. Complutense. 28040 Madrid. Spain
* Dpto. Física Aplicada. E.U.I.T.T.. Un. País Vasco. 20010 San Sebastián. Spain
** Dpto. Electricidad y Electrónica. Facultad Ciencias. Un. País Vasco. 48080 Vizcaya. Spain
*** Dpto. Electrophysics. Technical Un. Denmark. 2800 Lyngby. Denmark

1. INTRODUCTION

It is known that ribbon shaped metallic glasses can exhibit magnetic macroscopic anisotropies after thermal treatments at the presence of either magnetic field or applied stress[1]. Furthermore, as it has been recently reported for some Co-based amorphous alloys, annealing under the simultaneous action of applied stress and magnetic field enhances the induced magnetic anisotropy[2]. In the present work, we have performed stress, field and stress-field annealing for various alloys and we have examined the results taking into account the different annealing parameters and compositions.

2. EXPERIMENTAL PROCEDURE

Metallic glass ribbons were obtained by the single-roller quenching technique and the compositions of the sutided samples were $(Co_{1-x}Fe_x)_{75}Si_{15}B_{10}$ ($0 \leq x \leq 0.12$), $(Fe_{1-x}Ni_x)_{80}B_{20}$ ($0.25 \leq x \leq 0.75$) and $(Co_{1-x}Ni_x)_{75}Si_{15}B_{10}$ ($0 \leq x \leq 0.4$). The experimental set-up allowed us to anneal at the presence of applied tensile stress and/or magnettic field transverse to the ribbon axis. In order to carry out the thermal treatments we used the heating produced inside the ribbon by an electric current flowing through the ribbon[3,4]. The induced magnetic anisotropy was evalued from the change of the longitudinal magnetization work by applying tensile stress at room temperature after annealing. The values of the transverse field and tensile stress applied during annealing were $510^5 Am^{-1}$ and 600 MPa, respectively.

3. EXPERIMENTAL RESULTS AND THEIR ANALYSIS

Fig. 1 shows the typical evolution of the induced magnetic anisotropy as a function of the annealing time, t_{ann}, for a range of annealing temperatures, T_{ann}. A feature of this figure, which is common for every annealing and composition, is that for each curve (i.e. for each T_{ann}) we can find a maximum induced anisotropy, $K_{\perp}^{max}$, achievable after increasing t_{ann} as T_{ann} decreases (in our measurements, the anisotropy reaches a maximum for t_{ann} around 20 h).

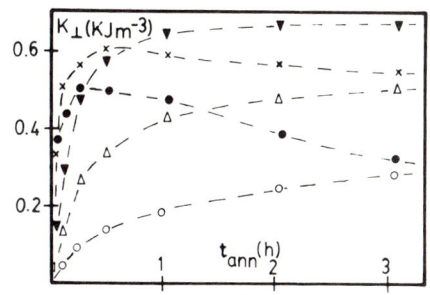 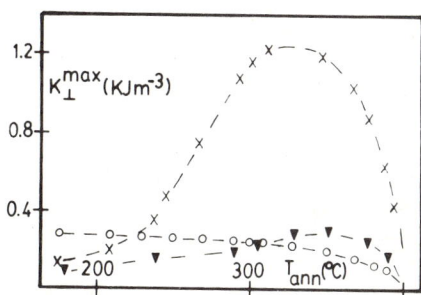

Fig. 1 $(Co_{0.95}Fe_{0.05})_{75}Si_{15}B_{10}$. stress-field induced magnetic anisotropy as a function of annealing time. T_{ann} = 260º C (o), 285 (△), 330 (▲), 350 (x) and 370 (●)

Fig. 2 $(Co_{0.92}Fe_{0.08})_{75}Si_{15}B_{10}$. The maximum induced anisotropy $K_{\perp}^{max}$ as a function of T_{ann} for stress (▼), field (o) and stress-field (x) annealing

This $K_{\perp}^{max}$ is found to be dependent on T_{ann} for all kinds of treatments that is stress, field and stress-field annealing as it can be observed in Fig. 2. In this figure we are able to remark the following points: a) stress-field induced anisotropy can not be considered as a simple arithmetic sum of single stress plus field induced anisotropies, and b) the T_{ann}-values for which $K_{\perp}^{max}$ reaches a maximum, $K_{\perp}^{MAX}$, are close in the cases of stress and stress-field annealings but differ from that of field annealing.

Finnaly, Fig. 3a and b show the composition dependence of the maximum $K_{\perp}^{MAX}$ for the sudied Co-Fe and Fe-Ni bases alloys. As it can clearly be observed, there is an evident dependence of $K_{\perp}^{MAX}$ on composition for these alloys. For the Co-Ni based alloys, induced anisotropies are rather small and depend in a confusing way on annealing temperature.

An interpretation of the results persented turns out to be rather difficult. Nevertheless, it seems to be clear that the combined action of the magnetic field and the stress enhances the orientational atomic arrangement during annealing. Owing to the many variables involved in this kind of thermal treatments, a quite systematic study is required to achieve a deeper understanding about the processes observed.

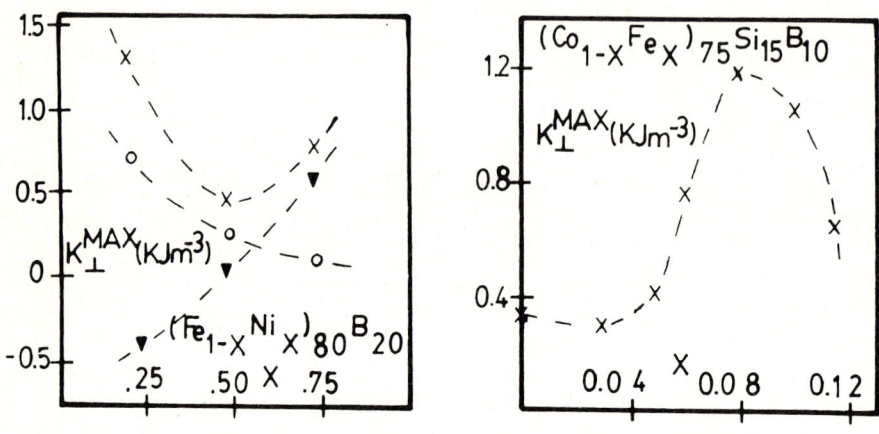

Fig. 3a   Fig. 3b

Maximum induceable anisotropy $K_\perp^{MAX}$ as a function of composition for stress (▼), field (o) and stress-field (x) annealing

REFERENCES

1.- M. Fujimori, "Amorphous Metallic Alloys", Ed. F.E. Luborsky (Butterworths Monographs in Materials, Londen, 1983) Chap. 16

2.- M. Vázquez, F. Ascasibar, A. Hernando and O.V. Nielsen. J. Magn. Mat. 66 (1987) 37

3.- M. Vázquez, J. González and A. Hernando. J. Magn. Magn. Mat. 53 (1986) 323

4.- J. González, M. Vázquez, J.M. Barandiarán, V. Madurga and A. Hernando, J. Magn. Magn. Mat 68 (1987) (in press)

TEMPERATURE DEPENDENCE AND CRITICAL EXPONENTS OF THE MAGNETISTRICTION OF METALLIC GLASS RIBBONS

M. Vázquez, C. Nuñez de Villacicencio*, V. Madurga, J.M. Barandiarán[+], A. Hernando and H. Kronmüller**

Laboratorio de Magnetismo. Facultad de Ciencias Físicas. Universidad Complutense. 28040 Madrid. Spain.
* Física Médica. Fac. Medicina. Universidad Complutense. 28040 Madrid. Spain.
+ Dpto. Electricidad y Electrónica. Facultad de Ciencias. Universidad País Vasco. 48080 Vizcaya. Spain.
** Max-Planck-Institut für Metallforschung Institut für Physik. Heinsenbergstr. 1. 7000 Stutgart 80. F.R. Germany.

INTRODUCTION

The saturation magnetostriction constant, λ_s, of metallic glasses decreases in most of the cases monotonically as the temperature increases, and approaches to negligeable values in the neighbourhood of the ferro-paramagnetic phase transition. In the present work, we have determined the thermal dependence of λ_s for a series of metallic glasses. Moreover, we were able to evaluate the critical exponents of λ_s for those alloys.

In order to measure the magnetostriction constant at different temperatures, we have made use of a very sensitive method which allowed us to detect very small values of λ_s either positive or negative. The method is based on the Wiedmann effect W.E. and has been recently developed[1]. The W.E. is associated with the appearance of a magnetostrictive torsional strain on an elongated sample when a magnetic field is applied in a helical direction. In our case, the magnetic field is obtained as the superposition of a longitudinal field, H_z, originated by a Helmholtz coil and a transverse one, H_y, originated by an electrical current flowing through the ribbon.

As calculated in reference 1, the magnotostriction can be evaluated through the expression

$$\lambda_s \, g(\gamma, \gamma') = \frac{a \, H_z}{3 \, H_y^{max}} \xi \tag{1}$$

where "a" is the half thickness of the ribbon, H_y^{max} is the maximum transverse field, that is the field at the surface of the ribbon for a given flowing current and ξ is the torsion angle per unit length magnetostrictively induced by the presence of both H_z and H_y fields. $g(\gamma, \gamma')$ is a function introduced to take into account the distribution of internal stresses, with γ and γ' being the ratio of the magnetoelastic anisotropy induced by applied stress and of the longitudinal field torque, respectively, to the mean value of the local anisotropy. The magnetostriction can be evaluated from eq.(1) providing a value close to 1 for the g-function which is achievable for $\gamma \to 0$ and $\gamma' \gg 1$, that is, for no applied stress and a longitudinal field high enough to overcome internal stresses. Furthermore, eq.(1) is valid only for the case $H_Y^{max} \ll H_Z$.

EXPERIMENTAL RESULTS

Ribbon-shaped samples of nominal compositions shown in Table I were obtained by the single-roller quenching technique. The experimental set-up allows us to apply the H_Z and H_Y fields and simultaneously to detect the angular deflection ξ. The increase of temperature within the sample is obtained with the help of a furnace inside which the ribbon is placed.

Fig. 1 shows the temperature evolution of the magnetostriction for some samples. Usual values of applied fields during measurement were $H_Z \sim 1.10^3$ Am^{-1} and $H_Y^{max} \sim 5$ Am^{-1} (which corresponds to 10 mA d.c. flowing along the ribbon).

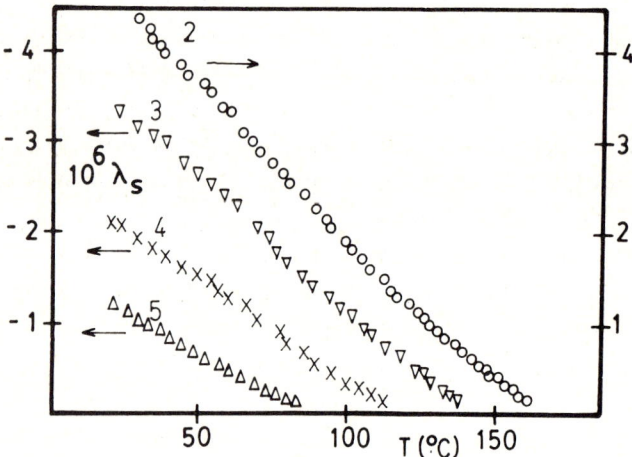

Fig. 1 Evolution of λ_S with temperature for some composition given in Table I.

Due to the sensitivity of the method we were able to evaluate the critical exponent κ defined recently[3] as:

$$\lambda_S(T) \propto (T_c - T)^\kappa \qquad (2)$$

where T_c and T are the Curie and measuring temperature respectively. The critical region embraces a temperature range around T_c which usually is given by $2\varepsilon = \dfrac{T - T_c}{T_c} \stackrel{<}{\sim} 0.1$. From eq.(2), the following equation is derived:

$$\lambda_s^{1/\kappa} = AT_c - AT \qquad (3)$$

"A" being a proportionality constant. Through the experimental pairs of values (λ_s, T), both κ and T_c are calculated from the value of $1/\kappa$ which gives the best fitting to a linear behavior between $\lambda_s^{1/\kappa}$ and T.

Table 1 shows the critical exponents, κ, the Curie temperature T_c, and the range ε for fit of the different investigated compositions.

Table I

	Composition	$T_c(K)$	κ	ε
1	$Co_{72.5}Si_{16.5}B_{11}$	482.6	1.15	0.062 - 0.019
2	$Fe_{20}Ni_{60}B_{20}$	445.7	1.03	0.052 - 0.009
3	$Co_{54}Ni_{21}Si_{15}B_{10}$	414.2	1.10	0.041 - 0.004
4	$Co_{70}Si_{18}B_{12}$	391.4	1.13	0.041 - 0.012
5	$Co_{47.3}Ni_{27.7}Si_{15}B_{10}$	365.0	1.08	0.048 - 0.007

The critical exponents exhibit an almost constant value close to 1 for all the samples. Although previously κ exponents have been reported[3-4] for a few alloys, we think that the present way of determining them is more accurate for no applied stress is present during measurement as in Ref. 4 which could have some influence on the value of λ_s. Also, the d.c. creating the $H\gamma$ field necessary to evaluate λ_s is very small, which avoids possible increases of the measuring temperature as in Ref. 3.

A new systematic experimental investigation of these critical exponents is in project. Also, a theoretical study would be necessary to complete such a result.

REFERENCES

/1/ C. Nuñez de Villacicencio, M. Vázquez, V. Madurga and A. Hernando. J. Magn. Magn. Mat. 59 (1986) 333

/2/ S. N. Kaul. J. Magn. Magn. Mat. 53 (1985) 5

/3/ M. Vázquez, A. Hernando and H. Kronmüller. Phys. Stat. Sol. (b), 133 (1986) 167

/4/ M. Vázquez, A. Hernando and O.V. Nielsen. J. Magn. Magn. Mat. 61 (1986) 390

MAGNETOSTRICTION OF a-RARE EARTH RANDOM MAGNETIC ANISOTROPY SPIN GLASSES

A. del MORAL and J.I. ARNAUDAS

ICMA and Dpto. Física de Materia Condensada, CSIC and Universidad de Zaragoza, 50009 Spain.

A model of magnetostriction for random magnetic anisotropy (RMA) spin glasses (SG) is developed. The obtained bulk magnetostriction is proportional to the average quadrupolar moment, which is dependent upon the ferromagnetic uniform exchange, J_o, and the RMA, D_o, strengths. Anisotropic magnetostriction, λ_t, has been measured on the RMA spin glasses a-$RE_{40}Y_{23}Cu_{37}$ (RE=Tb,Dy, and Er). The present model explains remarkably well the thermal variation of λ_t, the signs being in agreement with single-ion crystal field (CEF) origin.

1. INTRODUCTION

Random magnetic anisotropy (RMA) magnets is a subject of growing interest[1-2] Amorphous RE alloys are expected to present RMA with random local easy axes[1]. A feature of RMA magnets is the absence of spontaneous magnetization, RMA destroying long range order for spatial dimensionality d< 4[3]. Besides the initial susceptibility should diverge[3] or become very large ($\chi_i \sim (J_o/D_o)^4$)[4] at the critical isotherm. (When $D_o/J_o \gtrsim 1$ these systems become frozen in a SG state below a temperature T_{SG}[5]). Those features have been observed on the present alloys[6]. This paper deals, with a model for magnetostriction in RMA spin glasses, and, with magnetostriction measurements on amorphous $RE_{40}Y_{23}Cu_{37}$ (RE=Tb,Dy and Er) alloys.

2. MODEL OF MAGNETOSTRICTION IN RMA SPIN GLASSES

The starting Hamiltonian is the HPZ[1] one, plus the magnetoelastic contribution, i.e,

$$H = -\sum_{\vec{x},\vec{\delta}} J(\vec{x},\vec{x}+\vec{\delta})\vec{S}(\vec{x}) \cdot \vec{S}(\vec{x}+\vec{\delta}) - D_o \sum_{\vec{x}} (\hat{a}(\vec{x}) \cdot \vec{S}(\vec{x}))^2 - g\mu_B \vec{H} \cdot \vec{S}(\vec{x}) + H_{ms} \quad (1)$$

, where $\vec{S}(\vec{x})$ is the spin at site $\vec{x}$ (with z n.n. at distances $\vec{\delta}$), $\vec{H}$ the applied magnetic field, $\hat{a}$ the local anisotropy axis (EA), J the exchange interaction and H_{ms} the magnetoelastic coupling Hamiltonian, which reads

$$H_{ms} = -M_2 \sum_{\vec{x}} (\hat{a}(\vec{x}) \cdot \vec{S}(\vec{x}))^2 \varepsilon_{aa}(\vec{x}) + \frac{1}{2} C_e \sum_{\vec{x}} \left[\varepsilon_{aa}(\vec{x})\right]^2 \quad (2)$$

, where M_2 is the magnetoelastic coupling parameter, ε_{aa} the local strain component along $\hat{a}$ coupled to the local axial CEF and C_e some average elastic constant.

Assuming now an isotropic distribution of EA, with probability $p(\hat{a})$ and using replica technique one obtain the effective average Hamiltonian [5,7]

$$\beta H_{eff.} = -\beta J_0 \sum_{\alpha,\vec{x},\vec{\delta}} \vec{S}^\alpha(\vec{x}) \cdot \vec{S}^\alpha(\vec{x}+\vec{\delta}) - \beta^2 \frac{D_0^2}{m(m+2)} \sum_{\alpha,\beta} \sum_{\vec{x}} \sum_{i,j} S_i^\alpha S_i^\beta S_j^\alpha S_j^\beta$$

$$-\beta g \mu_B \sum_{\vec{x}} \vec{S}(\vec{x}) \cdot \vec{H} + \beta \tilde{H}_{ms} \qquad (3)$$

, where m is the spin dimensionality, α,β replica indexes and where i,j are spin component indexes ($\beta^{-1} \equiv K_B T$). The RMA part of Hamiltonian (3) is formally identical to the SK one for random exchange Ising SG, by interchanging site by spin component indexes. We, in the usual way, define an order parameter $q^{\alpha\beta}$ and a quadrupolar one, p^α. We assume now an uniform strain ε_{zz}, along the applied field ($\hat{z}$ axis), and project it along the local $\hat{a}$ axes to couple to the local CEF. Averaging over the distribution $p(\hat{a})$ and making use of the usual approximations[7,9], we obtain the effective magnetoelastic Hamiltonian,

$$\beta \tilde{H}_{ms} = \frac{1}{2} \beta b C_e \varepsilon_{zz}^2 - \beta M_2 b \varepsilon_{zz} \sum_{\vec{x},\alpha} \left[S_z^\alpha(\vec{x}) \right]^2$$

$$-\beta^2 D_0 M_2 b' \varepsilon_{zz} \sum_{\vec{x}} \sum_{\alpha,\beta} \sum_{i,j} S_i^\alpha S_i^\beta S_j^\alpha S_j^\beta \qquad (4)$$

, where b and b' are averaging constants. To progress we assume replica symmetry below T_{SG}, and from (3) and (4) we obtain for the free energy,

$$-(N^{-1}\beta)F = \frac{\beta^2}{4} \Delta^2 (q^2 - p^2) - \frac{1}{2} \beta b C_e \varepsilon_{zz}^2 - \frac{1}{2} \beta z J_0 M^2 +$$

$$\int_{-\infty}^{\infty} \frac{dx}{\sqrt{2\pi}} e^{-x^2/2} \ln \mathrm{Tr} \exp(\alpha S_z^2 + (\gamma x + \beta J_0 z M + \beta g \mu_B H) S_z) \qquad (5)$$

, where N is the ion number, $\Delta^2 = 2(m+2)^{-1} D_{ef}^2$, $\alpha = \beta^2(m+2)^{-1} D_{ef}^2 (p-q) + \beta M_2 b \varepsilon_{zz}$, $\gamma \equiv \beta \sqrt{q} \ (2(m+2)^{-1})^{1/2} D_{ef}$, with $D_{ef}^2 \equiv D_0^2 \left[1 + \frac{m(m+2)}{b} \frac{M_2}{D_0} \varepsilon_{zz} \right]$. The extremal of F gives an equilibrium strain $\varepsilon_{zz} = (M_2/C_e)p$, where p is the average quadrupolar moment. Then the anisotropic magnetostriction $\lambda_t = \lambda_\parallel - \lambda_\perp$ ($\lambda_\parallel$ and $\lambda_\perp$, respectively being the strains measured parallel and perpendicular to $\vec{H}$) becomes,

$$\lambda_t = \frac{M_2}{C_e} \int_{-\infty}^{\infty} \frac{dx}{\sqrt{2\pi}} e^{-x^2/2} \frac{\mathrm{Tr}\left[0_2^0 \exp(\ldots) \right]}{\mathrm{Tr} \exp(\ldots)} \qquad (6)$$

, where $(\ldots)$ is the argument of (5) and $0_2^0 \equiv 3 J_z^2 - J(J+1)$ the Stevens operator. Expressions for the SG parameter q and the induced magnetization M are similar to the SK model[8,9].

3. EXPERIMENT AND RESULTS

Measurements of $\lambda_\parallel$ and $\lambda_\perp$ have been performed up to 7T, from below T_{SG} (=36, 23.5 and 7 K respectively for RE=Tb,Dy, and Er)[6] up to r.t., and the thermal variation of λ_t is presented in Figs. 1.a-c. λ_t is large, remaining so well above T_{SG}, and besides has same sign for Tb and Dy compounds ($\alpha_J < 0$) and opposite

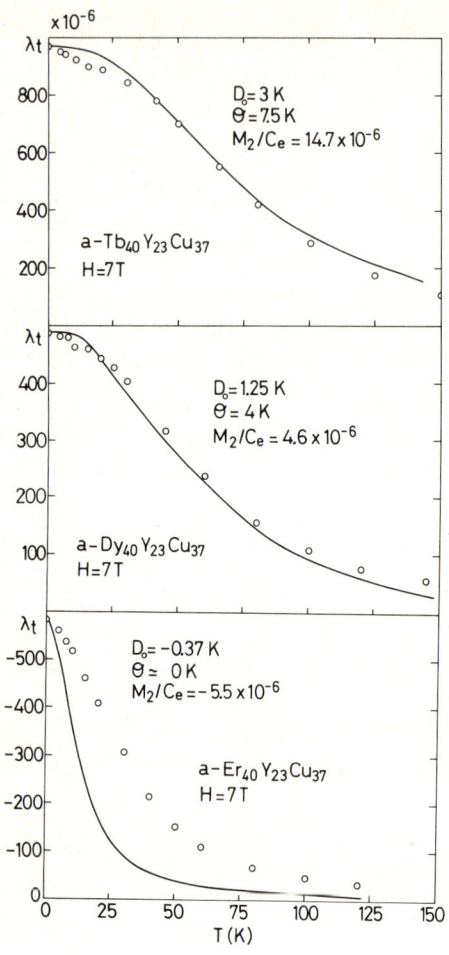

FIGURE 1. Thermal variation of λ_t for a - $RE_{40}Y_{23}Cu_{37}$. ((o), experiment; (——), theory.

for Er one ($\alpha_J > 0$), showing a single-ion CEF origin. On the same figures are shown the theoretical calculations of λ_t using eq. 6 and m=3. The values, of $\theta \equiv zJ_0$, D_0 and M_2/C_e are displayed on the Figs. 1a-c. The agreement is remarkable, except for the Er compound. Indeed the signs of M_2 are opposite for Tb and Dy and Er.

ACKNOWLEDGEMENTS

We are indebted to Prof. E.W. Lee and Dr. C. Cornelius for the experimental facilities at the University of Southampton and to the Spanish CAICYT for the grant n°789/84

REFERENCES

1) R. Harris, M. Plischke and M.J. Zuckermann, Phys. Rev. Lett., 31 (1973), 160.

2) D.J. Sellmyer and S. Nafis, Phys. Rev. Lett., 57, (1986), 1173.

3) A. Aharony and E. Pytte, Phys. Rev. Lett., 45, (1980), 1583.

4) E.M. Chudnovsky, W.M. Saslow and R.A. Serota, Phys. Rev. B33, (1986), 251.

5) J.H. Chen and T.C. Lubensky, Phys. Rev. B, 16, (1977), 2106.

6) A. del Moral, J.I. Arnaudas and K.A. Mohammed, Sol. St. Comm., 58, (1986), 395.

7) A. Aharony, Phys. Rev. B, 12, (1975), 1038

8) D. Sherrington and S. Kirkpatrick, Phys. Rev. Lett., 35, (1975), 1792

9) S.K. Ghatak and D. Sherrington, J. Phys. C., 10, (1977), 3149

SPECIFIC HEAT AND ELECTRONIC STRUCTURE OF FERROMAGNETIC $Fe_{90-x}Co_xZr_{10}$
AMORPHOUS ALLOYS.

M.Rosenberg and R.Wernhardt

Ruhr-Universität Bochum, Fakultät für Physik und Astronomie, NB03-34
P.O.B. 102148, D-4630 Bochum, BR Deutschland

The electronic contribution to the specific heat of $Fe_{90-x}Co_xZr_{10}$ amorphous alloys has been analyzed, starting from Fujiwara's band structrure of amorphous Fe and the compositional dependence of the Fe moments.

1. INTRODUCTION

Obi et al.[1] have reported an anomalously high value of 23.2 $mJmol^{-1}K^{-2}$ of the γ coefficient of the specific heat for the amorphous $a-Fe_{90}Zr_{10}$, in contrast to the 'normal' value of 5.3 $mJmol^{-1}K^{-2}$ for $a-Co_{90}Zr_{10}$ and $a-Co_{86}Zr_{14}$. This trend has been confirmed by Mizutani et al.[2] for $Fe_{90-x}Zr_x$ amorphous alloys with $8 \leq x \leq 11$ where γ ranges between about 22 and 26 $mJmol^{-1}K^{-2}$.

On the other hand, the metallic glasses $Fe_{90-x}M_xZr_{10}$ with M = Co, Ni and $0 \leq x \leq 4$ exhibit peculiar magnetic properties such as a rather low magnetic moment of about 1.6 μ_B/Fe-atom for $Fe_{90}Zr_{10}$ and a sharp increase of the average magnetic moment in the range of compositions $x \leq 10$. Previous Mössbauer studies have shown that this strong enhancement is mainly due to a strong increase of the Fe magnetic moment up to 2.4 μ_B for x = 20 whereas the Co moment has a composition independent value of 1.5 μ_B.

In order to better understand the electronic structure and the origin of the anomaly in the specific heat C_p of the Fe-rich $Fe_{90-x}Co_xZr_{10}$ glassy alloys, a study of the dependence of C_p on composition was undertaken.

2. EXPERIMENTAL AND RESULTS

$Fe_{90-x}Co_xZr_{10}$ with x = 0, 4, 10, 30, 60, 90 glassy alloys were prepared by the single-roller quenching technique in 0.2 atm Ar gas atmosphere. The C_p measurements have been carried out in an improved nonadiabatic calorimeter using the relaxation method.

Fig. 1 shows the specific heat results for $Fe_{90-x}Co_xZr_{10}$ in the temperature range 1.5 - 20 K plotted in the standard form C_p/T vs T^2. The dependence of the specific heat has been analyzed in terms of electron, lattice, nuclear and extra magnetic contributions. The complete analysis and the separation of all these terms will be presented elsewhere. In this paper emphasis will be put on the electronic contribution.

From the linear part in C_p/T plotted against T^2 a term γT and a Debye term

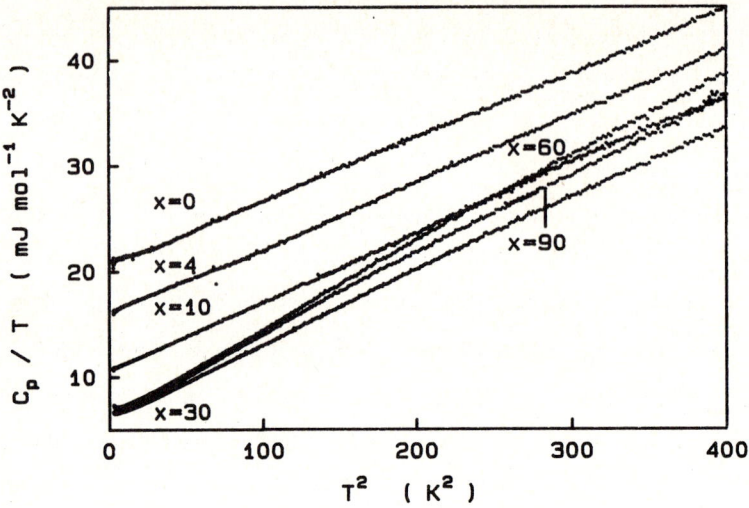

Figure 1. C_p/T against T^2 for a-$Fe_{90-x}Co_xZr_{10}$

βT^3 allowed us to determine the Debye temperature θ_D given in the third row of Table 1. The values of the coefficient γ are given in the second row of the same Table. One sees that γ takes a large value for $Fe_{90}Zr_{10}$ decreasing sharply in the concentration range $x \leq 10$ and reaching a rather constant value for $x \geq 30$.

If one tries to use the γ values for a determination of the dressed density of states $N^\gamma(E_F)$ one obtains for $x \leq 10$ unphysically large values. On the other side, γ values of 6.0 - 6.3 mJmol^{-1}K^{-2} are close to the ones found for Co-rich crystalline Fe-Co alloys with fcc structure.

3. DISCUSSION

Disregarding the influence of 10 at.% Zr on the band structure of $Fe_{90}Zr_{10}$ one can consider this glassy alloy as a kind of Fe stabilized in amorphous form in the presence of Zr. Based on the model of Yamamoto and Doyama[3] for the amorphous structure of relaxed densely randomly packed atoms, Fujiwara[4] made a band structure calculation for a-Fe and obtained for $N(E_F)$ a value of 2.27

Table 1

x	γ_{exp} mJmol^{-1}K^{-2}	θ_D K	$N(E_F)$ states At.$^{-1}$eV^{-1}	γ_E mJmol^{-1}K^{-2}
0	20.7	318	1.92	9.96
4	16.1	315	1.54	7.99
10	10.6	309	1.34	6.95
30	6.0	301	1.00	5.20
60	6.3	287	1.03	5.34
90	6.1	292	1.17	6.07

states $eV^{-1}at^{-1}$ which is substantially larger than that of α-Fe in the spin-polarized states (only 1.00 state $eV^{-1}at^{-1}$). An important point is the absence of structure in the dependence of the density of states on the electron energy E, so that even by increasing the number of electrons in the band from 5 to 9, $N(E_F)$ changes with ±10% only.

Using the concept of projected density of states at the Fe and Co sites and Fujiwara's density of states for spin up and spin down for both Fe and Co, and filling the Fe and Co bands with 7.4 electrons for the former and 8.4 for the latter, one can spin-polarize the bands by exchange splitting in order to get the values of the Fe and Co moments obtained previously from magnetic and Mössbauer studies. In such a manner we obtained for each value of x the projected density of states for Fe and Co electrons with spin up and spin down and finally the corresponding value of the bare density of states $N(E_F)$ of the alloys given in the fourth row of Table 1. One can see that by reaching its highest moment of $2.4\mu_B$ at x = 20 Fe changes from a weak ferromagnet according to the Stoner criterion to a strong one, as is already the case with Co. An important point is the quasi-constancy of $N(E_F)$ for $30 \leq x \leq 90$ in good agreement with the constancy of γ_{exp} in the same range of compositions.

The dressed density of states obtained from $N^\gamma(E_F) = 0.424*\gamma_{exp}$ implies an enhancement $1+\lambda$ of about 2.2 for the bare density in good agreement with similar results for bcc Fe and Fe-Co crystalline alloys. Taking the enhancement factor into account one obtains the values for the pure electronic coefficient γ_E given in the fifth row of Table 1. One can see that for $x \leq 10$, with decreasing Co content an extra magnetic contribution to γ_{exp} becomes more and more important. This contribution is due to the magnetic inhomogeneity of $Fe_{90}Zr_{10}$ and the alloys with low Co content and will be analyzed elsewhere.

ACKNOWLEDGEMENT

The authors are deeply indebted to Dr. K. Fukamichi for providing the samples and to U. Hardebusch and A. Schöne-Warnefeld for preforming a part of the measuring program. Financial support of the Deutsche Forschungsgemeinschaft is gratefully acknowledged.

REFERENCES

1) Y. Obi, L.C. Wang, R. Motsay, D.G. Onn, M. Nose, J. Appl. Phys 53 (1982) 2304

2) U. Mizutani, M. Matsuura, K. Fukamichi, J. Phys. F: Met. Phys. 14 (1984) 731.

3) R. Yamamoto and M. Doyama, J. Phys. F: Met. Phys. 9 (1979) 617.

4) T. Fujiwara, J. Phys. F: Met. Phys. 12 (1982) 661.

RELAXATION AND EMBRITTLEMENT OF Fe-Ni-P AMORPHOUS ALLOYS STUDIED
BY SMALL-ANGLE NEUTRON SCATTERING.

A.R. YAVARI*, D. BIJAOUI*, M.C. BELLISSENT[+], P. CHIEUX[§] and M. HARMELIN[°]

* LTPCM (CNRS UA 29), Institut National Polytechnique de Grenoble, BP 75, Domaine Universitaire, 38402 Saint Martin d'Hères Cédex, France.
[+] Laboratoire Léon Brilloin, C.E.A. Saclay, 91191 Gif sur Yvette, France.
[§] Institut Laue Langevin, BP 156X, 38042 Grenoble Cédex, France.
[°] C.E.C.M., CNRS, 94400 Vitry sur Seine, France.

We report small-angle neutron scattering measurements during in-the-beam annealing of FeNiP glassy tapes. The alloy becomes brittle but the scattering intensity remains unchanged during annealing. Measurements taken both above and below the Curie temperature indicate that compositional heterogeneities such as P-rich zones are present in the as-quenched tapes but no new P-rich zones form during annealing.

1. INTRODUCTION.

Most soft-magnetic glasses become brittle during relaxation annealing at temperature $T > T_b$ well below crystallisation at T_x and $T_b \simeq 150$ to 200 C for phosphorous containing alloys such as FeNiP and FeNiPB. Walter et al[1] using small angle X-ray scattering (SAXS), attributed thermal embrittlement in glassy $Fe_{40}Ni_{40}P_{14}B_6$ to the formation of P-rich zones. We have prepared a series of $(Fe_xNi_{1-x})_zP_{1-z}$ glassy tapes by planar-flow-casting[2,3] to minimize difficulties due to narrow irregular samples. Neutron scattering was used to detect formation of segregation zones during in-the-beam annealing. We present here the results concerning the composition $Fe_{35.5}Ni_{42.5}P_{19}$ $(Cr-V)_3$, hereafter referred to as FeNiP. The residual Cr and V are present in our industrial grade Fe-P mother alloy and were detected by chemical and X-ray microanalysis. Fig.1 shows DSC thermograms of relaxation and crystallisation at $T = 440$ C of the glassy tape and Figure 2 the ferromagnetic to paramagnetic transition at $T_c = 57$ C as measured by a magnetic balance. Thus $T_c \simeq 57$ C $< T_b \simeq 200$ C $\ll T_x = 440$ C and we can measure neutron scattering before and after embrittlement at T_b in both the para- and ferromagnetic range without crystallisation.

2. SMALL-ANGLE NEUTRON SCATTERING MEASUREMENTS : RESULTS AND DISCUSSION.

The various scattering experiments were performed on the PAXY and PACE spectrometers at CEA-Saclay and on D16 at ILL. Several wavelengths λ and sample to detector distances, L, were used all of which were in the ranges $4.6 \leqslant \lambda \leqslant 12.2$ A and $1m \leqslant L \leqslant 5m$, this allowed measurements in a wide scattering vector, $Q = 4\pi \sin \theta/\lambda$ range. Here we present data in the range

$0.01 \ll Q \ll 0.25$ Å^{-1}. 1 x 2 cm^2 sheets of the 30 μm thick tape were piled-up to thickness ≃ 0.5 mm and placed perpendicular to the beam. The intensities were corrected for background due to sample-holders and furnaces under vacuum or inert gas and for incoherent scattering and absorption. Normalized coherent scattering cross sections I(Q) in barns were obtained using the incoherent scattering of water. For full details see[4]. Figure 3 shows room temperature I(Q) for the FeNiP glass compared to I(Q) for amorphous $Fe_{40}Ni_{40}B_{20}$ at low Q and the large difference indicates the presence of heterogeneities in the former[5]. However magnetic and nuclear scattering of neutrons are both present at T < T_c and the former is dominant[6]. Furthermore STEM observations with X-ray microanalysis on chemically thinned films of the FeNiP tape showed evidence of local variations in P, Ni and Fe in the as-quenched sample. These heterogeneities which formed during the quench are not related to annealing effects.

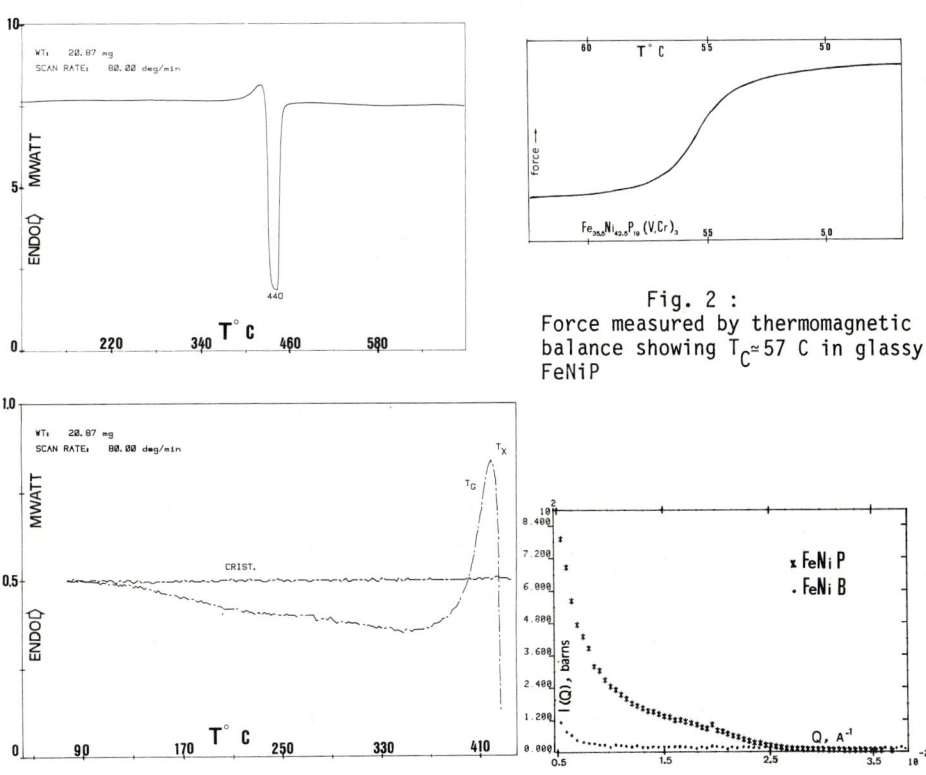

Fig. 1 :
DSC thermograms showing crystallisation at T 440 C and relaxation at 100<T<400 C in glassy FeNiP

Fig. 2 :
Force measured by thermomagnetic balance showing T_c ≈ 57 C in glassy FeNiP

Fig. 3 :
Very low angle neutron scattering in glassy FeNiP and FeNiB.

Figure 4 shows I(Q) at T = 20 C < T_c and T = 220 C >> T_c. It can be seen that most of the intensity is of magnetic origin. Figure 5 shows this in more detail between 200 and 60 C during cooling. The magnetic contribution persists at T >> T_c indicating presence of zones with $T_c' > T_c$ = 57 C which would be the case for zones with higher Fe or lower P content but again being already present in the as-quenched tape.

Walter et al[1] concluded from their SAXS that P-rich zones with radii 20 > r > 15 A formed during annealing resulting in embrittlement of FeNiPB.

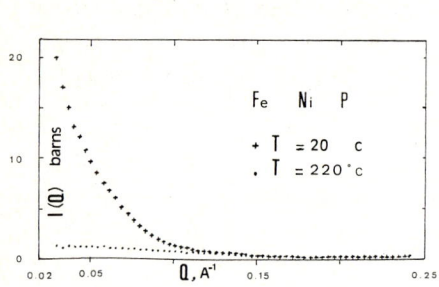

Fig. 4
Scattering intensity at T=20C<T_c and T = 220C>T_c before annealing.

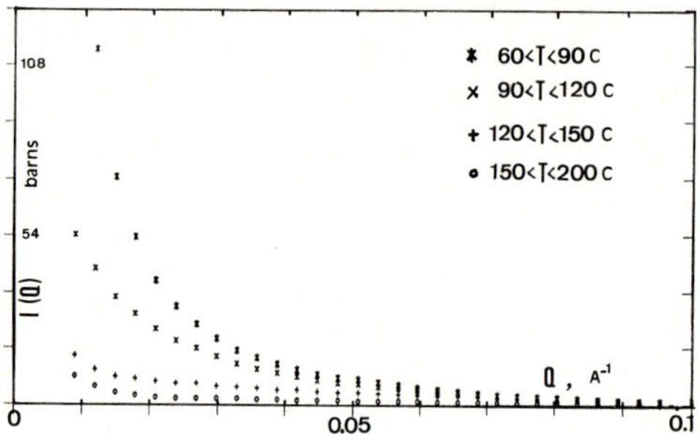

Fig. 5 :
Increase intensity during cooling showing dominance of magnetic scattering

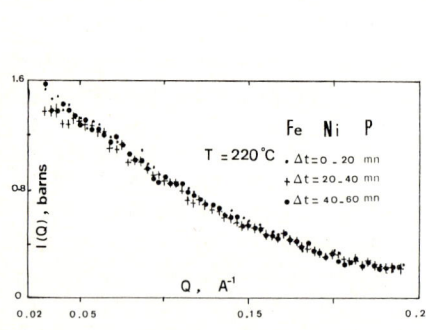

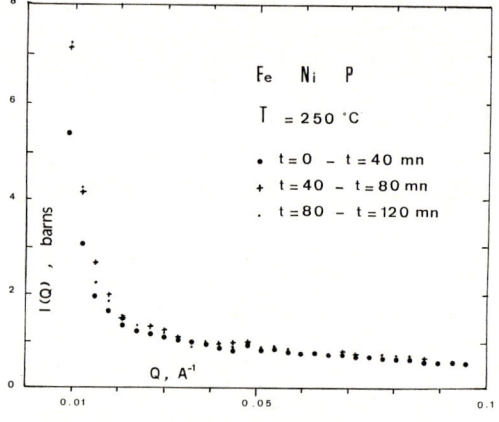

Fig. 6 :
Scattering intensity measured during annealing at T = 220 C>>T_b > T_c showing time independence in the range 0.03 < Q < 0.25 A^{-1}.

Fig. 7
Intensity during annealing at T≈250 C > T_c showing no change for 0.02 <Q≤0.1 A^{-1}.

Suppose that formation of P-rich zones with P-content C_2 results in depletion of P in the matrix from C_0 to C_1. however C_1 cannot be too low as to result in crystallisation at T_b such that in our case $C_1 \gg 10$ at %. If $0.25 < C_2 < 0.50$ (like in $(FeNi)_3P$ and $(FeNi)_2P$), the interzone distance, l, for P-rich zones of radius r will be approximately :

$$l = [(4\pi/3)(C_2 - C_1)/(C_0 - C_1)]^{1/3} r_1 \qquad (1)$$

and $3r_1 > l > 2r_1$. l may give rise to an interference peak or hump at $0.2 > Q = \frac{2\pi}{l} > 0.1$ A^{-1} for $20 > r > 15$ A. Furthermore development of P-rich zones with radius r should result in increased $I(Q)$ and $\partial \ln I/\partial Q^2$ should yield r for such emerging zones[7] in the wave vector range $Q < (1.5/r) \approx 0.1$ A^{-1} for $r \approx 15$ A. Such P effects should be present in the paramagnetic T range where coherent scattering from Fe and Ni atoms are nearly identical and only P redistribution results in a neutron scattering contrast. Fig.6 shows $I(Q)$ in the range $0.03 < Q < 0.25$ A^{-1} during annealing at $T = 220$ C $\gg T_b > T_c$ where only nuclear scattering occurs. Although the glass becomes brittle during annealing, $I(Q)$ does not change with time. No change in slope of $I(Q)$ occurs and no hump or interference peak appears. Figure 7 shows $I(Q)$ during annealing at 250 C at lower angles and again for $Q \approx 0.02$ A^{-1}, $I(Q)$ remains unchanged with time. Below 0.02 A^{-1} an increase occurs but $I(Q)$ in this range has been attributed to surface effects such as oxidation[5] although large precipitates would also contribute. However, we do not observe any formation and growth of such precipitates or zones which would have to be formed during the quench with only enrichment occuring during annealing. Walter et al[1] did not use in-situ annealing and variations in irradiated volumes at different stages may have occured (although their different results may also be due to the presence of boron). In our case, SAXS data using the powerful synchrotron radiation source of Tsukuba (Japan) with in-situ heating[8] also failed to show any time evolution. A more general discussion of results of mechanical testing as well as neutron, X-ray and Auger electron spectroscopy in $(Fe_xNi_{1-x})_4P$ glasses has been given elsewhere[4].

REFERENCES
1. J.L. Walter, D.G. Legrand and F.E. Luborsky, Mat. Sci. Eng. 29 (1977) 161.
2. M.C. Narasimhan, U.S. Patent n+ 4142571, 1979.
3. A.R. Yavari, this conference.
4. D. Bijaoui, Doctoral thesis, INPG, 1987.
5. A.R. Yavari, P. Desré, P. Chieux, "Atomic Transport and Defects in Metals by Neutron Scattering", eds. C. Janot et al, Springer Verlag, 1985, pp.842-8.
6. A.R. Yavari, Rapidly Quenched Metals V, eds : S. Steeb and H. Warlimont, Elsevier Science Publishers, 1985, pp. 495-500.
7. A. Guinier and G. Fournet, Small Angle Scattering of X-rays", Wiley, New York, 1955.
8. A.R. Yavari and K. Osamura, Rev. Sci. Inst. : in press 1987.

GALVANO-MAGNETIC PROPERTIES OF ION-BEAM MIXED Fe-Zr

N. KARPE[+], K.V. RAO[+], B. TORP[§] and J. BØTTIGER[§]
+ Department of S.S.P., Royal Institute of Technology, Stockholm, Sweden
§ Institute of Physics, Aarhus University, Aarhus, Denmark

We have ion-beam mixed multi-layers of Fe and Zr to produce well-characterized 'TEM-amorphous' and magnetically homogeneous alloys in the extended conc. range 15 to 80 at. % Zr, i.e. where 'glassy' alloys cannot be formed with melt-spinning. In addition to Hall and resistivity studies, we have determined the conc. dependence of the Curie temperature and the effective susceptibility exponent $\gamma(T)$ from Hall effect data. Our results suggest that previous data reported for sputtered films are subject to possible effects of dissolved gases arising from the environment in which they were prepared.

1. INTRODUCTION

The considerable interest for binary amorphous Fe-Zr and its magnetic and electrical properties has been manifested in the large number of recent publications on this subject. For studies of amorphous magnetic systems Fe-Zr offers the temptations of a rich variety of magnetic states - 'spin-glass', para- and ferro-magnetism and also superconductivity in the Zr rich region.

With melt-spinning,'glassy' $Fe_{100-x}Zr_x$ can only be formed near the eutectica in two regions of composition, $8 \leqslant x \leqslant 11$ and $60 \leqslant x \leqslant 80$. Ion beam mixing (IBM) provides a tool to form 'TEM-amorphous' alloys over almost the entire composition range, $15 \leqslant x \leqslant 80$ /1/. Sputtering techniques can also cover most of this range, but such films are subject to contaminations which can severely effect their physical properties /2/.

One aim of this paper is to compare Hall effect and electrical resistivity and other parameters derived from such measurements for IBM:ed $Fe_{100-x}Zr_x$, $15 \leqslant x \leqslant 43$, with results from melt-spun ribbons /3-4/ and rf-sputtered films /5-6/. Our results on the spontaneous Hall coefficient R_s, the electrical resistivity and the Curie temperatures differ systematically from those obtained from sputtered films. On the other hand our results are in resonable agreement with those from melt-spun ribbons, although a direct comparison is restricted because the range of compositions of our films discussed here and that of melt-spun ribbons do not overlap. We attribute the poor agreement between our results and those from rf-sputtered films to the higher homogeneity and less contaminations with IBM. Also, it is intersting to point out that the sputtered films were only confirmed to be 'x-ray amorphous'.

2. EXPERIMENTAL

High purity elements of Fe and Zr were deposited in a preassure of $\sim 10^{-5}$ Pa (during evaporation) into a sandwich-structure on sapphire substrates. The thicknesses of the individual layers (max. 150Å) were adjusted to obtain films of overall compositions of 15, 20, 24, 27, 34 and 43 at.% Zr with total thicknesses of 800Å. The mixing was performed by bombardment with 500 keV Xe^+ ions with a fluence of $1,5 \times 10^{16}$ ions/cm² at RT. We want to emphasize that the films were thin enough to allow essentially all the incident ions to penetrate through the film into the substrate. Film composition and uniformity with thickness after mixing were determined with Rutherford Backscattering Spectroscopy (RBS).

The amorphous structure was confirmed with transmission electron microscopy (TEM) for all films. Our ribbon of $Fe_{90}Zr_{10}$, produced by melt-spinning, was confirmed to be 'x-ray amorphous'.

Electrical measurements were performed with the 'double ac-technique', which enables simultaneous measurements of Hall effect and electrical resistivity.

3. RESULTS

The electrical resistivity and the spontaneous Hall coefficient of IBM:ed Fe-Zr show a monotonic increase with Zr content as shown in fig 1. A negative TCR (α) below T_c is observed for all alloys, fig 2, with a pronounced curvature around T_c. In amorphous magnetic metals the Hall effect can be explained in terms of the dominating spontaneous Hall effect with the coefficient R_S of magnetic origin. The Hall constant i. e. the observed Hall effect can thus be expressed as

$$R_H = \partial \rho_H / \partial H_{H \to 0} \approx R_S / (1 + 1/\chi) \qquad (1)$$

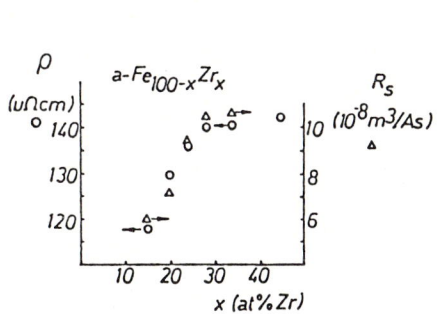

Fig 1. Electrical resistivity at RT and spontaneous Hall coefficient of Ion beam mixed Fe-Zr.

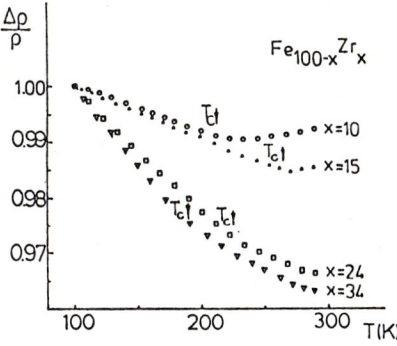

Fig 2. Temperature dependence of 'glassy' Fe-Zr, normalized to 100K

where H is the applied field and χ the magnetic susceptibility for amorphous magnetic materials in the conventional Hall effect geometry. From eqn. (1) the susceptibity above T_c and R_s considered to be essentially temperature independent can both be evaluated from the temperature dependence of R_H in fig. 3. The χ evaluated in this way agrees well this that obtained from other methods /7/.

From the evaluated susceptibility above T_c a determination of the Curie temperature is possible within less than $\pm 0{,}5K$. Using this determined T_c, the effective susceptibility exponent $\gamma(T)$, defined /8/ as

$$\gamma(T) = (T-T_c) \chi \, \partial \chi^{-1}/\partial T \qquad (2)$$

can be determined in the 'truly' critical region and above. To solve the derivative in eqn. (2) numerically with good precision accurate data are necessary. The result of this procedure is shown in fig.4 for melt-spun $Fe_{90}Zr_{10}$.

4. DISCUSSION

For the composition dependence of the resistivity of 'glassy' Fe-Zr, data from different techniques of production are compiled in fig. 5. This gives little support for the strong Zr dependence /6/ reported earlier from studies on rf-sputtered films and suggests that the observed descrepancy arises mostly because of the technique used in producing the films.

The T_c's derived from our Hall data were also verified with ac-susceptibility measurements and are among the lower ones in the wide range of reported values. Our preliminary ac-susceptibility investigations /11/ indicate a significantly extended region of 'spin-glass-like' behaviour in the Magnetic Phase Diagram of 'glassy Fe-Zr, fig 6.

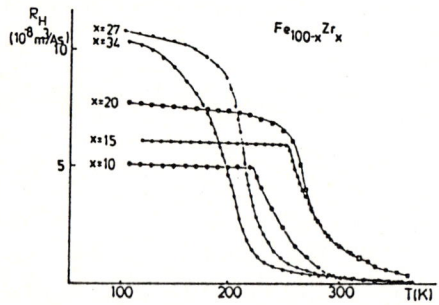
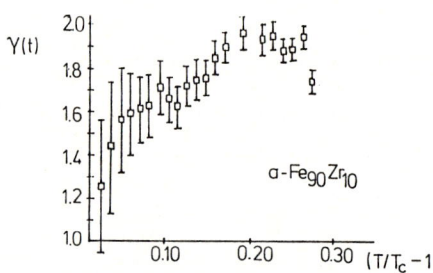

Fig 3. Temperature dependence of the Hall constant $R_H = \partial \rho_H / \partial H |_{H \to 0}$

Fig.4. The effective exponent $\gamma(T)$ derived from Hall data at ~ 100 Oe ac-field. The error bars show the maximal systematic error from an assumed uncertainty of $\pm 0{,}5K$ in T_c and 2% in $R_s(T_c)$. From static magn. measurements /9/ γ near T_c has been evaluated to be 1.38.

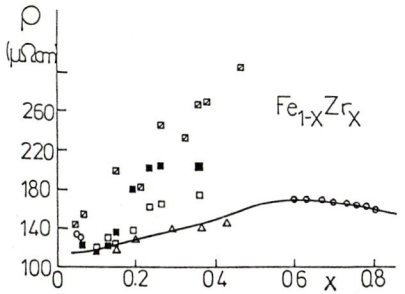
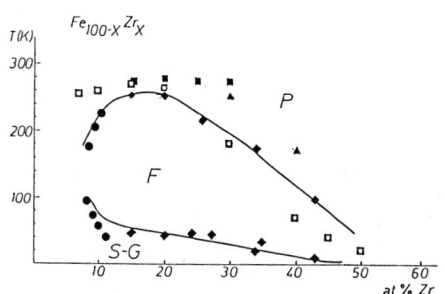

Fig 5. Compiled resistivity data for Fe-Zr made with three techniques;
1. △ IBM, RT (This work)
2. ○ Melt-spinning, RT ref. 4
3. ■ Rf-sputtering, ⎧ RT ref. 5
   ▫   ⎨ 4K ref. 5
   ▨   ⎩ 80K ref. 6
Note that the sputtered films tend to give higher values (line= method 1&2)

Fig 6. 'Magnetic Phase Diagram' for amorphous Fe-Zr. The lines correspond to boundaries obtained from ◆ IBM. Other symbols are ● /3/ melt-spun, ■ /5/ ▫ /13/ rf-sputtered and ▲ /10/ vapour deposited films.

A non-monotonic temperature dependence of $\gamma(T)$, as exemplified here for $Fe_{90}Zr_{10}$, has been observed in amorphous and crystalline disordered alloys. In recent Monte Carlo simulations /12/ this behaviour has been explained as an effect of site-disorder. Our results are in good agreement with this model's predictions of an increase in $\gamma(T)$ from the critical value to a broad maximum above the critical region.

In summary we have tried to show that many physical properties of 'glassy' Fe-Zr are sensitive to the method of preparation and that well-characterized samples should be studied to obtain reliable results. Ion-beam Mixing offers such a possibility.

4. REFERENCES

1. J Bøttiger et al (to be published)
2. T Stobiecki et al, J Mag Mag Mat, 23, 299 (1981)
3. N.Saito et al, J Phys, F16, 911 (1986)
4. W M Muir et al, J Non-cryst Solids, 61&62, 1115 (1984)
5. K Fukamichi, R J Gambino and T R McGuire, J Appl Phys, 53, 2310 (1982)
6. T Stobiecki et al, J Mag Mag Mat, 41, 199 (1984)
   T Stobiecki and M Przybyslki, Phys Stat Sol (b), 134, 131 (1986)
   A Paja and T Stobiecki, Phys Stat Sol (b), 134, 331 (1986)
7. R Malmhäll et al, Physica, 86B, 796 (1977)
8. J Kouvel and M Fisher, Phys Rev 136, A1626 (1964)
9. S N Kaul, A Hoffmann and H Kronmüller, J Phys, F16, 365 (1986)
10. J Büschow and P Smit, J Mag Mag Mat, 23, 85 (1981)
11. D-X Chen et al (to be published)
12. M Fähnle, J Phys, C18, 181 (1985)
13. K M Unruh and C L Chien, Phys Rev B , 30, 4968 (1984)

INVESTIGATION OF MAGNETIC PROPERTIES IN AMORPHOUS FeZr ALLOYS

A. FORKL, R. REISSER and H. KRONMÜLLER

Max-Planck-Institut für Metallforschung, Institut für Physik, Heisenbergstrasse 1, 7000 Stuttgart 80, Fed. Rep. Germany

The magnetization of melt-spun amorphous $Fe_{90}Zr_{10}$ alloys was investigated in the temperature range 4.2 K < T < 400 K and in fields 0 T < $\mu_0 H$ < 4.75 T. The temperature dependence of the polarization J is found to follow a T^2-law in a large field range, indicating itinerant ferromagnetism. The critical exponents determined by modified Arrott plots and by the method of Kouvel-Fisher were shown to agree with the predictions of the 3d Heisenberg model. Investigations of the domain structure prove the existence of ferromagnetic domains down to 4 K. Accordingly, it may be concluded that in Fe-rich FeZr alloys the ferromagnetic coupling dominates.

1. INTRODUCTION

Fe-rich amorphous FeZr alloys possess unusual magnetic properties. An irreversible behavior in thermomagnetic experiments at low temperatures appears. It is not clear whether this irreversibility has to be attributed to domain effects in a predominantly ferromagnetic matrix or to spin glass behavior[1]. Furthermore, it is an open question whether $Fe_{90}Zr_{10}$ belongs to the class of itinerant or localized spin materials. From the viewpoint of these questions the present paper deals with the temperature and field dependence of the dc-magnetization and with first results on the domain structure of $Fe_{90}Zr_{10}$.

2. EXPERIMENTAL PROCEDURE AND RESULTS

2.1. Experimental

The ribbons were produced by the melt-spin technique. Amorphicity was proved by transmission electron microscopy and polarization was measured in a SQUID magnetometer. The domain structure was investigated by the magnetooptical Kerr effect with the use of digital image processing. The domain structure is amplified by taking the difference of two pictures with and without domains.

2.2. Temperature and field dependence of the polarization

The temperature dependence of the polarization is shown in Fig.1 for several fields after cooling with and without applied fields (upper line for applied field). From 30 K to 170 K the polarization J can be described by a T^2-law (Fig.2) which holds for the band model in agreement with earlier measurements[2]. The critical temperature T_c = 208.0±0.5 K and the critical exponents β and γ were calculated from modified Arrott plots (Fig.3) and by the Kouvel-Fisher[3]

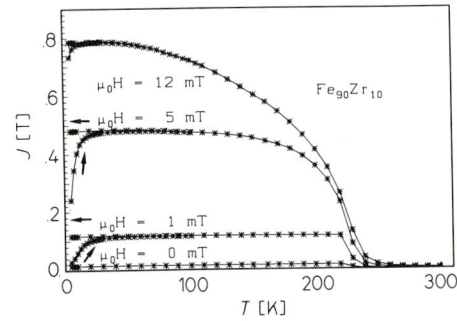
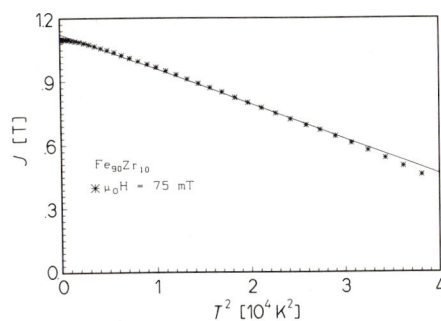

FIGURE 1
Temperature dependence of the polarization J

FIGURE 2
Polarization J as a function of T^2

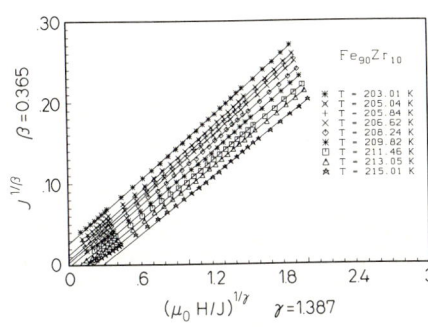
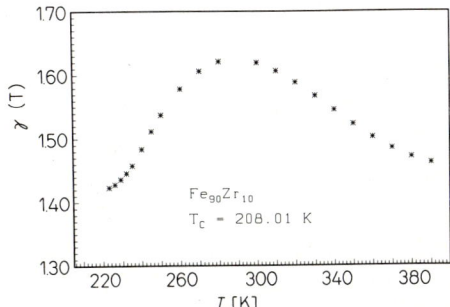

FIGURE 3
Modified Arrott plot

FIGURE 4
Temperature dependence of the Kouvel-Fisher exponent $\gamma(T)$

method. The results ($\gamma=1.39$, $\beta=0.38$ and $\delta=4.73$ all ± 10%) agree with the theoretical predictions for the 3d Heisenberg ferromagnet of localized spins. Furthermore it should be noted that the Kouvel-Fisher exponent $\gamma(T)$ (Fig.4) exhibits the same non-monotonic temperature dependence as observed for disordered ferromagnets with localised magnetic moments[4].

2.3. Domain structures

The following characteristics of the domain structure (ds) have been observed:

1. Broad domain stripes with the mean direction transversal to the ribbon direction (RD) as shown in Fig.5 a and b at 4 K after longitudinal and after transversal demagnetization. The width of the domains is about 80 to 250 μm and 30 to 50 μm after longitudinal and transversal demagnetization, respectively.

346 A. Forkl et al.

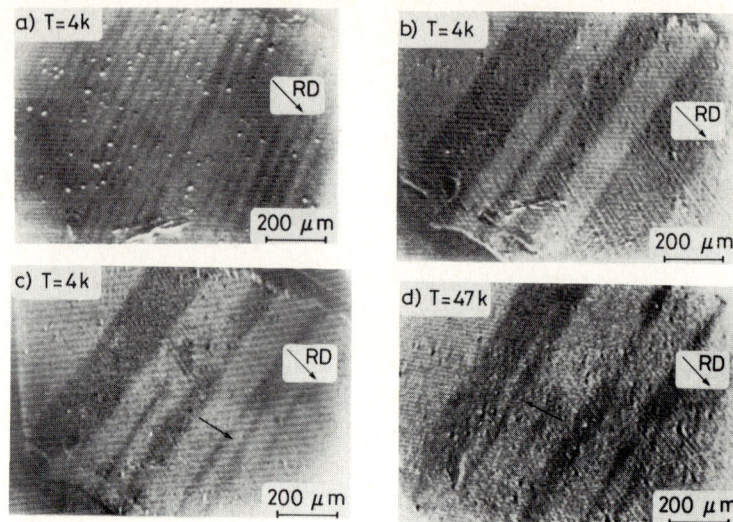

FIGURE 5
Domain structure after demagnetizing a) longitudinal (∥ RD) b) transversal (⊥ RD) and c) after applying a field of $\mu_0 H = 1$ mT and d) after heating up from 4 K to 47 K in the applied field of $\mu_0 H = 1$ mT

2. This behavior is found at temperatures from 4 K to 150 K.
3. Applying a magnetic field the domain walls move as shown in Fig.5 c. This ds emerges from Fig.5 b after applying a field of $\mu_0 H = 1$mT.
4. Rising the temperature in an applied field starting from 4 K the ds changes as indicated in Fig.5 d at 47 K (this ds results from Fig.5 c).

3. CONCLUSIONS

From the present measurements it is concluded that the apparent low temperature irreversibility (Fig.1) may be possibly attributed to a change in the domain structure. At temperatures between 30 K and 170 K the measurements (Fig.2) show itinerant band magnetism and around T_c the material may be described as a 3d Heisenberg ferromagnet.

REFERENCES

1) H. Hiroyoshi and K. Fukamichi, J. Appl. Phys. 53 (1982) 2226

2) W. Beck and H. Kronmüller, Phys. stat. sol. (b) 132 (1985) 449

3) W.-U. Kellner, T. Albrecht, M. Fähnle and H. Kronmüller, J. Magn. Magn. Mat. 62 (1986) 169

4) M. Fähnle, J. Magn. Magn. Mat. 65 (1987) 1

LOSSES, AFTEREFFECT AND DISACCOMODATIONS IN AMORPHOUS FERROMAGNETIC ALLOYS

P. ALLIA and F. VINAI

Ist.Elettr.Naz."Galileo Ferraris" -10125 Torino (Italy)
GNSM-CNR and CISM-MPI, Research Unit of Torino (Italy)

The permeability aftereffect in amorphous ferromagnetic alloys has been extensively studied from the experimental point of view, and different models have been proposed to explain the phenomenon. By extending the aftereffect measurements to very short times after impulsive sample demagnetization, it is possible to put into evidence both an entirely new effect, i.e. a relaxation related to a change of the energy dissipation of the system, and the short-time tail of the diffusive permeability disaccomodation. Preliminary analysis of the last effect, suggests the existence of correlations between the ordering defects which give rise to the aftereffect.

1. INTRODUCTION

The permeability aftereffect, or disaccomodation, is a reversible decrease with time of the initial permeability μ of a ferromagnet, after any rearrangement of the magnetic domain structure. This phenomenon is observed in many soft magnetic materials, and has been widely studied in crystalline alloys.

In amorphous ferromagnets the permeability decay follows a logarithmic law and can be observed at any temperature between 4K and the Curie temperature T_c. As a consequence, differently from the case of ferromagnetic crystals, in amorphous ferromagnets the experimental data support the hypothesis of the existence of a wide distribution of activation energies. In addition, the elementary events giving rise to the effect, pictured as structural defects able to re-orient according to the direction of M, [1] are generally supposed to be independent. [2]

We now analyse different aspects of the magnetic aftereffect, with special emphasis on the interaction between the ordering defects responsible for the permeability decay, and the magnetization. An evidence of correlations between events originating the aftereffect is also given.

CONVENTIONAL AFTEREFFECT

Let us call "conventional aftereffect" the permeability disaccomodation measured by conventional and impulsive methods[3] at times longer than 10^{-2} s after demagnetization. The permeability decay is well approximated by a law of the type a - b lnt. The aftereffect is present at any temperature below T_c and is a function of the measurement temperature. In fact, the reversible decay of the permeability observed after a rearrangement of magnetic domains, may be explained in terms of a superposition of relaxation times τ distributed at each temperature over an extended time interval. Usually, such a distribution is related to the presence of a spectrum of activation energies Q for the ordering processes responsible for the aftereffect, the relation between each τ and the corresponding Q being given, in the simplest model, by an Arrhenius law.[3] The disaccomodation is strictly connected with microscopic ordering processes taking place in the amorphous alloy. In fact, the phenomenon disappears with the crystallization of the material. In addition, structural relaxation, involving irreversible changes of the short-range order, induces irreversible modifications of the aftereffect intensity.[4] On the other hand, the different degree of atomic disorder frozen into the material by varying the quenching rate at which the glass is produced plays a role on the aftereffect intensity at fixed temperature.[5] These data indicate that the structural disorder plays a relevant role in determining the value of the aftereffect at a given temperature. Some models,[1,6,7] have been proposed in order to explain such results.

Another interesting experimental point is that the intensity of the aftereffect measured at room temperature is related to the square of the saturation magnetostriction constant λ_s, by a law of the type $\Delta\mu/\mu = A + B\lambda_s^2$.[6] Amorphous ferromagnets of the type $TM_{80}M_{20}$ can be divided in two families of alloys: Fe-based and Co-based alloys. The dependence of the aftereffect on the square of λ is observed for all materials, even if Fe-based and Co-based alloys are characterized by different values of the coefficients A and B.

Starting from this result, we made the hypothesis of a magnetostrictive interaction between local magnetization and "structural defects" defined in terms of local stresses and strains.[1,6] The model for the amorphous structure used in this theory is based on a description of the glassy metal structure in terms of fluctuations of atomic-level stress. These stresses can be shown to be roughly proportional to local strains, describing the degree of structural distortion of the environment of any atom in the metal.[8]

From this point of view, the defects responsible of the aftereffect are intrinsic to the structure. Under the hypothesis, that the magnetic aftereffect originates from atomic rearrangements occurring in order to minimize the magnetoelastic energy after a sudden variation of the local magnetization direction, we obtain[6]

$$\Delta \mu/\mu \propto \lambda_{eff}^2 <\tau^2> \qquad (1)$$

where $<\tau^2>$ is the second moment of the shear stress fluctuations (which is a constant for a given alloy), and λ_{eff}^2 is approximatively given by

$$\lambda_{eff}^2 = \lambda_s^2 + <m> \lambda_g^2 \qquad (2)$$

where λ_s is the saturation magnetostriction, λ_g is connected with the radial term of the local anisotropy, and $<m>$ is a parameter giving information on the degree of structural distortion of the elementary "cells" (defined by the nearest-neiglebors of each atom in the material).[9]

Eq 2 can be cast in the form

$$\Delta \mu/\mu = A + B\lambda_s^2 \qquad (3)$$

In this way, it is possible to explain the aftereffect in both Fe-based and Co-based alloys.

Measurements performed by means of different techniques seem to indicate the consistency of this model. Interesting results have been recently obtained[10] by studying a series of materials prepared by melt spinning under controlled conditions, in order to have a family of alloys with different composition but with a similar degree of disorder.

We point out the fact that measurements performed with different techniques confirm the results obtained by means of the impulsive technique. Measurements performed on $Co_{75-x} Fe_x Si_{15} B_{10}$ (x=0-50), with λ ranging from small negative to high positive values allow to study the transition of the behavior of the magnetic aftereffect from Co-based to Fe-based alloys.[7,10]

Recently, interesting results have been obtained by comparing the measurements of aftereffect and magnetostriction performed on samples plastically deformed at a constant strain rate.[11] (see Fig.1). The behavior of the aftereffect as a function of the plastic deformation is not in contrast with the assumption of a magnetostrictive origin of the permeability disaccomodation. These measurements indicate that the number of the structural defects responsible for the aftereffect, is significantly enhanced by sample deformation, in agreement with a model [12] suggesting that small regions of weakened resistance to certain deformations appear in an amorphous metal during plastic deformation, and are not annihilated upon unloading.

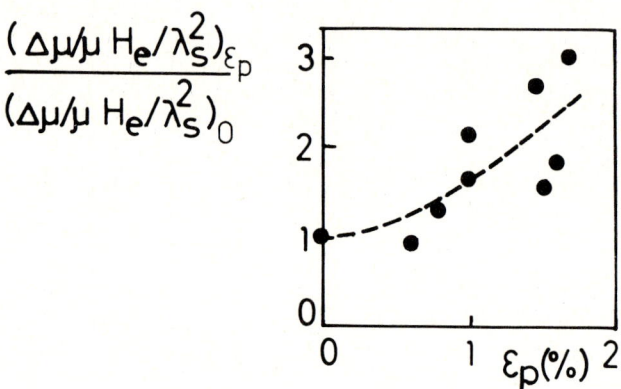

Fig. 1 - Behavior of $(\Delta\mu/\mu H_e/\lambda_s^2)_{\varepsilon_p}/(\Delta\mu/\mu H_e/\lambda_s^2)_0$ as a function of the plastic deformation.

PERMEABILITY AFTEREFFECT AT VERY SHORT TIMES

More information about the form of the energy distribution function, P(Q), and more details about the nature of the defects responsible for the aftereffect and their possible correlations, can be obtained in principle by extending the time interval of analysis down to very short times after impulsive demagnetization.

By using the experimental technique developed in our laboratory,[13] it is now possible to investigate the magnetic permeability decay from about 10^{-5} s after demagnetization. This new technique allowed us to put into evidence two aspects of the domain wall dynamics immediately after a sudden rearrangement of the magnetic domains. First, a strong permeability relaxation of dissipative type is observed in the interval $10^{-5} s \leqslant t \leqslant 10^{-3}$ s.

On the other hand, it is still possible to observe, even at very short times after demagnetization a permeability aftereffect of diffusive type similar to the ordinary long-time aftereffect.

Details about the dissipative relaxation may be found in Ref. (14).

By modifying the measuring procedure in order to make the dissipative contribution negligible, it is possible to separate the contributions of the two effects. It is possible in fact to obtain such a disentanglement by applying to the walls a square-wave field of suitable amplitude, in order to move each wall between two positions at a distance of the order of the wall thickness.[15]

On the other hand the proposed method for the measurement of the permeability aftereffect of diffusive type provides the first direct check of the adequacy of Neel's theory for the magnetic disaccomodation in amorphous ferromagnets. The time behavior of the short-time disaccomodation of diffusive type is again quasi-logarithmic, and similar to the conventional aftereffect mesured at longer time, see Fig. 2.

Such a behavior is typically interpreted in terms of the existence of a quasi-uniform activation energy distribution $p(Q)$. As mentioned previously, the events responsible for the aftereffect are considered to be independent. However, the study of the short-time disaccomodation allows one to conclude that this hypothesis has to be withdrawn is a more detailed description of the microscopic ordening processes giving rise to the aftereffect.

The presence of correlations between the microscopic processes is suggested by the analysis of a simple experiment. The nature of these correlations is still to be clarified. In this experiment, whose details ave reported elsewhere,[15] each domain wall is cyclically moved between two contiguous positions by making use of a suitable D.C. field, alternatively positive and negative in sign, with field inversion occurring every t_o seconds; t_o is then the time of permanence of a domain wall in each position. First of all, we perform a complete demagnetization of the sample in order to avoid any memory of the previous domain structure. A new domain pattern is created (at the time t=0) and the permeability is measured. A set of subsequent permeability decays is obtained. These decays start every t_o seconds, and it is possible to observe the time behavior of the permeability between two domain wall jumps. An interesting result is that the shape of the permeability decays varies with time, i.e. with the number of domain wall jumps. In particular, it is possible to observe: a) a decrease of the intensity of the decays; and b) a progressive

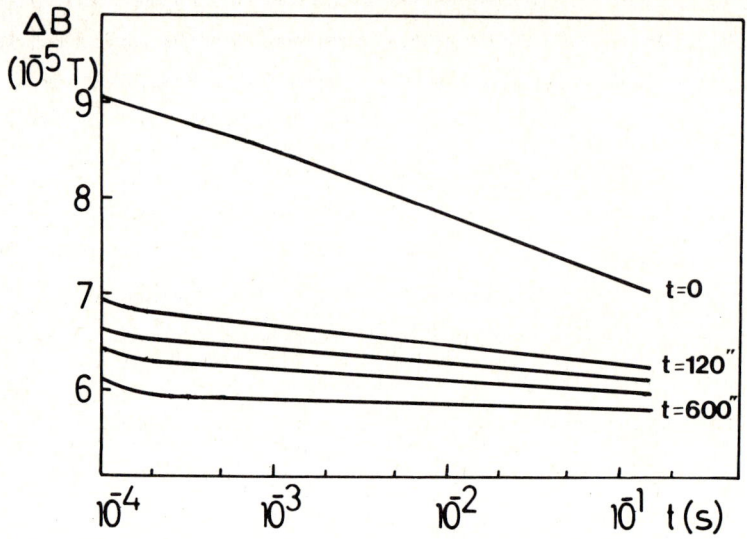

Fig. 2 - Behavior of the short-time diffusive aftereffect B(T) in $Fe_{81} B_{13.5} Si_{3.5} C_2$ at different times after initial demagnetization

flattening of the curve (see Fig. 2).

These experimental results are in contrast with the hypothesis of a broad distribution of time constants for <u>independent</u> events. In fact, the presence of non interacting events should imply, in principle, a complete reversibility of the processes. In other world, if some ordering processes have occurred when the wall was in one position for a time t_o, they should completely forget the previous situation when the wall is displaced to the other position (again for t_o seconds), in such a way that when the wall comes back to the first position, the ordering process starts always from the same initial condition.

The observed decrease of the intensity of the aftereffect can be explained within the framework of the existing models by taking into account the presence of time constants much longer than t_o. The domain wall stabilization proceeds, also when the walls are cycling between two contiguous positions, resulting in a deepening of both potential wells.

However, the second effect, i.e., the flattening of the curves, cannot be explained on the basis of independent events, as easily shown by a simple calculation. Details of this calculation are reported in a recent paper [15].

These preliminary results show that our understanding of the magnetic aftereffect of diffusive type in amorphous ferromagnets is far from being complete. A better knowledge of the structural defects is needed, in order to describe the permeability aftereffect of amorphous ferromagnets in a more realistic way.

REFERENCES

1) P.Allia and F.Vinai, Phys. Rev. B26, 6141 (1982)

2) H.Kronmüller, Philos.Mag. B48, 127 (1983)

3) P.Allia, P.Mazzetti and F.Vinai, J.Magn.Magn.Mat. 19, 281 (1980)

4) P.Allia and F.Vinai, J.Phisique 41, C8-654 (1980)

5) P.Allia, F.E.Luborsky, R.Sato Turtelli, G.P.Soardo and F.Vinai, IEEE Trans.on Magnetics MAG 17, 2615 (1981)

6) P.Allia and F.Vinai, Phys. Rev. B33, 422 (1986)

7) H.Kronmüller, N.Moser and F.Rettenmeier, IEEE Trans on Magnetics MAG 20, 1388 (1984)

8) T.Egami, K.Maeda and V.Vitek, Philos.Mag. 41, 883 (1980)

9) P.Allia and F.Vinai, J.Physique 46, C8-317 (1985)

10) A.J.de Rezende, R.Sato Turtelli and F.P.Missell, IEEE Trans. on Magnetics (1987) in press

11) P.Allia, C.Beatrice, E.Bonetti and F.Vinai, Zeitschr. Physik. Chemie (1987), in press

12) D.Srolovitz, V.Vitek and T.Egami, Acta metall. 31, 335 (1983)

13) P.Allia, C.Beatrice, P.Mazzetti and F.Vinai, J.Magn.Magn.Mat. 54-57, 273 (1986)

14) P.Allia, C.Beatrice, P.Mazzetti and F.Vinai, Appl.Phys.Letters (1987), in press

15) P.Allia, C.Beatrice, P.Mazzetti and F.Vinai, submitted to J.Appl.Phys.

INDUSTRIAL APPLICATIONS OF METALLIC GLASS RIBBONS

G. HERZER and H.R. HILZINGER

Vacuumschmelze GmbH, D-6450 Hanau, Federal Republic of Germany

A survey is given of some major commercial applications of ferromagnetic metallic glass ribbons in electronics industry. Topics to be discussed are magnetic cores with low losses and specifically designed hysteresis loops for inductive components in switched-mode power supplies or magnetic switches for pulse compression in power sources; magnetic heads; flexible magnetic shieldings and various types of magnetic sensors.

1. INTRODUCTION

Their promising soft magnetic properties, combined with their high electrical resistivity and their production-inherent low thickness have meanwhile opened the way for ferromagnetic glass ribbons to a variety of commercial applications in electronics.

Candidates for application are on the one hand the Fe-rich magnetic alloys which exhibit the highest saturation flux densities among the amorphous metals and at the same time are based on inexpensive raw materials such as Fe, B, Si and C. The Co-based metallic glasses, on the other hand, are distinguished by their low or vanishing magnetostriction leading to highest permeabilities and lowest losses.

Table 1:
Basic properties of some amorphous non-magnetostrictive alloys

VITROVAC® *	6025	6030	6150
Composition	$(CoFeMo)_{73}(SiB)_{27}$	$(CoMnFeMo)_{77}(SiB)_{23}$	$(CoMnFe)_{80}(SiB)_{20}$
B_s (T)	0.55	0.80	1.0
T_c (°C)	210	350	500
$|\lambda_s|$	$\leq 0.1 \times 10^{-6}$	$\leq 0.1 \times 10^{-6}$	$\leq 0.1 \times 10^{-6}$
ρ_{el} (μΩcm)	135	130	115

* VITROVAC® is the registered trademark of Vacuumschmelze GmbH for amorphous metals.

Both Co- and Fe-rich metallic glass ribbons are in current commercial use. The specific requirements of electronics industry hereby often demand for the near-zero magnetostrictive Co-based material on which we will focus in the following. Table 1 shows some of such commercially available alloys and their basic material parameters.

2. INDUCTIVE COMPONENTS FOR SWITCHED-MODE POWER SUPPLIES

In power electronics switched-mode power supplies operating at frequencies of 100 kHz and still more are replacing the standard power supplies to an increasing extent. Hence low iron losses of the magnetic core material are important for an efficient performance of the inductive components. Due to their high electrical resistivity and the low ribbon thickness (which both limit eddy current losses) and due to their low hysteresis losses, ferromagnetic metallic glass ribbons show exceptional low core losses in this frequency range. Fig. 1 compares the specific losses for different core materials. Amorphous alloys seem to be favorable even to recently developed ferrite cores. Tape wound

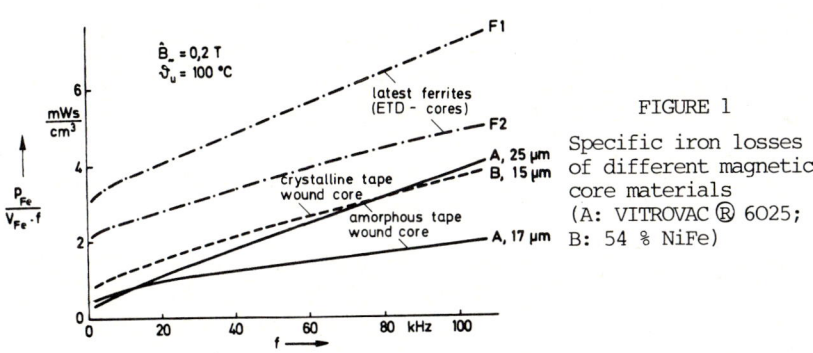

FIGURE 1
Specific iron losses of different magnetic core materials
(A: VITROVAC® 6025; B: 54 % NiFe)

cores of crystalline material would have similar losses only at a ribbon thickness below about 15 μm, which is generally not accessible at reasonable production costs. This is the reason why amorphous cores with low magnetic losses and specifically designed hysteresis loops are already in wide commercial use for various inductive components[1,2]:

For example <u>saturable reactors</u> with toroidal tape wound cores are used to regulate several output voltages independently of each other on purely magnetic basis[3]. Apart from low losses under premagnetizing conditions and extremely low coercivity a pronounced squareness of the hysteresis loop is required for this purpose. This is realized advantageously with amorphous non-magnetostrictive alloys, like VITROVAC® 6025 or VITROVAC® 6030, where a square loop can be obtained with a comparatively low induced magnetic anisotropy along the ribbon axis and thus low ac coercivity and low iron losses[4].

<u>Inverter transformers</u>, as another example, require a core material with sufficiently high saturation induction and a flat hysteresis loop with a low remanence. Lowest iron losses again are the prerequisit to obtain a maximum output power of the transformer[5]. All these requirements are well met by a non-magnetostrictive Co-based metallic glass like VITROVAC® 6030 annealed in a transverse field[6]. Since $\lambda_s \approx 0$ additional losses due to magnetostrictive vibrations of the core are avoided and losses are only determined by classical eddy current theory. The high unipolar flux density swing $\Delta B = 0.7$ T at 100 °C combined with the low losses thus allows a considerable reduction in size of the transformer.

3. MAGNETIC SWITCHES FOR PULSE COMPRESSION

Magnetic switches for pulse compression in power sources, like for example for pulsed excimer lasers, require a well squared hysteresis loop. Since the material is subjected to magnetization rates up to 10 T/μsec or more (i.e. MHz range), low core losses are of tremendous importance for an efficient performance[7]. This is why the thin metallic glass ribbons are already advantageously used for this application.

Fe-based metallic glasses like e.g. METGLAS® 2605 CO offer highest saturation induction and thus the most economical solution. Co-based non-magnetostrictive alloys like VITROVAC® 6030 or VITROVAC® 6150, however are preferred for higher repetition rates because of their significant lower anomalous eddy current losses (compare FIGURE 2). Moreover manufacturing of the magnetic switch cores, like for example electrical insulation of the individual layers, is facilitated because of the mechanical insensivity of the magnetic properties of this material.

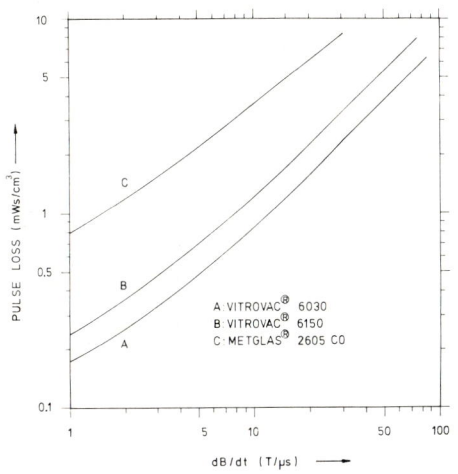

FIGURE 2
Pulse loss as a function of the magnetization rate dB/dt for several amorphous alloys annealed to a squared loop (Ribbon thickness: 22 μm).

4. MAGNETIC HEADS

Materials for magnetic heads should exhibit high initial permeability, adequate saturation flux density, thermal stability and good resistance to wear and corrosion[8,9]. These requirements have made the Co-based non-magnetostrictive amorphous alloys attractive for this application. Presently audio heads of Co-based amorphous alloys are manufactured for tape recorders in the top and medium prize level and for duplicating music tapes[8]. For example, VITROVAC ® 6025, annealed to a round hysteresis loop, is successfully used in audio heads for studio tape machines.

As for video-recording, amorphous Co-based metal ribbons in thin gauge of about 10 μm thickness and less are reported to show higher permeability and better output signal than Sendust and even Mn-Zn ferrite in the frequency range of 1 - 10 MHZ[8].

5. MAGNETIC SHIELDINGS

External magnetic fields may induce interfering voltages in cable systems. Since usual conductive shields do not operate below about 10 kHz, shields of soft magnetic material is required. Moreover on bending the cables the material should not be degraded magnetically or mechanically.
The combination of high yield strength and high, stress independent permeability makes a near-zero magnetostrictive Co-based amorphous alloy uniquely well suited because it can be bent in manufacturing and in service practically without loss of permeabi-

lity - unlike to soft magnetic crystalline ribbons. Best performance has been found with VITROVAC® 6025 X which is increasingly used for flexible cable shielding. Wound helically around the cable, shielding factors up to 100 are realized in a wide frequency (up to 10 kHz) and field range (several mA/cm up to 10 A/cm)[10]

Furtheron magnetic shields with near-zero magnetostrictive amorphous alloys are available as flexible shields either manufactured by braiding long strips[11] or embedding amorphous flakes[12] in between plastic foils.

6. MAGNETIC SENSORS

Sensors containing magnetic material as the principle functional elements to measure magnetic fields, electric currents, displacements, stresses, torques or forces have been in use for many years. The unusual mechanical and magnetic properties of amorphous metallic glass ribbons or wires have given fresh impetus to this field since they meet most of the particular sensor requirements such as mechanical robustness and relibability, temperature stability, high sensitivity, wide dynamic range, quick response and small size[13-15].

Examples from automobile industry using magnetometers with non-magnetostrictive amorphous cores are an electronic compass for traffic guidance and, combined with small permanent magnets, a speed sensor in which the magnetic flux of a permanent magnet is periodically affected by the cogs of a gear, or a displacement sensor which controls the fuel injection process in Diesel engines (cf. Ref 1).

The probably most advanced application of amorphous magnetometers is that as a theft detection device. Such a security device consists of an elongated ferromagnetic strip which is attached to the articles to be protected. When excited in an ac-field produced in the passageway, the non-linearity of the hyster--esis loop generates higher harmonics of the fundamental frequency which are detected by pick-up coils. Amorphous zero-magnetostrictive alloys are advantageous because of their better signal output already at small excitation. Moreover the performance is not affected by mechanical stress during production or handling.

An example for a magnetostrictive transducer using magnetostrictive amorphous material is a torque transducer which enables the contactless measurement of torques in a shaft[13].

7. CONCLUSIONS

Metallic glasses since long have made the step from laboratory curiosity to industrial application. Monthly production figures are being measured in tons now and are continuously raising. Substitution of conventional soft magnetic alloys and soft ferrites is to be expected to an increasing extent in existing applications. The superior properties of metallic glasses often bring about redesign and improvement of the device.

REFERENCES

1) H.R. Hilzinger, IEEE Trans. Magn. MAG 21 (1985) 2020

2) Vacuumschmelze GmbH, Amorphous Metals Data Sheet VC 004 (1984)

3) D. Grätzer, IEEE Trans. Magn. MAG 16 (1980) 922

4) V. Ungemach, H.R. Hilzinger and W. Kunz, J. Magn. Magn. Mat. 42 (1984) 363

5) D. Grätzer, J. Magn. Magn. Mat 9 (1978) 91

6) H.R. Hilzinger and W. Kunz, IEEE Trans. Magn. MAG 20

7) C.H. Smith, D. Nathasingh and H.H. Liebermann, IEEE Trans. Magn. MAG 20 (1984) 1320

8) Y. Makino, Proc. Conf. Rapidly Quenched Metals, RQ 5 (Würzburg 84), North Holland 1985, p. 1699

9) M. Takahashi, H. Fujimori and T. Miyazaki, in: "Recent Magnetics for Electronics", ed. Y. Sakurai, Ohmsha and North Holland, 1983, p. 137

10) L. Borek, Elektronik 4 (1982) 43

11) Vacuumschmelze GmbH, Amorphous Metals Data Sheet VC 001 (1982)

12) Riken Corp., Magnetic Shield Data Sheet

13) R. Boll and G. Hinz, Technisches Messen 52 (1985) 189

14) K. Mohri, IEEE Trans. Magn. MAG 20 (1984) 942

15) K. Mohri, Proc. Conf. Rapidly Quenched Metals, RQ 5 (Würzburg 84), North Holland 1985, p. 1687

AMORPHOUS METAL SENSOR

Kaneo MOHRI

Recent advances in the field of application of amorphous metals to high-performance sensors for mechatronics and medical electronics are reviewed. Various new sensors which have robustness, high resolution (μG, μA, μm, μKg, μrad), quick response (0 - 10 kHz) and microlization in size are constituted, in which sensors using conventional semiconductor, crystalline magnetic material and other materials are difficult to operate stably.

I. INTRODUCTION

Sensors are one of the most important devices for establishment of the intelligent electronic control systems for mechatronics field such as industrial robots, automobiles, motor drive systems, machining equipments, and industrial measurements systems. The requisite conditions of these sensors are as follows,

(i) high resolution such as 10^{-6} G, 10^{-6} A, 10^{-6} m, 10^{-6} Kg, and 10^{-6} rad,

(ii) quick response or wide dynamic range such as 0 - 10 kHz (f_{-3dB} = 10 kHz),

(iii) robustness with which sensors stable operate under severe disturbances such as - 50 to 180 °C temperature variation, mechanical shock stresses, and humidity,

(iv) microlization in size for constitution of "built-in sensor", in which the diameter of the sensor element is less than 1 mm,

(v) good matching with microcomputers, and

(vi) good value of cost-performance ratio.

Amorphous cores are the most suitable sensor material accompanied with all these conditions. The basic principle of amorphous sensor constitution and the properties of amorphous material as the high-performance sensor material have been summarized in 1984 [1][2].

Basic structure and properties of new amorphous sensors are presented, in this paper, about (a) sensors for field and current detections using a two-core type multivibrator bridge with a negative feedback, (b) motor flux sensors for induction motors and brush-less small motors, (c) a thin and light-weighted rotary encoder with a million pulses per rotation, (d) torque sensors detecting a twisted angle of the shaft, (e) mechanocardiogram sensors, and (f) new pulse generator element utilizing the large Barkhausen and Matteucci effects of cold-drawn then stress annealed amorphous wires. A table for amorphous sensors developed in Japan is also shown.

Kyushu Institute of Technology, Sensui-cho 1-1, Tobata, Kitakyushu 804, Japan. This work is supported by the Grant-in-Aid for Developmental Scientific Research (2) of Ministry of Education in Japan and the New Technology Development Foundation in Japan.

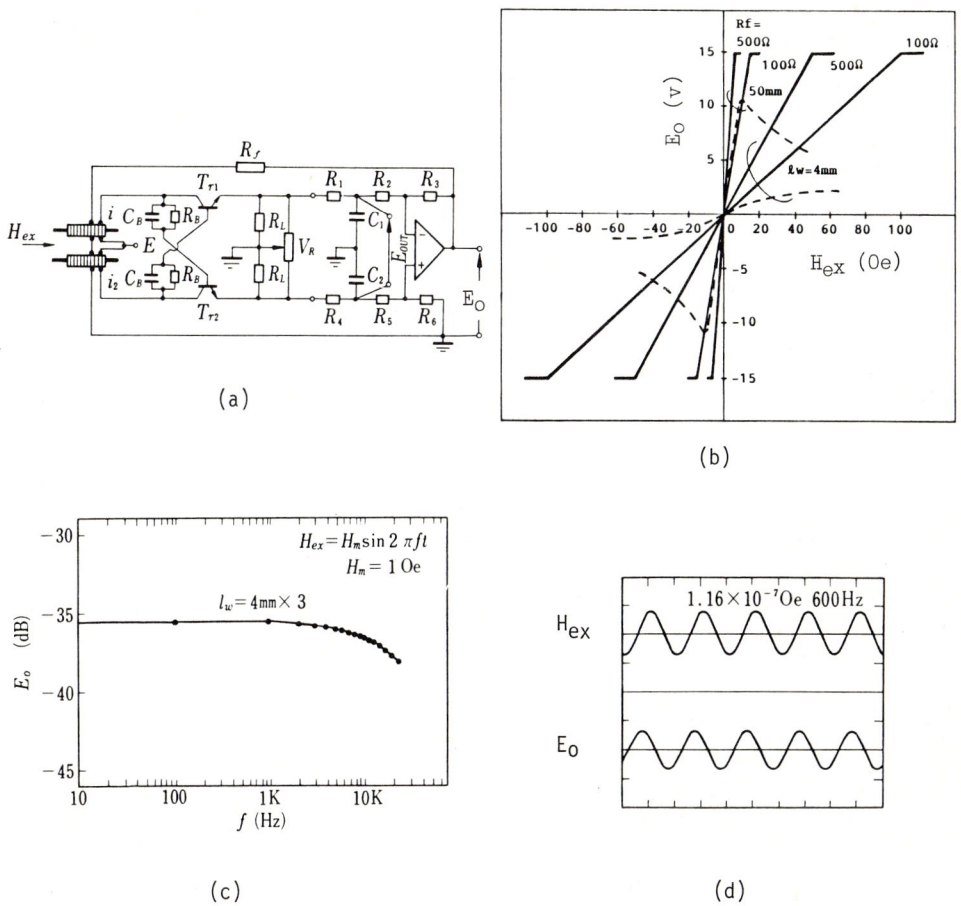

Fig.1. Amorphous field sensor: a two core type multivibrator bridge in (a), E_o vs. H_{ex} characteristics in (b), frequency characteristics of E_o in (c), and an example to detect small ac field in (d).

II. FIELD AND CURRENT SENSORS
[II - 1] Field sensor

Figure 1 shows a field sensor using an amorphous two core type multivibrator bridge with a negative feedback loop in (a) and its some properties in (b)(c) and (d). Two inductances of the straight-form amorphous cores and the two dummy resistances R_ls constitute a bridge circuit. The output voltage, E_{out}, of the multivibrator is amplified through a differential amplifier, and a part of the amplifier output, E_o, is fedback through a feedback resistance, R_f, to generate a negative field in the cores against the external field, H_{ex}.

The field detection gain is expressed as follows when $KN_f \gg R_f \ell_w$,

$$\frac{E_o}{H_{ex}} = \frac{R_f \ell_w}{N_f} \qquad (1)$$

where ℓ_w is the length of zer-magnetostrictive amorphous wire cores, N_f the feedback winding turn, and K the total gain of the multivibrator and the amplifier. The dotted curves are E_{out} versus H_{ex} characteristics for $R_f = \infty$.

The field detection characteristics and the frequency characteristics are shown in (b) and (c). The resolution of the field sensor is 10^{-7} Oe for ac field and 10^{-6} for dc field as shown in (d). The sensor cores stably work in the temperature variation from -50 °C to 200 °C. The maximum detectable field is about 100 Oe. The cut-off frequency for ac H_{ex} is about 20 kHz.

[II - 2] Inverter current sensor

Accuracy of direct drive (DD) industrial robots is greatly improved by introducing the high performance current sensor. Figure 2 illustrates a non-contact type accurate current sensor using n pairs of amorphous small cores (5-mm long) set around a copper wire. An inverter current with 2 kHz switching frequency is precisely detected due to the linear and quick response field sensor [3].

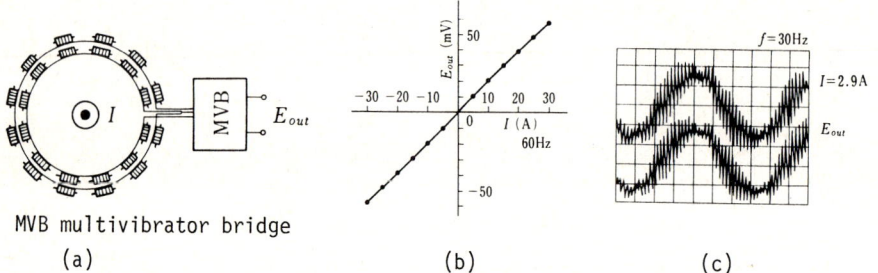

MVB multivibrator bridge
(a)   (b)   (c)

Fig.2. Non-contact current sensor: core arrangement in (a), E_{out} vs. I characteristics in (b) and inverter current sensing in (c).

[II - 3] Secondary current sensor for induction motors

AC servo motors using squirrel-cage induction motors are the most reliable control motors. However, they have been required to have more accuracy by establishment of true vector control systems, which can be realized by direct sensing of the secondary current, I_2. Figure 3 shows a new I_2 sensing method setting 8 pairs of small zero-magnetostrictive amorphous cores (4-mm long, 0.8-mm diameter) facing both endrings of a 1.5 kW, 4 pole induction motor as shown in (a). Stable I_2 sensing was done as in (b) by cancelling irregular mechanical vibrations in the 8-pair cores, which were connected with a multivibrator bridge.

The electromagnetic torque is detected by multiplying I_2 and the main flux which is detected using a search coil around a stator tooth. New vector control system becomes possible using a torque feedback system [4].

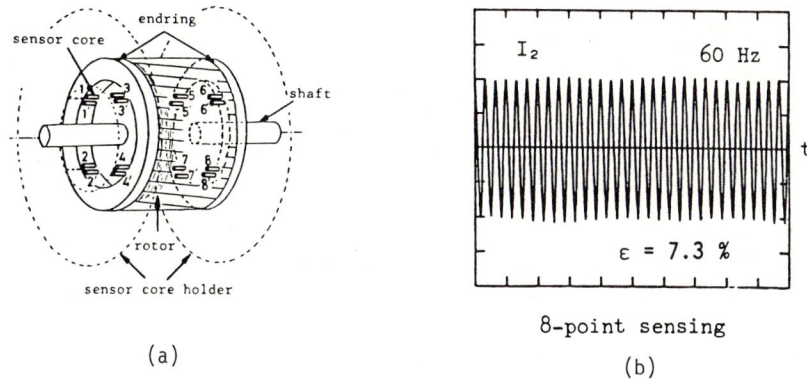

Fig.3. Secondary current sensing: amorphous core arrangement in (a), and I_2 sensing results for 1.5 kW 4 pole induction motor.

[II - 4] Brush-less motor sensor

Brush removing from the magnet-rotor type motors improve their reliability and suppress electromagnetic-noise generation. These brush-less motors need non-contact rotor position sensors. Figure 4 represents a 300 W brush-less motor with an outer magnet rotor and 3 pairs of amorphous small cores (4-mm long, 1-mm diameter). Each pair cores are connected with a multivibrator bridge, and generate a three phase voltage as the input of the transistor inverter circuit. Measured results show a stable operation of the amorphous field sensors, while Hall elements are destroyed for temperature over 70 °C in (b)[5].

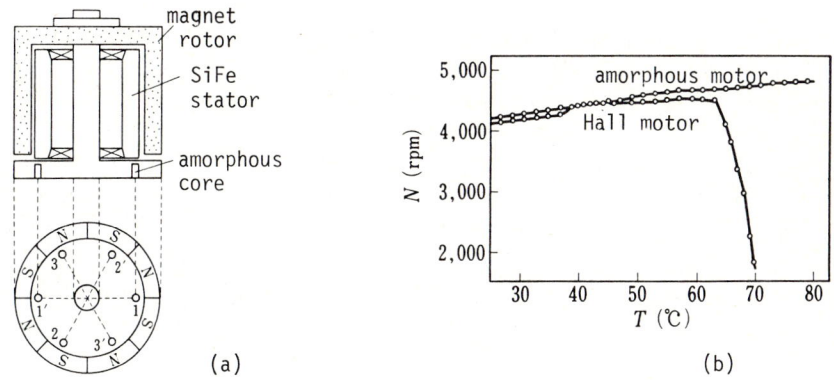

Fig.4. Amorphous brush-less magnet motor: amorphous core position in (a), and rotation vs. motor temperature characteristics in (b).

III. SENSORS USING AMORPHOUS STAR-SHAPED CORES

[III - 1] Amorphous 10^6 pulse rotary encoder

Conventional rotary encoders generate 10^6 pulses per shaft rotation are the photo-encoder and the resolver type encoder. The accuracy of the photo-encoder is rather sensitive to off-center rotation of the shaft, while the resolver is heavy in weight and big in size. A flat, light-weighted and robust rotary encoder with 10^6 pulses is constituted using the amorphous star-shaped core, as illustarated in Figure 5. The n pairs of small zero-magnetostrictive amorphous wire cores (4-mm long, 0.8-mm diameter) are set along the outer surface of a n-pole ring magnet, and each n-pair cores are connected serially with each other. These two n-pair cores are the cores in a multivibrator bridge.

The star-shaped amorphous core field sensor detecting the leakage flux from the magnet has the four advantages:

(i) cancellation of irregular distribution of the field strength of each pole in the n series cores,

(ii) cancellation of external disturbance fields in the radially arranged n series cores (the resolution of angle detection is about 0.002° for n = 40 [5]),

(iii) cancellation of off-center motion of the rotating shaft in the circularly arranged n series cores, and

(iv) positioning of n-pair cores is flexible and the difference of magnetic properties in each core is averaged in n series cores.

A leakage flux waveform for n = 40 is shown in (b) [5]. 10^6 pulses per shaft rotation is estimated for a ripple of the flux waveform is less than 0.3 % FS, which was realized in a star-shaped core with n = 100 [6].

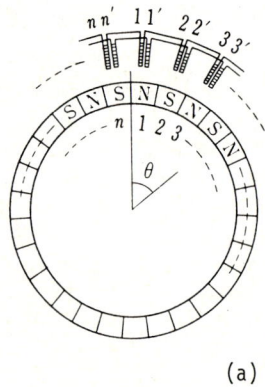
(a)

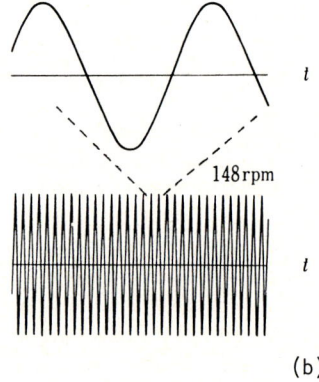
(b)

Fig.5. Core arrangement of a star-shaped core rotary encoder in (a), and detected leakage flux waveform from a ring magnet in (b).

[III - 2] Torque sensor

Torque sensors detecting the twisted angle, $\Delta\theta$, proportional to applied torque T ($\Delta\theta \cong 0.05°/cm$ for T = 100 kg-m in automobile engine shaft) are independent of problems in the torque sensors utilizing magnetostrictive effects such as the difference of temperature coefficient between the magnetostrictive layer and the shaft, temperature gradient of the shaft, and corrosiveness and aging of the magnetostrictive layers.

Figure 6 illustrates a torque sensor with a pair of star-shaped cores in (a) which resolution is about 0.002° in (b)[7].

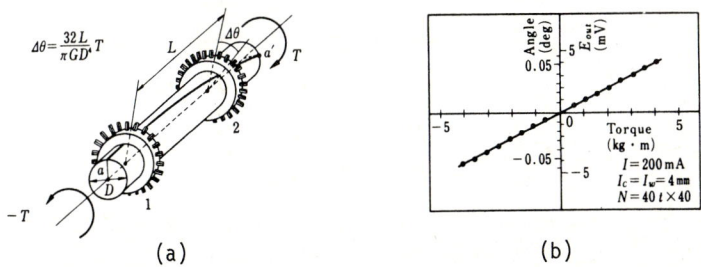

Fig.6. Torqu sensor in in (a) and torque detection results in (b).

[III - 3] Mechanocardiogram (micro-vibration) sensor

Microvibrations (0.1 μm - 200 μm) of the human body surface present important diagnosis informations for heart and circulatory organs, brain, eyes, teeth, and phonation organs. Conventional microvibration sensors are difficult for operation and for accurate measurements.

Figure 7 shows an accurate microvibration sensor using a star-shaped amorphous core multivibrator bridge detecting a small magnet displacement in (a). A small magnet is adhered on the body surface. Some detected waveforms are shown in (b). A resolution of the sensor is 0.02 μm [8].

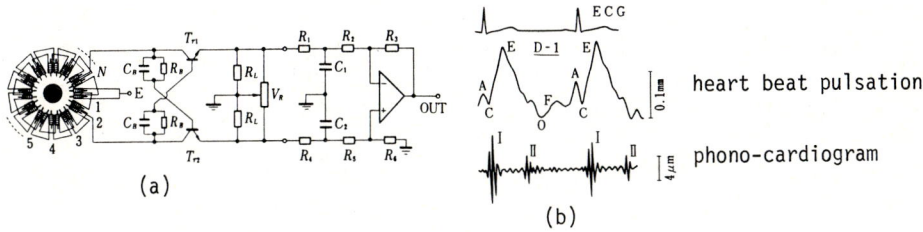

Fig.7. Mechanocardiogram sensor in (a) and detected waveforms in (b)

IV. SENSORS UTILIZING LARGE BARKHAUSEN AND MATTEUCCI EFFECTS

Amorphous magnetostrictive wires (made by UNITIKA Co.) are sensitive and stable pulse generator elements for ac fields with 0.01 Hz to 50 kHz. Critical values of flux-reversal field, H^*, and minimum wire length, ℓ_w^*, are 0.08 Oe and 70 mm for as-prepared 120-μm diameter wires, and 1 Oe and 15 mm for cold-drawn tension-annealed 30-μm diameter wires [9].

SENSORS DEVELOPED IN JAPAN

Security sensor, rotation sensor, current sensor, rotary encoder (UNITIKA Co.), data tablet (WACOM Co.), cartridge (SONY Co.), displacement sensor (K-TEC Co.), mechanocardiogram sensor (TDK Co.), field and current sensors (Mishima Time Co.), pressure sensor, stroke sensor, liquid He level sensor (AISIN Co.)

REFERENCES

1) K. Mohri, T.IEEE MAG-20, No.5, 942 (1984).
2) K. Mohri, Proc. RQ-5 Conf., 1687 (1984).
3) K. Mohri et al., Digest 9th Ann. Conf. on Magnetics in Japan, 28pD-9 (1985).
4) K. Mohri et al., T.IEEE MAG-22, No.5, 397 (1986).
5) K. Mohri et al., 39th Conf. IEE Kyushu Div. No.554 (1986).
6) K. Mohri et al. 10th Ann Conf. Magnetics in Japan (1987).
7) K. Mohri et al., T.IEEE MAG-23, No.5 (1987).
8) K. Mohri et al., ibid. (1987)
9) F.B. Humphrey, this symposium full paper.

SWR LINEWIDTH IN THIN FILMS OF Fe-B AND LONG RANGE FLUCTUATION OF EXCHANGE PARAMETER [x]

R. S. ISKHAKOV [xx], L. J. MAKSYMOWICZ, D. TEMPLE, R. ŻUBEREK [xxx]

Solid State Physics Department, Academy of Mining and Metallurgy, al. Mickiewicza 30, 30-059 Cracow, Poland

SWR linewidths in Fe-B metallic glasses have been studied by means of microwave spectroscopy at X-band. The spin wave dispersion relation was found to have the form predicted by the model of Ignatchenko-Iskhakov in which fluctuations of the exchange parameter in amorphous films are taken into account. The influence of the fluctuation on the linewidth of SWR modes is a topic of interest. Two sets of experimental data which could be a source of information about relaxation processes of magnons were studied:
1) dependence of the linewidth ΔH on the spin wave vector k - it exhibits a characteristic inflection for the same value of k for which an inflection is seen in the dispersion relation,
2) dependence of $\Delta H_{\parallel}$ on temperature (4.2 K to 295 K) - this gives an opportunity to study reentrant phenomena.

1. INTRODUCTION

An amorphous material is characterized by a set of spin system parameters. These parameters are not constants as in the case of crystalline materials but they are random functions of spatial coordinates (Fig. 1).

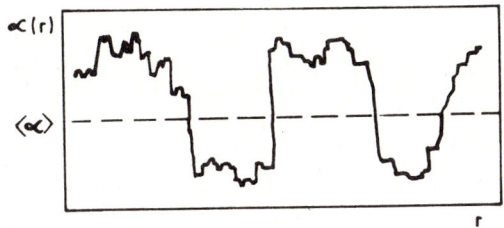

FIGURE 1
A spatially fluctuating value of the exchange parameter in an amorphous ferromagnet as an example of random function.

---

x   Supported by the central Research Program 01.08B and Government Project PR-3
xx Institute of Physics, Siberian Branch AS USSR, Krasnoyarsk
xxx Institute of Physics, Polish Academy of Sciences, Warsaw

Basic characteristics of a random function are its stochastic moments: 1) the average value of the parameter and 2) the correlation function that determines both the average amplitude of the fluctuation and the correlation radius. The fluctuations of the parameters in the spin system lead to a modification of the spin wave dispersion relation[1] which can be studied by means of the spin wave resonance (SWR) method[1,2]. One can also expect that the fluctuations will influence the relaxation processes of magnons, thus the linewidth of SWR modes.

2. RESULTS AND DISCUSSION

Thin amorphous films of Fe-B metallic glasses obtained by rf sputtering have been studied using microwave spectroscopy. It was found[2] that the spin wave dispersion relation had the form predicted by the model of Ignatchenko and Iskhakov[1] in the case when the fluctuation of the exchange parameter prevails over the fluctuations of other parameters of the spin system. Fig. 2 presents the logarithmic dependence of the SWR linewidth ΔH_n on the SWR mode number n ($k = n\pi/L$, L - film thickness) for $Fe_{46}B_{54}$ film at the temperature of 4.2 K. This relationship exhibits the inflection at $k=k_f$, where k_f is approximately the value of k for which the inflection in the dispersion curve is observed. Yet there are no theoretical predictions of the dependence of linewidth on k in amorphous materials (particularly when the long - range fluctuations of magnetic parameters are present).

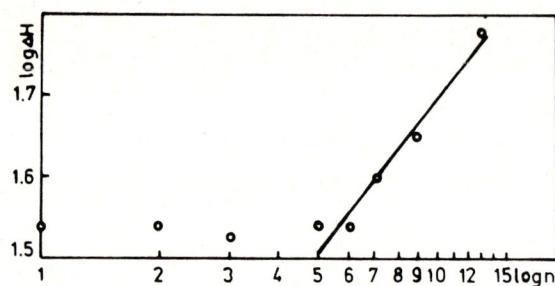

FIGURE 2

The log ΔH_n vs log n at T = 4.2 K for $Fe_{46}B_{54}$ film (ΔH_n is the SWR linewidth in the perpendicular geometry).

It is seen from these experimental data that $\Delta H_n \sim n^m (m \approx 0.8)$ and there is no significant temperature dependence of m for this sample. However, in the case of the samples richer in Fe some temperature dependence is observed (while temperature decreases, m decreases also).

In the investigated amorphous material, apart from the long range fluctuation of the exchange parameter, some intrinsic magnetic inhomogeneity is also observed in the concentration range corresponding to the neighborhood of the so-called percolation limit in the crystalline state where strongly coupled spin clusters (with cluster size about 20 - 25 Å) embedded in a weakly coupled "infinite cluster" can be formed [3]. The phase diagram for Fe_xB_{100-x} indicates four distinct temperature and concentration regimes: ferromagnetic (FM), reentrant (REE), spin glass (SG) and paramagnetic (PM). The REE regime occours in the concentration range $40 \lesssim x \lesssim 50$.

Besides the standart set of experiments which characterize the magnetic behavior of the REE and SG regimes the spin wave resonance data are also a source of important information:
1) the temperature dependence of the SWR mode resonance field in the parallel geometry $(H_{\parallel})$ shows a maximum for $T \approx T_f$ (T_f - the critical temperature for the REE regime) due to the unidirectional anisotropy energy in the SG phase (for the $Fe_{46}B_{54}$ film $T_f \approx 50K$)
2) there is a maximum in the temperature dependence of $\Delta H_{\parallel}$ (parallel geometry) at $T \approx T_f$. This reflects the relaxation in the strongly coupled spin clusters immersed in a weakly coupled infinite cluster with slow relaxation effects.

3. CONCLUSIONS

Inhomogeneities in the amorphous metallic glasses can be considered in two aspects: the long range correlations with fluctuations of the magnetic parameters with the correlation radius on the order of 100 Å and the intrinsic magnetic inhomogeneity, formation of spin clusters 20 - 25 Å in size, can be expected in the concentration range corresponding to the REE regime.

REFERENCES
1) V. A. Ignatchenko, R. S. Iskhakov, Proceedings of the 2nd Int. Conf. on Physics of Magn. Mat., Jadwisin, Poland 1985, p. 527
2) L. J. Maksymowicz, D. Temple, R. Żuberek
   J. Magn. Magn. Mat. 58 (1986) 303
3) D. J. Webb, S. M. Bhagat, K. Moorjani, T. O. Poehler,
   F. G. Satkiewicz, M. A. Manheimer,
   J. Magn. Magn. Mat. 44 (1984) 158

AUTHOR INDEX

Abe, S.   212
Adeva, P.   265
Aeschlimann, M.   70
Alameda, J.M.   259, 262
Alegria, A.   157
Allia, P.   347
Altounian, Z.   80
Amado, M.M.   182
Amaral, V.S.   182
Armstrong, J.V.   309
Arnaudas, J.I.   330
Arnold, J.   312
Aroca, C.   221, 281, 297
Ascasibar, E.   294, 321

Baczewski, L.T.   41, 170
Barandiarán, J.M.   142, 145, 185, 306, 324, 327
Barbara, B.   182
Bayreuther, G.   188
Bellissent, M.C.   336
Bellissent-Funel, M.C.   77
Bijaoui, D.   132, 336
Birchenough, A.D.   315
Bona, G.L.   70
Bøttiger, J.   340
Boucher, B.   77
Briones, F.   262
Brzózka, K.   61

Carrizo, J.   54
Celasco, M.   274
Celotta, R.J.   221
Charles, S.W.   86
Chen, D-X   95
Chien, C.L.   104
Chieux, P.   336
Coey, J.M.D.   309
Collins, A.   215
Colmenero, J.   157
Conde, A.   291
Contreras, M.C.   259, 262
Criado, A.   291
Cristina, M.C.   265

Day, R.K.   58, 256
De Pedro, L.   281

Del Moral, A.   330
Delcroix, P.   41, 64
Desre, P.   6
Dieny, B.   182
Duhaj, P.   206
Dunlop, J.B.   58, 256
Durand, J.   41

Fdez-Gubieda, M.L.   38
Fernandez, A.   51
Fernandez Barquin, L.   185
Flores, J.   197
Foley, C.P.   256
Forkl, A.   344
Fuertes, F.   262
Fujimori, H.   212

García, J.A.   54
Garcia, M.   297
García, N.   83
García Escorial, A.   265, 321
Ghafari, M.   58
Givord, D.   262
Glazer, A.A.   48
Gomez Sal, J.C.   185
González, J.   324
Gonzalez, J.M.   294, 318
Goscianska, I.   35
Greer, A.L.   145, 265
Grössinger, R.   203
Grundy, P.J.   315
Gruütter, P.   83
Guilmin, P.   64
Guimaraes, D.   77
Güntherodt, H.-J.   83
Gutierrez, J.   142
Guyot, P.   64
Guz, J.   278

Hadjipanayis, G.   123
Hagiwara, M.   117
Handstein, A.   312
Han Yongzhu   45
Han Zhenge   148
Harmelin, M.   336
Hartmann, M.   176

Heinzelmann, H.   83
Heitmann, H.   176
Hembree, G.   221
Hernando, A.   321, 324, 327
Hernando, B.   54
Herzer, G.   203, 354
Hibder, H.R.   83
Hidalgo, C.   318
Hilzinger, H.R.   354
Hirata, T.   92
Hoffmann, H.   188
Hoshi, Y.   160
Ho Su Nam   139
Humphrey, F.B.   89, 95, 110

Ignatchenko, V.A.   250
Inoue, A.   117, 200
Iskhakov, R.S.   250, 367

Jezuita, K.   61

Kaczkowski, Z.   136, 139
Kaczmarek, W.A.   67
Kamarad, J.   312
Kamigaki, K.   212
Kaneyoshi, T.   303
Karpe, N.   340
Kawamura, H.   89, 95, 110
Kawamura, h.   110
Ke Cheng   148
Kitamura, N.   92
Klahn, S.   176
Komatsu, T.   74
Konc, M.   35
Kraus, L.   206
Krauss, W.   15
Kronmüller, H.   164, 327, 344
Krzywinski, J.   173

Lacey, E.T.M.   315
Lachowicz, H.K.   232
Lagunas, A.R.   259
Lanotte, L.   28, 129
Leake, J.A.   238
Lienard, A.   262
Linderoth, S.   318
Liniers, M.   294, 318
Livet, F.   132
López, E.   281, 297
Lord, D.G.   315
Lucinski, T.   268
Lunney, J.G.   309

Madurga, V.   321, 324, 327
Maksymowicz, L.J.   367
Malmhäll, R.   89, 95, 110
Manabe, T.   89, 95
Marchal, G.   41
Márquez, R.   291
Ma Ruzhang   45
Masoero, A.   274
Masumoto, T.   117
Matras, G.   278
Matusita, K.   74
Matyja, H.   191
McHenry, M.E.   215
Meier, F.   70
Meyer, E.   83
Millán, M.   291
Mohri, K.   89, 95, 110, 360
Moreira, J.M.   182
Mørup, S.   1
Muñoz, C.   281
Muñoz, J.S.   287

Nafalski, A.   278
Naoe, M.   92, 160
Nazareth, A.   123
Nielsen, O.V.   38, 324
Nuñez de Villacicencio, C.   327

Ogasawara, I.   110, 117
O'Handley, R.C.   215
Okamoto, K.   227
Otero, T.F.   179

Pérez Frias, M.T.   194
Piecuch, M.   41
Pierce, D.T.   221
Pierna, A.R.   179
Pietrzak, J.   67
Pilpczuk, E.   191
Plazaola, F.   142, 185
Politis, C.   15
Pönninger, A.   203
Pont, M.   200
Porreca, F.   28, 129
Potapov, A.P.   48
Puchalska, I.B.   21
Puźniak, R.   287

Rao, K.V.   200, 287, 340
Ratajczak, H.   35, 67
Reisser, R.   344
Rivas, J.   164
Rivero, G.   271, 294

Author Index

Roberts, R.B. 256
Rodriguez Fernandez, J. 185
Rosenberg, M. 333
Rosenthaler, L. 83
Rubio, H. 51
Ryan, D.H. 244

Sáenz, J.J. 83
Sánchez, M.C. 281, 297
Sánchez, P. 281, 297
Sanquer, M. 77
Sarmento, E.F. 303
Sassik, H. 32
Schneider, J. 312
Sellmyer, D.J. 98
Serebryakov, A.V. 284
Shulika, V.V. 48
Siegmann, H.C. 70
Siemko, A. 173
Škorvanek, I. 154
Soiński, M. 154
Sousa, J.B. 182
Spisak, P. 35
Stampanoni, M. 70
Startsewa, I.E. 48
Stepanescu, A. 274
Stobiecki, F. 151
Stobiecki, T. 188
Szlanta, J. 61
Szymański, B. 154
Szymczak, H. 232

Tejedor, M. 51, 54
Telleria, I. 145, 157
Temple, D. 367
Thompson, J.R. 15
Torp, B. 340

Tourbot, R. 77
Twarowski, K. 167

Unguris, J. 221

Van Wonterghem, J. 1
Vaterlaus, A. 70
Vázquez, M. 321, 324, 327
Vicent, J.L. 194, 197, 318
Vinai, F. 347
Voitanik, P. 38
Vojtaník, P. 209

Wac-Włodarczyk, A. 278
Wadas, A. 300
Walz, F. 164
Wang Xinlin 148
Wegrzyn, A. 170
Wells, S. 86
Wenda, J. 167
Wernhardt, R. 333
Wezulek, R. 32
Wiesendanger, R. 83
Witter, K. 176

Xu Zuxiong 45

Yamasaki, J. 89, 95, 110
Yavari, A.R. 6, 132, 336

Zabala, I. 142
Zabala, N. 306
Zaluski, L. 173
Zárubová, N. 206
Závěta, K. 206
Żuberek, R. 367

SUBJECT INDEX

activation energy 74, 200
— spectrum 145, 157, 238
after-effect, *see* magnetic after-effect
amorphization
— by mechanical alloying 15
— by solid state reaction 64
amorphous iron 244
amorphous particles 1
amorphous superconductors 104, 340
anelastic deformation 173
anisotropy
— induced, *see* magnetic anisotropy induced by
— magnetic, *see* magnetic anisotropy
— of magnetoresistance 191, 197
— perpendicular, *see* magnetic anisotropy
— random magnetic, *see* random magnetic anisotropy
applications 89, 95, 278, 354, 360
Arrott plot – modification 344
asperomagnetism 244
atomic
— bond reorientation 321
— pair distribution function 250, 321
— short-range order, *see* short range order
— size difference 104

Barkhausen effect 274
band structure 188, 215
binding energy 215
Bloch's law 104, 244
Bloch lines memories 21
Bloch walls 21, 221

chemical short-range order (CSRO) 238
— mechanism 74
coating 160
coercive field 32, 35, 123, 129, 154
coercivity
— degradation 176
— temperature dependence 123
compositional fluctuations 336
concentration depth profile 151
corrosion 179
critical
— exponents 61, 98, 327, 340, 344

— phenomena 67
— thickness 6
Crystal Electric Field (CEF) effects 330
crystallization
— kinetics 123, 256, 268, 284, 287, 291
— rules 256, 265, 284
— of surface 45, 80
— temperature 67, 80, 185
Curie–Weiss law 77

Debye temperature 185, 333
density of states 215, 330, 333
Differential Scanning Calorimetry (DSC) 80, 145, 256
diffusion 151
directional
— ordering 28
— chemical bonding 74
disaccomodation 347
distribution of
— activation energy 74
— hyperfine field 58, 61, 309
— internal stresses 51
domains
— cylindrical 21
— magnetic, *see* magnetic domains
— mono- 77
— stripe 21, 227
— structure 206, 262
— zig-zag 45
domain wall 21, 221, 281, 347
— velocity 95, 110
— width 221
ductility 132

eddy current losses 154
effect
— ΔE 294, 312
— galvano-magnetic 340
elastic properties 136, 212
elastoresistivity 321
electrical properties 188
electrical resistivity 35, 48, 67, 268
— irreversible changes 238
— kinetics 74, 340
— magnetic contribution 182, 185

—reversible changes 74
—temperature dependence 35
electron–phonon interaction 300
electronic structure 215, 333
embrittlement 132, 336
exchange interaction
—constant 215, 250, 268
—fluctuations 244, 250

fatigue strength 117
Fermi surface 191
ferromagnetic glasses 336, 354
Ferromagnetic Resonance (FMR) 67, 300
ferromagnets
—isotropic 67
—itinerant-electron 300
field theory, effective 303
flash annealing 89, 303, 306
fluctuation
—of exchange constant 215
—of local stresses 347
—spin 188
forced volume magnetostriction 170, 312
free volumes 148, 318

galvano-magnetic effect 340
glass transition 256, 344
glassy tapes 6, 132

Hall effect 188, 194
—anomalous 200
—spontaneous 35
hybridization 215
hydrogenation 244
hyperfine field 58, 61, 333
hysteresis loop 41, 48, 154, 271

induced anisotropy, see magnetic anisotropy induced by
interaction, electron–phonon 300
internal
—friction 139, 212
—stresses 51
Invar effect 312
Ising model 303
itinerant-electron model 300, 344

Kerr effect, see magneto-optical Kerr effect
kinetics of crystallization 123, 256, 268, 284, 287, 291
Kissinger plot 123

laser surface treatment 297, 309
layer
—multi- 41, 64, 98, 151, 194
—oxide 179, 259
linear magnetostriction 167, 191
log-time kinetics 238
losses of power 48
low angle diffuse X-ray scattering 132

magnetic after-effect 209, 330, 347
—irreversible 164, 274
—reversible 164
magnetic alloys
—Invar 312
magnetic anisotropy 28
—bulk 51, 70
—circumferential 110
—coherent 98
—constants – evaluation 54
—perpendicular 41, 45
—planar 315
—radial 110
—random, see random magnetic anisotropy
—surface 51, 70
magnetic anisotropy field 227, 294
magnetic anisotropy induced by:
—creep 206
—directional short range order 28
—laser surface treatment 297, 309
—(magnetic) field 38, 48
—stress 38, 173, 303, 324
—stress-field 38
—thermomagnetic treatment 45
magnetic domain
—cylindrical 21
—single 227
—stripe 21, 227
—structure 83, 206, 262
—orientation 70
magnetic microstructure 221
magnetic moment 188, 215, 244
magnetic phase diagram 200, 303, 340
magnetic properties 41, 77, 244, 344
—bistable 89, 95
magnetic relaxation 157, 367
magnetization
—field dependence 312
—ripple 21, 262
—saturation – evaluation 54
—temperature dependence 312
—near surface 51
magnetoelastic coupling 142
magnetoelastic

—anisotropy field   232
—coupling constants   28
—properties   232, 294
magnetomechanical coupling   136, 139
Magneto-Optical Kerr Effect (MOKE)   51, 70, 259, 315
—magneto-optical recording media   92, 160
magnetoresistance   182, 197
—anisotropy   197
magnetostriction   32, 315
—critical exponent   327
—forced volume   170, 312
—linear   21, 129, 167, 191
—model   330
—saturation   28, 167, 232
—single-ion model   203
—stress dependence   232
—temperature dependence   327
—two-ion contribution   232
—volume   170, 312
—zero   203
magnetostrictive wires   89, 95, 110
magnetovolume effect   312
Matteucci effect   110, 117
mean field approximation   303
mechanical alloying   15
modulated solids   104
Mössbauer effect   1, 45, 58, 61, 64, 309
multilayers   41, 64, 98, 151, 194

Néel walls   21
non-magnetostrictive alloys   164, 203, 232, 360
nucleation   265

oxidation   176, 179

pair distribution function   250, 321
particles
—single domain   123
—size   1, 86
—ultrafine amorphous   1
perminvar   209
perpendicular anisotropy   41, 45
phase diagram
—magnetic   200, 244, 303, 340
—structural   6, 15
photoemission spectroscopy   70
piezomagnetic properties   136
planar flow casting   1, 6, 32, 336
Poisson's coefficient   321
positron annihilation   318
precipitation   256
pressure dependence of

—Curie temperature   312
—forced volume magnetostriction   170
—magnetization   170
protective coatings   160

quenching   256
quench rate   80

random magnetic anisotropy   98, 123, 182, 330
rapid solidification   15
rare earth iron alloys   123, 315
rare earth transition metal alloys   41, 70, 98, 123, 176, 315, 330
relaxation
—structural, see structural relaxation
—magnetic, see magnetic relaxation
—non-exponential   157
—of stress   238
resistivity, see electrical resistivity

scattering
—low angle diffuse X-ray   132
—small angle, neutron   77, 336
sensors   89, 95, 354, 360
short range order (SRO)   142, 185, 203
single-ion magnetostriction   203, 330
size difference   104
small angle neutron scattering   77, 336
soft
—amorphous ribbons   32, 271
—magnetic ribbons   32, 271
—metglasses   98
specific heat   244, 333
speromagnetism   98, 244
spin-polarized photoemission   70
spin fluctuations   188
spin glass   340
spin-glass phase transition   98, 244
spin-wave
—constant   104
—resonance (SWR)   250, 367
stress relaxation   238
stripe domain   21, 227
structural relaxation   74, 145, 318, 336
—enthalpy   80, 145
—irreversible   80, 238
—reversible   80, 238
structure
—magnetically inhomogeneous   333
—spin-glass   244
superconductivity   104, 340
superlattice   104
superparamagnetism   1

surface
— amorphization 303
— crystallization 45, 80
— effect 259
— magnetic properties – measurements 83
— magnetism 51, 70
— oxidation 45, 259
susceptibility
— critical exponents 340
— transverse-biased initial 259

temperature dependence
— of coercivity 123
— of electrical resistivity 35, 327
— of magnetostriction 327
— of spontaneous Hall effect 35
tensile stress 321
thermal expansion 148, 256
thermal stability 58, 148, 200
thermoelectric power 268
thermomagnetic analysis 256
topological short-range order 238
transducers 354, 360

transition
— amorphous–crystalline 86, 256
— ferromagnetic–paramagnetic 61
— ferromagnetic–spin glass 98, 344
— paramagnetic–ferromagnetic 340
trigonal prisms 74
two-ion magnetostriction 232
two-level systems 164, 232, 238

ultrafine particles 1
ultrasonic transducers 136
ultrasound velocity 139

Vickers hardness 117
viscosity field 274
volume magnetostriction 170, 312

wires 89, 95, 110
— mechanical properties 117

Young's modulus 28, 117, 321

zero-magnetostrictive alloys 203, 321

MATERIALS INDEX

The elements within each alloy are arranged so that the transition metals Co, Fe and Ni, are mentioned first and the metalloids B, Be, C, P and Si, last. Within these groups the ordering is alphabetical.

The letter M denotes unspecified transition metals. R denotes unspecified rare-earth metals.

Au–R   123

Co–Au   21
Co–B   200, 209
Co–B–C   200
Co–B–Si   32, 110, 117, 327
Co–Cr   21, 227
Co–Cr–B   191
Co–Cr–B–Si   117, 206
Co–Er–Gd   98
Co–Fe   74, 86, 312
Co–Fe–B   1, 32, 265, 321
Co–Fe–B–Si   28, 38, 48, 74, 170, 312, 321, 324, 347
Co–Fe–Mn–B–Si   360
Co–Fe–Mn–Mo–B–Si   203, 360
Co–Fe–Mo–B–Si   278, 321, 360
Co–Fe–Ni   185
Co–Fe–Ni–B–Si   164, 173, 206
Co–Fe–Ta–B–Si   148
Co–Fe–Tb   92
Co–Fe–Zr   170, 333
Co–Gd   21, 227
Co–Gd–Er   98
Co–Gd–Mo   21
Co–Gd–Tb   98
Co–Ho   21
Co–M–B–Si   104
Co–Mn–B   191
Co–Nb–B   306
Co–Nd   123
Co–Ni–B   321
Co–Ni–B–Si   185, 300, 324, 327
Co–P   21, 129, 271, 294
Co–Pr   160
Co–R   123
Co–Sn   64
Co–Tb   123
Co–V–B   191
Co–Y   259

Co–Zr   188, 215, 333
Cu–Dy–Y   333
Cu–Er–Y   330
Cu–Hf   15
Cu–Nb   104
Cu–Pd–Ti   15
Cu–Tb   77
Cu–Tb–Y   330
Cu–Ti   15
Cu–Zr   117, 215

Fe–B   1, 54, 80, 104, 142, 151, 167, 209, 215, 244, 312
Fe–B–C–Si   6, 51
Fe–B–Si   6, 89, 95, 110, 117, 136, 139, 154, 167, 197, 212, 318
Fe–C   1, 104
Fe–C–P   117, 212
Fe–Cr–B   142
Fe–Cr–B–Si   117, 206, 312
Fe–Dy–B   98
Fe–Dy–Tb   315
Fe–Er   98
Fe–Ga–Tb   98
Fe–Gd   21
Fe–Gd–Tb   176
Fe–Hf   104, 244, 312
Fe–M–B–Si   117
Fe–M–C–P   117
Fe–Mn–B   142
Fe–Mn–B–Si   312
Fe–Mo   104
Fe–Mo–B   142
Fe–Nd   41, 98, 123
Fe–Nd–B   262
Fe–Ni   110, 312
Fe–Ni–B   1, 132, 157, 179, 324, 327
Fe–Ni–B–P   164, 312, 336
Fe–Ni–B–Si   61, 312, 321
Fe–Ni–B–C–Si   45

Fe–Ni–Cr–P 336
NiDy 98
Fe–Ni–Mo–B 145, 309
Fe–Ni–Mo–B–Si 321
Fe–Ni–P 287
Fe–Ni–V–P 336
Fe–Ni–Zr 170, 333
Fe–P 312
Fe–Pd–B 142
Fe–R 123
Fe–Sb 35
Fe–Sc 256
Fe–Sc–B 58, 256
Fe–Si 167, 215, 268
Fe–Sm–B 67
Fe–Ta 98, 104
Fe–Tb 70, 123
Fe–Ti 104
Fe–Y 244
Fe–Zr 104, 170, 188, 244, 312, 333, 340, 344
Fe–Zr–B 142

Ni–Al–B–P 117
Ni–Al–B–Si 117
Ni–Al–Cr–B–Si 117
Ni–Al–M–B–P 117
Ni–Al–M–B–Si 117
Ni–B 32, 104
Ni–B–Si 32
Ni–Cr–B–C 291
Ni–Cr–P 291
NiDy 98
Ni–Dy–Gd 182
Ni–Hf 15
Ni–Nb 104, 194
Ni–Pd–P 117, 265, 321
Ni–Pt–P 117
Ni–Zr 215

Pd–Ti 15
Pd–Zr 215

Rh–Zr 15